Biopolymer Composites

Also of interest

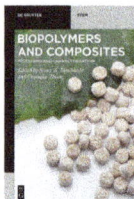

Biopolymers and Composites
Processing and Characterization
Madbouly, Zhang (Eds.), 2021
ISBN 978-1-5015-2193-5, e-ISBN 978-1-5015-2194-2

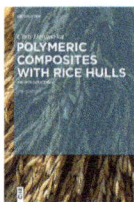

Polymeric Composites with Rice Hulls
An Introduction
Defonseka, 2019
ISBN 978-3-11-063968-1, e-ISBN 978-3-11-064320-6

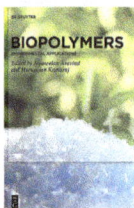

Biopolymers
Environmental Applications
Aravind, Kamaraj (Eds.), 2023
ISBN 978-3-11-099872-6, e-ISBN 978-3-11-098718-8

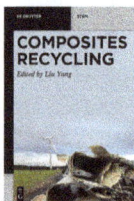

Composites Recycling
Yang (Ed.), 2023
ISBN 978-3-11-075441-4, e-ISBN 978-3-11-075443-8

Physical Sciences Reviews.
e-ISSN 2365-659X

Biopolymer Composites

Production and Modification from Tropical Wood and
Non-Wood Raw Materials

Edited by
Salit Sapuan, Syeed SaifulAzry Osman Al Edrus,
Ahmad Adlie Shamsuri, Aizat Abd Ghani and
Khalina Abdan

DE GRUYTER

Editors

Prof. Dr. Salit Mohd Sapuan
Universiti Putra Malaysia
Department of Mechanical and Manufacutring
Engineering
43400 UPM Serdang Selangor
Malaysia

Dr. Syeed SaifulAzry Osman Al Edrus
Universiti Putra Malaysia
Institute of Tropical Forestry and Forest Products
Universiti Putra Malaysia
43400 Serdang
Malaysia

Dr. Ahmad Adlie Shamsuri
Universiti Putra Malaysia
Institute of Tropical Forestry and Forest Products
Universiti Putra Malaysia
43400 Serdang
Malaysia

Dr. Aizat Abd Ghani
Universiti Putra Malaysia
Institute of Tropical Forestry and Forest Products
Universiti Putra Malaysia
43400 Serdang
Malaysia
and
Faculty of Tropical Forestry
Universiti Malaysia Sabah
Jalan UMS
88400 Kota Kinabalu, Sabah
Malaysia

Prof. Khalina Abdan
Universiti Putra Malaysia
Institute of Tropical Forestry and Forest Products
Universiti Putra Malaysia
43400 Serdang
Malaysia

ISBN 978-3-11-162237-8
e-ISBN (PDF) 978-3-11-076922-7
e-ISBN (EPUB) 978-3-11-076931-9

Library of Congress Control Number: 2022948639

Bibliographic information published by the Deutsche Nationalbibliothek
The Deutsche Nationalbibliothek lists this publication in the Deutsche Nationalbibliografie;
detailed bibliographic data are available on the internet at http://dnb.dnb.de.

© 2024 Walter de Gruyter GmbH, Berlin/Boston
This volume is text- and page-identical with the hardback published in 2023.
Cover image: ronstik / iStock / Getty Images Plus
Typesetting: TNQ Technologies Pvt. Ltd.

www.degruyter.com

About the editors

Mohd Sapuan Salit is an "A" Grade Professor of composite materials at Department of Mechanical and Manufacturing, Universiti Putra Malaysia (UPM) and a Head of Laboratory of Biocomposite Technology, INTROP, UPM. He has a BEng in Mechanical Engineering from University of Newcastle, Australia, an MSc in Engineering Design from Loughborough University, UK and PhD in Material Engineering from De Montfort University, UK. He is a Professional Engineer, a Society of Automotive Engineers Fellow, an Academy of Science Malaysia Fellow, a Plastic and Rubber and Institute Malaysia Fellow, a Malaysian Scientific Association Fellow, an International Biographical Association Fellow and an Institute of Material Malaysia Fellow. He is an Honorary Member and immediate past Vice President of Asian Polymer Association based in IIT Delhi and Founding Chairman and Honorary Member of Society of Sugar Palm Development and Industry, Malaysia. He is the co-editor-in-chief of Functional Composites and Structures, and member of editorial boards of more than two dozen journals. To date he has produced more than 1800 publications including over 860 journal papers, 50 books, and 175 chapters in book. He has delivered over 50 plenary and keynote lectures, and over 150 invited lectures. He organized 30 journal special issues as a guest editor, presented over 650 technical articles in conferences and seminars, reviewed over 1300 journal papers and has 8 patents.

Dr. S.O.A. SaifulAzry received his BSc in Forestry (Wood Science & Technology) from Universiti Putra Malaysia (UPM) in 2002, his MSc (Biocomposite Technology) in 2007, and his PhD (Bicomposite Technology) in 2017 also from UPM. Currently he is a Research Officer in Laboratory of Biocomposite Technology (Biocomposite), Institute of Tropical Forestry and Forest Products (INTROP), Universiti Putra Malaysia. His research interests include wood composites, nanocellulose and nano biocomposite. He has published several articles on these topic areas.

https://doi.org/10.1515/9783110769227-201

Ahmad Adlie Shamsuri received his BSc (Hons) in Chemistry from Universiti Kebangsaan Malaysia (UKM) in 2006, his MSc (Chemistry) in 2009, and his PhD (Chemistry) in 2015 also from UKM. His career began as a research officer at the Institute of Bioscience, Universiti Putra Malaysia (UPM), in 2008. UPM and government agencies primarily fund his research projects. Later, he was promoted to senior research officer at the Laboratory of Biocomposite Technology, Institute of Tropical Forestry and Forest Products (INTROP), UPM, in 2013. His areas of research include polymer blend, polymer composite, ionic liquid, and surfactant. He has about 13 years of experience in research and development. So far, he has published more than 40 journal papers, and he is also a reviewer of several WoS indexed journals.

Aizat Ghani received his Bachelor degree at Universiti Malaysia Sabah on 2011 in Wood Fiber Technology and Industry. He obtained his Master Science and Doctoral of philosophy from University Putra Malaysia at 2015 and 2019, respectively in field of Wood Composite Technology and Bioresource and Technology. He just started his career on 2020 as Research Officer at Laboratory of Biocomposites, INTROP, UPM. In October 2022, he started his career as senior lecturer in Universiti Malaysia Sabah. His researches interest is in wood and non-wood forest product, wood composite and modification, wood treatment and also wood quality enhancement. On 2019, he has been selected as one of the Japan Society for the Promotion of Science (JSPS) Fellow.

A. Khalina is a Professor of biocomposite manufacturing technology, polymer Biocomposite, disaster management research, and postharvest engineering at Faculty of Engineering, Universiti Putra Malaysia (UPM). She earned her B.Eng degree in Agricultural Engineering in 1996 and Doctor of Philosophy in Biocomposite Technology from Universiti Putra Malaysia in 2006. Currently she is a Director of Institute of Tropical Forestry and Forest Products (INTROP) from Mac 2020. She was officially appointed as a lecturer in 2006 and promoted to Associate Professor in 2010. She was appointed as Head of Biomaterial Processing Laboratory in 2007, and Head of Biocomposite Technology Laboratory in 2010. In April 2017, Khalina was appointed as Deputy Dean, Research & Innovation of Faculty of Engineering. She has been appointed as visiting scientist at Universiti Tun Hussein Onn (UTHM) Malaysia, Istanbul Technical University (Turkey), University of Queensland (Australia), and Seoul National University (South Korea). To date, she has authored and published more than 170 articles in citation-indexed journals. She also has authored or co-authored more than 22 chapters in books and 81 national/international conference papers. She was also invited to be a keynote and invited speaker at national and international conferences. She has been appointed as a technical assessor to evaluate projects for the Industrial Grant Scheme, MOSTI, and Ministry of Plantation Industries and Commodities. She has been appointed as a consultant in the biocomposite roadmap, steering, and technical committee on the kenaf roadmap, and several projects.

Preface

Biopolymer composites: Production and Modifications from Tropical Wood and Non-wood Raw Materials delivers a novel and systematic collection of producing and modifying biopolymer composites from the valuable bioresource of tropical wood and non-wood species. Lately, the use of biopolymer composites in the composite industry has turned out to be the accountability of the manufacturer due to the environmental awareness and rising of the biopolymer composites industry.

Moreover, the sustainability of biopolymer composites supplies has also contributed to the development of the industry. In addition, the utilization of tropical wood and non-wood species in biopolymer composites has increased due to their excellent properties and environmental friendliness. Herein we focus on the preparation and properties of green composites, including the material selection and conceptual design in natural fiber composites, as well as the new molecular design concept of high-performance bioplastics. On top of that, the application of versatile natural fiber like kenaf and woven natural fiber in advanced sustainable polymer composites were discussed as well. The under-utilized species and non-wood species like bamboo, oil palm trunk, was explained as potential raw materials for manufacturing polymer composites. This book is suitable for students, academicians, researchers, and industry players. It also includes the current development, possible applications, and future insights of biopolymer composites.

We are incredibly grateful to all of the authors who contributed chapters to this edited book and shared their insightful thoughts and expertise. We make an effort to compile all the various information from authors in many fields who have written about wood and bio-composites in order to successfully complete this endeavor. We are very grateful for the dedication of the contributors who have helped us put our ideas into practice. Finally, we would like to express our gratitude to the De Gruyter team for their kind assistance during the entire book-production process.

https://doi.org/10.1515/9783110769227-202

Contents

Abdan Khalina, Ching hao Lee and Aisyah Humaira

Fathin Sakinah Mohd Radzi, Anuar Abu Bakar, Mohd Azman Asyraf, Nik Adib Nik
Abdullah, and Mat Jusoh Suriani

Mohd Khairun Anwar Uyup, Lee Seng Hua, Alia Syahirah Yusoh, Asniza Mustapha, and
Siti Norasmah Surip

Alia Syahirah Yusoh, Mohd Khairun Anwar Uyup, Paridah Md Tahir, Lee Seng Hua, and
Ong Chee Beng

Nurul Ain Maidin, Salit Mohd Sapuan, Mastura Mohammad Taha and Zuhri Mohamed
Yusoff Mohd

Nor Amira Izzati Ayob, Nurul Fazita Mohammad Rawi, Azniwati Abd Aziz, Baharin
Azahari and Mohamad Haafiz Mohamad Kassim

Asniza M, Alia Syahirah Y, Mohd Fahmi A, Lee SH, and Anwar UMK

Moustafa Alaa, Khalina Abdan, Lee Ching Hao, Ammar Al-Talib, Muhammad Huzaifah
and Norkhairunnisa Mazlan

Muhammad Aizuddin Mohamad, Aidah Jumahat and Napisah Sapiai

Mohd Azrul Jaafar, Shahrul Azam Abdullah, Aidah Jumahat, Mohamad Asrofi Muslim,
Napisah Sapiai and Raymond Siew Teng Loy

Mohamad Asrofi Muslim, Aidah Jumahat, Shahrul Azam Abdullah, Mohd Azrul Jaafar,
Napisah Sapiai and Raymond Siew Teng Loy

List of contributing authors

Khalina Abdan
Laboratory of Biocomposite Technology
Institute of Tropical Forestry and Forest Products
(INTROP)
Universiti Putra Malaysia
43400 Serdang, Selangor
Malaysia
and
Department of Biological and Agricultural
Engineering
Faculty of Engineering
Universiti Putra Malaysia
43400 Serdang, Selangor
Malaysia
E-mail: khalina@upm.edu.my

Nik Adib Nik Abdullah
MSET Inflamable Composit Sdn. Bhd.
PT 7976K Gong Badak Industrial Zone
21030 Kuala Nerus, Terengganu
Malaysia

Norihan Abdullah
Laboratory of Biocomposite
Institute of Tropical Forestry and Forest Products
(INTROP)
Universiti Putra Malaysia
43400 Serdang, Selangor
Malaysia
E-mail: GS60118@student.upm.edu.my

Shahrul Azam Abdullah
School of Mechanical Engineering
College of Engineering
Universiti Teknologi MARA
40450, Shah Alam, Selangor
Malaysia
E-mail: shahrulazam@uitm.edu.my

Nabilah Ahamad
Department of Wood and Fiber Industries
Universiti Putra Malaysia
43400 UPM Serdang, Selangor
Malaysia

Humaira Alias Aisyah
Laboratory of Bio composite Technology
Institute of Tropical Forestry and Forest Products
(INTROP)
Universiti Putra Malaysia
43400 Serdang, Selangor
Malaysia

Malaysia Ammar Al-Talib
Department of Mechanical Engineering
Faculty of Engineering
Technology & Built Environment
UCSI University
Cheras, 56000
Malaysia

Moustafa Alaa
Department of Biological and Agricultural
Engineering
Faculty of Engineering
Universiti Putra Malaysia
Serdang 43400
Malaysia
E-mail: Moustafaalaa1@outlook.com

Isah Aliyu
Advanced Engineering Materials and Composites
Research Centre (AEMC)
Department of Mechanical and Manufacturing
Engineering
Universiti Putra Malaysia
43400 UPM Serdang, Selangor
Malaysia
and
Department of Metallurgical Engineering
Waziri Umaru Federal Polytechnic
Brinin Kebbi
Nigeria
https://orcid.org/0000-0003-3564-3500

https://doi.org/10.1515/9783110769227-203

UMK Anwar
Forest Products Division
Forest Research Institute Malaysia (FRIM)
52109 Kepong
Malaysia
E-mail: mkanwar@frim.gov.my
https://orcid.org/0000-0002-3787-9907

Zaidon Ashaari
Department of Wood and Fiber Industries
Universiti Putra Malaysia
43400 UPM Serdang, Selangor
Malaysia

M. Asniza
Forest Products Division
Forest Research Institute Malaysia (FRIM)
Kepong, Selangor
Malaysia

Mohd Azman Asyraf
Faculty of Ocean Engineering Technology and
Informatics
Universiti Malaysia Terengganu
21030 Kuala Nerus, Terengganu
Malaysia
E-mail: asyrafazman23@yahoo.com

Fatimah A'tiyah
Faculty of Biotechnology and Biomolecular
Sciences
Universiti Putra Malaysia
43400 UPM Serdang, Selangor
Malaysia

Nor Amira Izzati Ayob
School of Industrial Technology
Universiti Sains Malaysia
Gelugor
11800 Penang
Malaysia

Rafiqah Shafi Ayu
Laboratory of Bio composite Technology
Institute of Tropical Forestry and Forest Products
(INTROP)
Universiti Putra Malaysia
43400 Serdang, Selangor
Malaysia

Baharin Azahari
School of Industrial Technology
Universiti Sains Malaysia
Gelugor
11800 Penang
Malaysia

Azniwati Abd Aziz
School of Industrial Technology
Universiti Sains Malaysia
Gelugor
11800 Penang
Malaysia

Anuar Abu Bakar
Faculty of Ocean Engineering Technology and
Informatics
Universiti Malaysia Terengganu
21030 Kuala Nerus, Terengganu
Malaysia
and
Marine Materials Research Group
Faculty of Ocean Engineering Technology and
Informatics
Universiti Malaysia Terengganu
21030 Kuala Nerus, Terengganu
Malaysia
E-mail: anuarbakar@umt.edu.my

Ong Chee Beng
Forest Products Division
Forest Research Institute Malaysia
Kepong, Selangor
Malaysia

A. Mohd Fahmi
Forest Products Division
Forest Research Institute Malaysia (FRIM)
Kepong, Selangor
Malaysia

Nur Diyana Ahmad Fazil
Laboratory of Bio Composite Technology
Institute of Tropical Forestry and Forest Products
(INTROP)
Universiti Putra Malaysia,
43400 Serdang
Malaysia
E-mail: nur.diyana.ahmad.fazil@gmail.com

Aizat Ghani
Institute of Tropical Forestry and Forest Product
Universiti Putra Malaysia
43400 UPM Serdang, Selangor
Malaysia
and
Faculty of Tropical Forestry, University
Malaysia Sabah
Jalan UMS
88400 Kota Kinabalu, Sabah
Malaysia
E-mail: aizatabdghani@gmail.com;
muhammad.aizat@upm.edu.my

M. S. Gilbert
Faculty of Tropical Forestry, University
Malaysia Sabah
Jalan UMS
88400 Kota Kinabalu, Sabah
Malaysia

Aisyah Humaira
Department of Mechanical and Manufacturing
Engineering
Faculty of Engineering
Universiti Putra Malaysia
43400 Serdang, Selangor
Malaysia
and
Universiti Pertahanan Nasional Malaysia
Sungai Besi
Malaysia

Muhammad Huzaifah
Department of Crop Science
Universiti Putra Malaysia
Serdang 43400
Malaysia

Che Nor Aiza Jaafar
Advanced Engineering Materials and Composites
Research Centre (AEMC)
Department of Mechanical and Manufacturing
Engineering
Universiti Putra Malaysia
43400 UPM
Serdang, Selangor, Malaysia

Mohd Azrul Jaafar
School of Mechanical Engineering
College of Engineering
Universiti Teknologi MARA
40450, Shah Alam, Selangor
Malaysia

Aidah Jumahat
School of Mechanical Engineering
College of Engineering
Universiti Teknologi MARA (UiTM)
40450, Shah Alam, Selangor
Malaysia
E-mail: aidahjumahat@uitm.edu.my

Ridhwan Jumaidin
Fakulti Teknologi Kejuruteraan Mekanikal dan
Pembuatan
Universiti Teknikal Malaysia Melaka
Hang Tuah Jaya
76100 Durian Tunggal, Melaka
Malaysia
and
Institute of Tropical Forestry and Forest Products
Universiti Putra Malaysia
Serdang 43400
Malaysia
E-mail: ridhwan@utem.edu.my

Mohamad Haafiz Mohamad Kassim
School of Industrial Technology
Universiti Sains Malaysia
Gelugor
11800 Penang
Malaysia

Ching Hao Lee
Laboratory of Biocomposite Technology
Institute of Tropical Forestry and Forest Products
(INTROP)
Universiti Putra Malaysia
43400 Serdang, Selangor
Malaysia
and
Department of Mechanical Engineering
School of Computer Science and Engineering
Taylor's University
Subang Jaya
Malaysia
leechinghao@upm.edu.my

Chuan Li Lee
Institute of Tropical Forestry and Forest Product
Universiti Putra Malaysia
43400 UPM Serdang, Selangor
Malaysia

Seng Hua Lee
Laboratory of Biocomposite Technology
Institute of Tropical Forestry and Forest Products
(INTROPS)
Universiti Putra Malaysia
Serdang, Selangor
Malaysia
and
Institute of Tropical Forestry and Forest Product
Universiti Putra Malaysia
43400 UPM Serdang, Selangor
Malaysia

Raymond Siew Teng Loy
Carbon Tech Global Sdn Bhd
PT-1361
Jalan Kesidang 5
Kampung Mohd Taib Kawasan Industri
Kampung Sungai Choh,
48000, Rawang, Selangor
Malaysia

Nurul Ain Maidin
Department of Mechanical and Manufacturing
Faculty of Engineering
Universiti Putra Malaysia
43400 UPM Serdang, Selangor
Malaysia
and
Department of Manufacturing Engineering
Technology
Faculty of Mechanical and Manufacturing
Engineering Technology
Universiti Teknikal Malaysia Melaka
76100 Durian Tunggal, Melaka
Malaysia

Norkhairunnisa Mazlan
Department of Aerospace Engineering
Universiti Putra Malaysia
Serdang 43400
Malaysia

Siti Nurul Ain Md. Jamil
Department of Chemistry
Faculty of Science
Universiti Putra Malaysia
UPM Serdang 43400, Selangor
Malaysia
and
Centre of Foundation Studies for Agricultural
Science
Universiti Putra Malaysia
UPM Serdang 43400, Selangor
Malaysia
E-mail: ctnurulain@upm.edu.my

Muhammad Aizuddin Mohamad
School of Mechanical Engineering
College of Engineering
Universiti Teknologi MARA (UiTM)
40450, Shah Alam, Selangor
Malaysia

Mohamad Asrofi Muslim
Faculty of Mechanical Engineering
Universiti Teknologi MARA
40450, Shah Alam, Selangor
Malaysia

Asniza Mustapha
Forest Products Division
Forest Research Institute Malaysia
52109 Kepong, Selangor
Malaysia

Mohd Nurazzi Norizan
Laboratory of Bio composite Technology
Institute of Tropical Forestry and Forest Products
(INTROP)
Universiti Putra Malaysia
43400 Serdang, Selangor
Malaysia

I. Palle
Faculty of Tropical Forestry
University Malaysia Sabah
Jalan UMS
88400 Kota Kinabalu, Sabah
Malaysia
E-mail: isspalle@ums.edu.my

Fathin Sakinah Mohd Radzi
Faculty of Ocean Engineering Technology and
Informatics
Universiti Malaysia Terengganu
21030 Kuala Nerus, Terengganu
Malaysia
E-mail: fathinsakinah96@gmail.com
https://orcid.org/0000-0002-5348-3101

Mohd Nazren Radzuan
Department of Biological and Agricultural
Engineering
Faculty of Engineering
Universiti Putra Malaysia
43400 Serdang, Selangor
Malaysia

Amirul Hazim Abdul Rahman
Fakulti Teknologi Kejuruteraan Mekanikal dan
Pembuatan
Universiti Teknikal Malaysia Melaka
Hang Tuah Jaya
76100 Durian Tunggal, Melaka
Malaysia

Nurul Fazita Mohammad Rawi
School of Industrial Technology
Universiti Sains Malaysia
Gelugor
11800 Penang
Malaysia
E-mail: fazita@usm.my

Muhammad Huzaifah Mohd Roslim
Department of Crop Science
Faculty of Agricultural Science and Forestry
Universiti Putra Malaysia Bintulu Campus
97008 Bintulu, Sarawak
Malaysia

Ahmad Ilyas Rushdan
Sustainable Waste Management Research Group
(SWAM)
School of Chemical and Energy Engineering
Faculty of Engineering
Universiti Teknologi Malaysia
81310 UTM Johor Bahru, Johor
Malaysia
and
Centre for Advanced Composite Materials (CACM)
Universiti Teknologi Malaysia
81310 UTM Johor Bahru, Johor
Malaysia

Syeed SaifulAzry
Institute of Tropical Forestry and Forest Product
Universiti Putra Malaysia
43400 UPM Serdang, Selangor
Malaysia

Malaysia Mohd Sapuan Sali
Department of Mechanical Engineering
Universiti Putra Malaysia
43400, Selangor, Serdang
Malaysia

Napisah Sapiai
School of Mechanical Engineering
College of Engineering
Universiti Teknologi MARA
40450, Shah Alam, Selangor
Malaysia

Salit Mohd Sapuan
Laboratory of Biocomposite Technology
Institute of Tropical Forestry and Forest Products
Universiti Putra Malaysia
43400 UPM Serdang, Selangor
Malaysia
and
Department of Mechanical and Manufacturing
Faculty of Engineering
Universiti Putra Malaysia
43400 UPM Serdang, Selangor
Malaysia
E-mail: sapuan@upm.edu.my

Ayu Rafiqah shafi
Laboratory of Biocomposite
Institute of Tropical Forestry and Forest Products
(INTROP)
Universiti Putra Malaysia
43400 Serdang, Selangor
Malaysia

Ahmad Adlie Shamsuri
Laboratory of Biocomposite Technology
Institute of Tropical Forestry and Forest Products
Universiti Putra Malaysia
UPM Serdang 43400, Selangor
Malaysia
E-mail: adlie@upm.edu.my

Mat Jusoh Suriani
Faculty of Ocean Engineering Technology and
Informatics
Universiti Malaysia Terengganu
21030 Kuala Nerus, Terengganu
Malaysia
and
Marine Materials Research Group
Faculty of Ocean Engineering Technology and
Informatics
Universiti Malaysia Terengganu
21030 Kuala Nerus, Terengganu
Malaysia
E-mail: surianimatjusoh@umt.edu.my

Siti Norasmah Surip
Faculty of Applied Sciences
Universiti Teknologi MARA (UiTM)
40450 Shah Alam, Selangor
Malaysia

Y. Alia Syahirah
Forest Products Division
Forest Research Institute Malaysia (FRIM)
Kepong, Selangor
Malaysia

Mastura Mohammad Taha
Department of Manufacturing Engineering
Technology
Faculty of Mechanical and Manufacturing,
Engineering Technology
Universiti Teknikal Malaysia Melaka
76100 Durian Tunggal, Melaka
Malaysia

Paridah Md Tahir
Institute of Tropical Forestry and Forest Product
(INTROP)
Universiti Putra Malaysia
Serdang, Selangor
Malaysia

Mohd Khairun Anwar Uyup
Forest Products Division
Forest Research Institute Malaysia
52109 Kepong, Selangor
Malaysia
E-mail: mkanwar@frim.gov.my
https://orcid.org/0000-0002-3787-9907

Ridwan Yahaya
Science and Technology Research Institute for
Defence (STRIDE)
Kajang, Selangor
Malaysia

A. A. Mohd Yunus
Faculty of Tropical Forestry, University
Malaysia Sabah
Jalan UMS
88400 Kota Kinabalu, Sabah
Malaysia

Mohd Zuhri Mohamed Yusoff
Advanced Engineering Materials and Composites
Research Centre (AEMC)
Department of Mechanical and Manufacturing
Engineering
Universiti Putra Malaysia
43400 UPM Serdang, Selangor
Malaysia
and
Laboratory of Biocomposite Technology
Institute of Tropical Forestry and Forest Products
(INTROP)
Universiti Putra Malaysia
43400 UPM Serdang, Selangor
Malaysia

Alia Syahirah Yusoh
Forest Products Division
Forest Research Institute Malaysia
Kepong, Selangor
Malaysia
E-mail: aliasyahirah@frim.gov.my

Edi Syams Zainudin
Advanced Engineering Materials and Composites
Research Centre (AEMC)
Department of Mechanical and Manufacturing
Engineering
Universiti Putra Malaysia
43400 UPM Serdang, Selangor
Malaysia
and
Laboratory of Biocomposite Technology
Institute of Tropical Forestry and Forest Products
(INTROP)
Universiti Putra Malaysia
43400 UPM Serdang, Selangor
Malaysia

W. L. Zen
Faculty of Tropical Forestry, University
Malaysia Sabah
Jalan UMS
88400 Kota Kinabalu, Sabah
Malaysia

Ahmad Adlie Shamsuri*, Khalina Abdan and
Siti Nurul Ain Md. Jamil

1 Polybutylene succinate (PBS)/natural fiber green composites: melt blending processes and tensile properties

Abstract: An increase in the environmental consciousness at present has enhanced the awareness of researchers in utilizing biodegradable materials for the production of environmentally friendly products. Currently, biodegradable polymers, for example, polylactic acid, polybutylene succinate, polycaprolactone, etc., can be utilized as matrices to produce green composites. Meanwhile, natural fibers have been used as fillers for green composites as they are biodegradable and renewable. In this brief review, the physicochemical properties of selected biodegradable polymer, specifically polybutylene succinate, are demonstrated. Moreover, examples of natural fibers that are usually used to produce green composites are also shown. Additionally, practical methods employed for the preparation of green composites were exposed. The tensile properties of green composites, such as the tensile strength, tensile modulus, and elongation at break at different loadings of natural fibers, are also briefly reviewed. The information obtained in this review provides detailed differences in the preparation methods of green composites. In addition, this brief review supplies a clearer comprehension of the tensile properties of green composites for the usage of semistructural and packaging applications.

Keywords: biocomposite; bioplastic; green composite; natural fiber; polybutylene succinate.

1.1 Introduction

It is well known that there are three types of polymer composites. The first is a typical polymer composite, the second is a polymer biocomposite, and the third is a green

---

*Corresponding author: Ahmad Adlie Shamsuri, Laboratory of Biocomposite Technology, Institute of Tropical Forestry and Forest Products, Universiti Putra Malaysia, UPM Serdang 43400, Selangor, Malaysia, E-mail: adlie@upm.edu.my
Khalina Abdan, Laboratory of Biocomposite Technology, Institute of Tropical Forestry and Forest Products, Universiti Putra Malaysia, UPM Serdang 43400, Selangor, Malaysia, E-mail: khalina@upm.edu.my
Siti Nurul Ain Md. Jamil, Department of Chemistry, Faculty of Science, Universiti Putra Malaysia, UPM Serdang 43400, Selangor, Malaysia; and Centre of Foundation Studies for Agricultural Science, Universiti Putra Malaysia, UPM Serdang 43400, Selangor, Malaysia, E-mail: ctnurulain@upm.edu.my

As per De Gruyter's policy this article has previously been published in the journal Physical Sciences Reviews. Please cite as: A. A. Shamsuri, K. Abdan and S. N. A. Md. Jamil "Polybutylene succinate (PBS)/natural fiber green composites: melt blending processes and tensile properties" *Physical Sciences Reviews* [Online] 2022. DOI: 10.1515/psr-2022-0072 | https://doi.org/10.1515/9783110769227-001

composite. Typical polymer composites are made of synthetic polymers, such as polyethylene (PE), polypropylene (PP), polystyrene (PS), polyvinyl chloride (PVC), polyamide (PA), and so forth. The polymers have acted as matrices for the polymer composites. Another component in polymer composites is fillers, in which synthetic materials, for instance, glass, aramid, carbon, etc., are incorporated into the polymer matrix. Besides that, polymer biocomposites can be made from synthetic polymers, as mentioned above; however, the fillers are from natural resources, such as natural fibers [1, 2]. In addition, green composites are made of natural or synthetic biodegradable polymers, for example, polylactic acid, polybutylene succinate, polycaprolactone, etc., and natural fibers are used as fillers. Figure 1.1 shows the components of polymer composites, as stated before. On the other hand, typical polymer composites are not biodegradable compared to polymer biocomposites. Polymers with carbon backbones in their structure, such as vinyl polymers require an oxidation process for biodegradation, and hydrolysis cannot occur [3]. Nevertheless, green composites are easily biodegradable because they are comprised of biodegradable matrices and fillers. Therefore, green composites are suitable for the production of environmentally friendly products.

Recently, the use of synthetic biodegradable polymers in the preparation of green composites has been extended due to the green consciousness and growth of the many synthetic biodegradable polymer industries. The utilization of synthetic biodegradable polymer like polybutylene succinate (PBS) for the preparation of green composites is a promising way because it can be processed with conventional processing machines and its mechanical properties are comparable to PP [4]. Table 1.1 indicates the physicochemical properties of PBS. PBS is a type of thermoplastic produced through melt condensation polymerization. PBS is a polyester which contains the ester functional group in every repeat unit of the main chain. PBS also has an opaque color like other amorphous thermoplastics. The melting point of PBS is moderate, around 115 °C like low-density polyethylene. Furthermore, PBS can be dissolved in a nonpolar solvent, such as chloroform. At the moment, PBS can be relatively expensive compared to polylactic acid because it is synthesized from petrochemical sources. Nonetheless, the main advantages of PBS are its excellent biodegradability and high toughness suitable for producing green composites in semistructural and packaging applications. Figure 1.2 demonstrates the chemical structure of PBS.

In order to decrease the production cost of biodegradable products made from PBS, natural fibers can be utilized as fillers for the preparation of green composites [5, 6]. Natural fibers are fibers obtained from plants or animals, whereas synthetic fibers are completely man-made. There are so many types of natural fibers in the world. Figure 1.3 exhibits the examples of natural fibers, namely kenaf, bamboo, wood, cotton, miscanthus, and hemp. Table 1.2 displays the comparison between natural fiber and synthetic fiber. It can be observed that natural fibers have a lower density than synthetic fibers [7]. However, natural fibers have moderate mechanical properties compared to synthetic fibers. This is because they have different microstructures and biochemical

(a) Typical polymer composite = Synthetic polymer + Synthetic filler

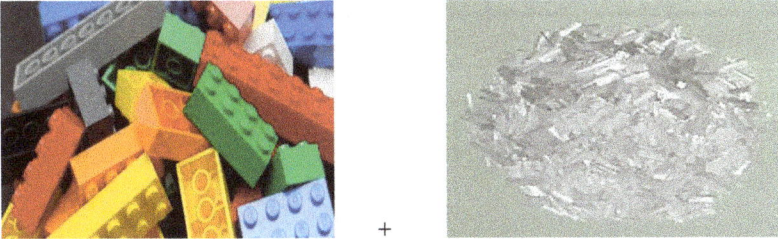

 +

(b) Polymer biocomposite = Synthetic polymer + Natural fiber

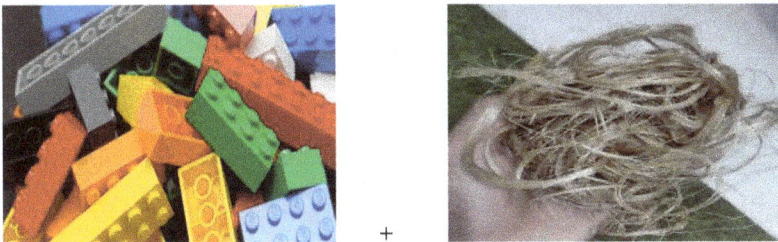

 +

(c) Green composite = Biodegradable polymer + Natural fiber

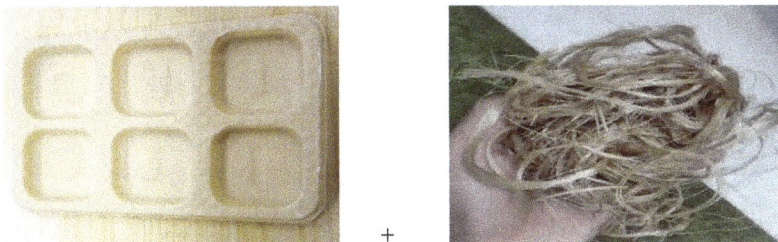

 +

Figure 1.1: Components of polymer composites (a) typical polymer composite, (b) polymer biocomposite, and (c) green composite.

compositions [8]. In addition, natural fibers have low thermal stability; hence, they only require moderate processing temperatures to produce green composites [9]. On the other hand, the best thing about natural fibers is that they have infinite resources in comparison to the synthetic fibers [10, 11]. They can also be completely biodegradable, which can conserve the environment. However, natural fibers cannot be recycled as they are comparatively less durable than synthetic fibers, but they are recognized as renewable and sustainable materials. Another exceptional element about natural

Table 1.1: Physicochemical properties of polybutylene succinate.

Physicochemical properties	Polybutylene succinate
Type	Thermoplastic
Family	Polyester
Appearance	Opaque
Melting point	115 °C
Solubility	Chloroform
Physical	Tough
Source	Petrochemical
Cost	High
Advantage	Biodegradable

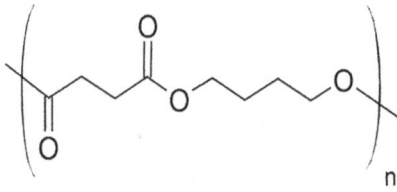

Figure 1.2: Chemical structure of polybutylene succinate.

Figure 1.3: Examples of natural fibers (a) kenaf, (b) bamboo, (c) wood, (d) cotton, (e) miscanthus, and (f) hemp.

fibers is that they are cheaper than synthetic fibers, which can reduce the production costs of green composites.

In the last few years, various natural fibers have been used in the production of green composites with the aim of expanding the usage of natural fibers and protecting

Table 1.2: Comparison between natural fiber and synthetic fiber.

Characteristics	Natural Fiber	Synthetic Fiber
Density	Low	High
Mechanical properties	Moderate	High
Thermal stability	Low	High
Resources	Infinite	Limited
Biodegradability	Yes	No
Recyclability	No	Yes
Cost	Low	High

the environment. As previously mentioned, the employment of natural fibers as primary fillers provides an advantage. The incorporation of natural fibers can reduce the utilization of PBS for green composites and decrease their density. This leads to an increase in using natural fibers and subsequently decreases the price of biodegradable products [7, 12], in addition to giving several choices of natural fiber resources. On top of that, to the best knowledge of the authors, no brief review has been made concentrating on the melt blending processes and tensile properties of PBS/natural fiber green composites. That is the purpose of making an organized review in this paper. Furthermore, this brief review is limited, and although not comprehensive, it is related to other correlated studies.

1.2 Preparations of green composites

There are two stages for the preparation of green composites: the first stage involves melt blending, and the second is composite molding. These stages are similar to the preparation of typical polymer composites. Nevertheless, as stated earlier, natural fibers have low thermal stability; therefore, moderate processing temperatures are required to avoid the thermal degradation of natural fibers during the preparation of green composites. Moreover, the usage of appropriate methods or techniques for preparing green composites is an important matter that must be taken into account.

1.2.1 Melt blending

Green composites can be prepared via solution blending and melt blending methods. Nonetheless, melt blending is a practical option for industrial scale compared to solution blending. In addition, the solution blending method requires a large amount of solvent, and it is also limited to producing films [13] and fibers [14]. The melt blending method using extruders is an excellent way to produce green composites in the form of pellets for the molding process. Besides that, internal mixers can also be used to produce green composites. Nevertheless extruders have a higher production capacity than internal mixers [15]. There are two types of extruders that are generally used for the production of polymer composites, specifically twin-screw extruder and single-screw extruder. Before extruding the polymer matrix and natural fiber, they must be premixed using a high-speed mixer [4] or mixed manually to facilitate the preparation of well-mixed composites [16]. The PBS/natural fiber green composites can be prepared using a twin-screw extruder at a temperature of 140 °C with a speed of 100 rpm for 2 min [17]. The green composites can also be prepared using a single-screw extruder at 150 °C with 100 rpm for 2 min [18]. Figure 1.4 presents the schematic illustrations of twin-screw extruder and single-screw extruder.

Table 1.3 shows the comparison between twin-screw extruder and single-screw extruder. It can be seen that even though they have an almost similar mechanism for preparing green composites, however, they have different capabilities; whereby a twin-screw extruder has a better mixing capability than a single-screw extruder [19]. In addition, a twin-screw extruder has high energy efficiency with a self-cleaning function compared to the single-screw extruder [20]. Nonetheless, the price of a single-screw extruder is relatively lower than that of a twin-screw extruder; therefore, it has a lower maintenance cost compared to the twin-screw extruder. Furthermore, the benefit of using a twin-screw extruder for the preparation of green composites is that the composite components will stay within a short time within its barrel [21], which can prevent the thermal decomposition of natural fibers. Moreover, a twin-screw extruder has higher productivity than a single-screw extruder [22]. Thus, from the comparison, it can be perceived that the twin-screw extruder is the ideal machine for preparing green composites.

1.2.2 Composite molding

Green composites molding is an essential process for shaping them for many applications. There are two types of molding techniques for structural and semistructural products, in particular, injection molding and compression molding techniques [23]. Figure 1.5 exhibits

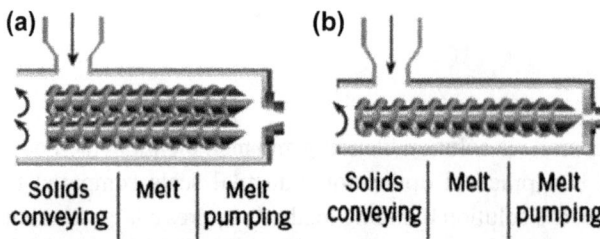

Figure 1.4: Schematic illustrations of (a) twin-screw extruder and (b) single-screw extruder.

Table 1.3: Comparison between twin-screw extruder and single-screw extruder.

Performance	Twin-Screw Extruder	Single-Screw Extruder
Mixing capability	Better	Poor
Energy efficiency	High	Low
Self-cleaning function	Yes	No
Price	High	Low
Maintenance	High	Low
Stay time	Short	Long
Productivity	High	Low

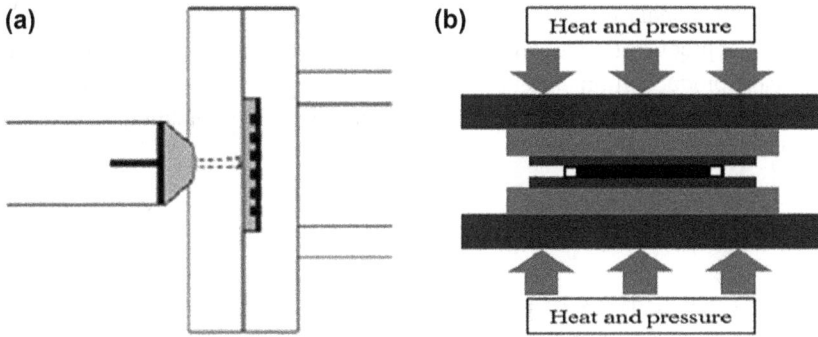

Figure 1.5: Schematic illustrations of (a) injection molding machine and (b) compression molding machine.

the schematic illustrations of injection molding machine and compression molding machine. The PBS/natural fiber green composites can be molded at a temperature of 145 °C with a pressure of 8 MPa using an injection molding machine [24], and the composites can also be molded by the same machine at 130 °C with 10 bar [17], and at 180 °C [6]. In addition, the green composites can be molded using a compression molding machine at 145 °C with 4 MPa for 6 min [25], whereas the composites can also be molded by the same machine at 150 °C with 10 MPa for 2 min [18, 26], and at 150 °C with 2000 kg for 5 min [27]. Table 1.4 indicates the comparison between injection molding machine and compression molding machine. The price of an injection molding machine is reasonably higher than that of a compression molding machine. Moreover, the initial set-up cost for an injection molding machine is high because it needs a special mold. Besides that, an injection molding machine can produce products with complex and small dimensions, whereas a compression molding machine can only produce large but simple products. However, no substantial wastage of green composites will be created by the compression molding machine during the molding process. Nevertheless, products molded by an injection molding machine are more consistent compared to the products molded by a compression molding machine; this is due to the orientation of composite for the injection molding

Table 1.4: Comparison between injection molding machine and compression molding machine.

Performance	Injection Molding Machine	Compression Molding Machine
Price	Expensive	Cheap
Set-up cost	High	Low
Product dimension	Complex and small	Simple and large
Composite wastage	Yes	No
Product consistency	More	Less
Production quantity	High	Low
Production time	Short	Long

machine is unidirectional [28]. Additionally, an injection molding machine can produce a higher quantity of products in comparison to the compression molding machine because it requires a short production time. Hence, from the comparison, it can be found that an injection molding machine is a good machine for green composites molding.

1.3 Tensile properties of green composites

1.3.1 Tensile properties of green composites at low natural fiber loading

Table 1.5 displays the tensile properties of PBS/natural fiber green composites at low natural fiber loading. Few preceding studies [29–31] have demonstrated that 30 wt% is still considered a low natural fiber loading because the PBS matrix can be incorporated up to 50 wt% of natural fibers. The polybutylene succinate/kenaf fiber (PBS/KF) green composites were prepared by Liang et al. using a single-screw extruder, and the composites were molded by a compression molding machine [18]. The tensile properties of the composites have been characterized using an EZ test machine. The tensile strength of the composite with a composition of 70/30 (wt/wt) decreased by 44% compared to the neat PBS. This was due to the insufficient adhesion between KF and PBS, which caused the weak interfacial interaction and debonding of the PBS matrix from the KF during the tensile deformation. However, the tensile modulus of the

Table 1.5: Tensile properties of PBS/natural fiber green composites at low natural fiber loading.

Sample	Composition	TS	TM	EAB	References
PBS/KF	70/30	↓	↑	–	[18]
PBS/CBS	70/30	↕	↑	↓	[5]
PBS/CF	70/30	↑	–	↓	[27]

KF, kenaf fiber; CBS, coffee beans skin; CF, cotton fiber; TS, tensile strength; TM, tensile modulus; EAB, elongation at break. The symbol '↓' corresponds to a decrease in the properties, and '↑' corresponds to an increase in the properties, while '↕' describes unchanged.

composite increased by up to 53% in comparison to the neat PBS. This was attributed to the stress transferred from the PBS matrix to the stiffer KF [18]. Therefore, the study indicated that the incorporation of the KF filler in the PBS matrix has provided economic green composites with a high tensile modulus and low tensile strength, suitable for fruit packaging application.

On the other hand, the polybutylene succinate/coffee beans skin (PBS/CBS) green composites were prepared by Totaro et al. using a batch mixer, and the composites were molded to a dog-bone shape [5]. The tensile properties of the composites have been characterized using an Instron 5966 series apparatus. The tensile strength of the composite with a composition of 70/30 (wt/wt) remains unchanged, which is similar to the neat PBS. Furthermore, the tensile modulus of the composite significantly increased by up to 136% compared to the neat PBS. This was ascribed to the good dispersion and wettability of the CBS filler in the PBS matrix. Nevertheless, the elongation at break of the composite decreased by 57% in comparison to the neat PBS. This phenomenon is normal for thermoplastic matrices incorporated with natural fibers [5]. Thus, the study suggested that the incorporation of the CBS filler in the PBS matrix has given green composites with a high tensile modulus and low elongation at break, suitable for packaging applications.

Besides that, the polybutylene succinate/cotton fiber (PBS/CF) green composites were prepared by Calabia et al. using a compression molding machine [27]. The tensile properties of the composites have been characterized using a Shimadzu Autograph AG-1000B. The tensile strength of the composite with a composition of 70/30 (wt/wt) increased by up to 73% compared to the neat PBS. This revealed the effectiveness of CF as a filler for PBS. Nonetheless, the elongation at break of the composite decreased by 56% in comparison to the neat PBS. This was because the presence of CF decreased mobility and enhanced the brittleness of the composite [27]. Hence, the study demonstrated that the incorporation of the CF filler in the PBS matrix has granted green composites with a high tensile strength and low elongation at break, suitable for food packaging application.

1.3.2 Tensile properties of green composites at high natural fiber loading

Table 1.6 exhibits the tensile properties of PBS/natural fiber green composites at high natural fiber loading. Previous studies [12, 32, 33] have shown that 50 wt% of natural fibers is the maximum percentage typically incorporated into the PBS matrix; therefore, it is considered a high natural fiber loading. The polybutylene succinate/miscanthus fiber (PBS/MF) green composites were prepared by Muthuraj et al. using a twin-screw extruder, and the composites were molded by an injection molding machine [17]. The tensile properties of the composites have been characterized using an Instron 3382 machine. The tensile strength of the composite with a composition of 50/50 (wt/wt)

Table 1.6: Tensile properties of PBS/natural fiber green composites at high natural fiber loading.

Sample	Composition	TS	TM	EAB	References
PBS/MF	50/50	↓	↑	↓	[17]
PBS/FF	50/50	↑	↑	–	[6]
PBS/RF	50/50	↑	–	↑	[26]

MF, miscanthus fiber; FF, flax fiber; RF, ramie fiber; TS, tensile strength; TM, tensile modulus; EAB, elongation at break. The symbol '↓' corresponds to a decrease in the properties, and '↑' corresponds to an increase in the properties.

decreased by 22% compared to the neat PBS. This was caused by the deficiency of interfacial interaction between the MF filler and the PBS matrix, as well as incompatibility between the components. However, the tensile modulus of the composite significantly increased by up to 488% in comparison to the neat PBS. This was due to the tensile modulus of the composites relying on the MF content, which was increased along with the increasing MF content. In contrast, the elongation at break of the composite was considerably lower than that of the neat PBS. This was because of the phase separation phenomenon and the decreased PBS chain entanglement in the presence of rigid MF [17]. Therefore, the study showed that the incorporation of the MF filler in the PBS matrix has provided green composites with a high tensile modulus and low elongation at break, suitable for biomedical applications.

On top of that, the polybutylene succinate/flax fiber (PBS/FF) green composites were prepared by Frackowiak et al. using an internal mixer, and the composites were molded by an injection molding machine [6]. The tensile properties of the composites have been characterized using a Lloyd LR10K test machine. The tensile strength of the composite with a composition of 50/50 (wt/wt) increased by up to 30% compared to the neat PBS. This implied that FF acted as effective filler for PBS. Additionally, the tensile modulus of the composite significantly increased by up to 343% in comparison to the neat PBS. This was attributed to an increase in the stiffness of the composite, which allowed for transferring stress from the PBS matrix to the FF filler and consequently contributed to the performance of the composite [6]. Thus, the study exhibited that the incorporation of the FF filler in the PBS matrix has given green composites with a high tensile strength and high tensile modulus, suitable for 3D printing application.

Meanwhile, the polybutylene succinate/ramie fiber (PBS/RF) green composites were prepared by Han et al. using a compression molding machine [26]. The tensile properties of the composites have been characterized using an Instron 5967 machine. The tensile strength of the composite with a composition of 50/50 (wt/wt) significantly increased by up to 216% compared to the neat PBS. This was ascribed to the reinforcement effect of RF fabric, which has excellent mechanical properties [26]. Moreover, the elongation at break of the composite slightly increased by up to 11% in comparison to the neat PBS. In most studies [34], it has described that when the tensile strength increases, the elongation will drastically decrease. This is caused by an increase in the strength of the composites. Nevertheless, the study showed an opposite trend; this is probably due to the employment of RF was in the form of fabric, which gave the pliability to the composite [35]. Hence, the study displayed that the incorporation of the RF filler in the PBS matrix has granted green composites with a high tensile strength and high elongation at break, suitable for household and car-interior applications.

1.4 Challenges and future recommendations

The PBS/natural fiber green composites have been developed as green materials for producing environmentally friendly products and have received much attention due to their biodegradable characteristic. However, several challenges seemed, such as the incompatibility and marine degradability of green composites that hindered the commercialization of some promising green products created in academic circles for a variety of applications. Future recommendations should focus on the enhancement of the interfacial adhesion between the PBS matrix and natural fibers, as well as on the assessment of the effect of natural fibers on the marine degradability of PBS in the green composites.

1.5 Conclusions

This paper briefly reviewed the types of polymer composites, physicochemical properties of PBS, examples of natural fibers, and types of machines for preparing and molding green composites. The significant tensile properties, such as the tensile strength, tensile modulus, and elongation at break of green composites, were also expressed in this brief review. The usage of natural fibers can reduce the utilization of PBS and lessen the cost of green composites. On top of that, the preparation of green composites using a twin-screw extruder is deemed to be the best option, as it has excellent mixing capability, high energy efficiency, short stay time, and high productivity. Besides that, green composites molding using an injection molding machine is a great way because it can produce products with complex dimensions, more consistent, high quantity, and short production time. On the other hand, the use of natural fibers at low loading can efficiently increase the tensile modulus of green composites. Additionally, the use of natural fibers at high loading can effectively improve the tensile strength and tensile modulus of green composites. This brief review may be valuable for commercializing green composites for semistructural and packaging applications.

References

1. Sanmuham V, Sultan MTH, Radzi AM, Shamsuri AA, Md Shah AU, Safri SNA, et al. Effect of silver nanopowder on mechanical, thermal and antimicrobial properties of kenaf/HDPE composites. Polymers 2021;13:3928.
2. Darus SAAZM, Ghazali MJ, Azhari CH, Zulkifli R, Shamsuri AA. Mechanical properties of gigantochloa scortechinii bamboo particle reinforced semirigid polyvinyl chloride composites. J Teknol 2020;82:15–22.
3. Vroman I, Tighzert L. Biodegradable polymers. Materials (Basel) 2009;2:307–44.
4. Shamsuri AA, Md. Jamil SNA. Compatibilization effect of ionic liquid-based surfactants on physicochemical properties of PBS/rice starch blends: an initial study. Materials 2020;13:1885.
5. Totaro G, Sisti L, Fiorini M, Lancellotti I, Andreola FN, Saccani A. Formulation of green particulate composites from PLA and PBS matrix and wastes deriving from the coffee production. J Polym Environ 2019;27:1488–96.
6. Frackowiak S, Ludwiczak J, Leluk K. Man-made and natural fibres as a reinforcement in fully biodegradable polymer composites: a concise study. J Polym Environ 2018;26:4360–8.
7. Kumar KP, Sekaran ASJ. Some natural fibers used in polymer composites and their extraction processes: a review. J Reinforc Plast Compos 2014;33:1879–92.
8. Le A, Correa D, Ueda M, Matsuzaki R, Castro M. A review of 3D and 4D printing of natural fi bre biocomposites. Mater Des 2020;194:108911.
9. Bourmaud A, Corre YM, Baley C. Fully biodegradable composites: use of poly-(butylene-succinate) as a matrix and to plasticize l-poly-(lactide)-flax blends. Ind Crop Prod 2015;64:251–7.
10. Darus SAAZM, Ghazali MJ, Azhari CH, Zulkifli R, Shamsuri AA, Sarac H, et al. Physicochemical and thermal properties of lignocellulosic fiber from gigantochloa scortechinii bamboo: effect of steam explosion treatment. Fibers Polym 2020;21:2186–94.
11. Sudari AK, Shamsuri AA, Zainudin ES, Tahir PM. Exploration on compatibilizing effect of nonionic, anionic, and cationic surfactants on mechanical, morphological, and chemical properties of high-density polyethylene/low-density polyethylene/cellulose biocomposites. J Thermoplast Compos Mater 2017;30:855–84.
12. Yan-hong F, Yi-jie L, Bai-ping X, Da-wei Z, Jin-ping Q, He-zhi H. Effect of fiber morphology on rheological properties of plant fiber reinforced poly (butylene succinate) composites. Compos B 2013;44:193–9.
13. Huang J, Cui C, Yan G, Huang J, Zhang M. A study on degradation of composite material PBS/PCL. Polym Polym Compos 2016;24:143–8.
14. Abudula T, Saeed U, Memic A, Gauthaman K, Hussain MA, Al-Turaif H. Electrospun cellulose Nano fibril reinforced PLA/PBS composite scaffold for vascular tissue engineering. J Polym Res 2019;26: 110–24.
15. Alavi SAR, Angaji MT, Gholami Z. Twin-screw extruder and effective parameters on the HDPE extrusion process. World Acad Sci Eng Technol 2009;25:204–7.
16. Nicharat A, Sapkota J, Foster EJ. Pre-mixing and masterbatch approaches for reinforcing poly(vinyl acetate) with cellulose based fillers. Ind Crop Prod 2016;93:244–50.
17. Muthuraj R, Misra M, Mohanty AK. Injection molded sustainable biocomposites from poly(butylene succinate) bioplastic and perennial grass. ACS Sustainable Chem Eng 2015;3:2767–76.
18. Liang Z, Pan P, Zhu B, Dong T, Inoue Y. Mechanical and thermal properties of poly(butylene succinate)/plant fiber biodegradable composite. J Appl Polym Sci 2010;115:3559–67.
19. Martin C. Twin screw extruders as continuous mixers for thermal processing: a technical and historical perspective. AAPS Pharm Sci Tech 2016;17:3–19.

20. Eitzlmayr A, Matić J, Khinast J. Analysis of flow and mixing in screw elements of corotating twin-screw extruders via SPH. AIChE J 2017;63:2451–63.
21. Wu H, Ning N, Zhang L, Tian H, Wu Y, Tian M. Effect of additives on the morphology evolution of EPDM/PP TPVs during dynamic vulcanization in a twin-screw extruder. J Polym Res 2013;20: 266–73.
22. Sakai T. Screw extrusion technology - past, present and future. Polimery/Polymers 2013;58: 847–57.
23. Bledzki AK, Faruk O. Wood fiber reinforced polypropylene composites: compression and injection molding process. Polym Plast Technol Eng 2004;43:871–88.
24. Kim HS, Lee BH, Kim HJ, Yang HS. Mechanical-thermal properties and VOC emissions of natural-flour-filled biodegradable polymer hybrid bio-composites. J Polym Environ 2011;19:628–36.
25. Hong G, Meng Y, Yang Z, Cheng H, Zhang S, Song W. Mussel-inspired polydopamine modification of bamboo fiber and its effect on the properties of bamboo fiber/polybutylene succinate composites. Bioresources 2017;12:8419–42.
26. Han Q, Zhao L, Lin P, Zhu Z, Nie K, Yang F, et al. Poly(butylene succinate) biocomposite modified by amino functionalized ramie fiber fabric towards exceptional mechanical performance and biodegradability. React Funct Polym 2020;146:104443.
27. Calabia BP, Ninomiya F, Yagi H, Oishi A, Taguchi K, Kunioka M, et al. Biodegradable poly(butylene succinate) composites reinforced by cotton fiber with silane coupling agent. Polymers 2013;5: 128–41.
28. Shamsuri AA. Is the polymer orientation can influence the mechanical properties of thermoplastic composites? Glob J Mater Sci Eng 2019;1:1–2.
29. Feng Y, Zhang D, Qu J, He H, Xu B. Rheological properties of sisal fi ber/poly (butylene succinate). Polym Test 2011;30:124–30.
30. Tan B, Qu J, Liu L, Feng Y, Hu S, Yin X. Non-isothermal crystallization kinetics and dynamic mechanical thermal properties of poly (butylene succinate) composites reinforced with cotton stalk bast fibers. Thermochim Acta 2011;525:141–9.
31. Gowman A, Wang T, Rodriguez-Uribe A, Mohanty AK, Misra M. Bio-poly(butylene succinate) and its composites with grape pomace: mechanical performance and thermal properties. ACS Omega 2018;3:15205–16.
32. Liminana P, Quiles-carrillo L, Boronat T, Balart R, Montanes N. The effect of varying almond shell flour (ASF) loading in composites with poly(butylene succinate) (PBS) matrix compatibilized with maleinized linseed oil (MLO). Materials 2018;11:2179.
33. Ogihara S, Okada A, Kobayashi S. Mechanical properties in a bamboo fiber/PBS biodegradable composite. J Solid Mech Mater Eng 2008;2:291–9.
34. Shamsuri AA, Jamil SNAM. Application of quaternary ammonium compounds as compatibilizers for polymer blends and polymer composites — a concise review. Appl Sci 2021;11:3167.
35. Shamsuri AA, Azid MKA, Ariff AHM, Sudari AK. Influence of surface treatment on tensile properties of low-density polyethylene/cellulose woven biocomposites: a preliminary study. Polymers 2014; 6:2345–56.

Nurul Ain Maidin, Salit Mohd Sapuan*, Mastura Mohammad Taha
and Mohd Yusoff Mohamed Zuhri

2 Material selection and conceptual design in natural fibre composites

Abstract: Material selection is the process of determining which material is best
suited to meet the needs of a specific application. Mechanical characteristics, chemical
properties, physical properties, electrical properties, and cost are all aspects that define
the selection requirements. During the material selection process, these must be
weighed. Materials selection is a process used by design engineers to choose the best
materials for a specific component. To find the best composite materials, a materials
selection system is used to find candidate materials from various composite materials
that meet all of the material selection criteria, such as strength, stiffness, cost, and
aesthetics. Similarly, a materials selection system will require candidate materials to
contain several forms of NFC to identify the best appropriate NFC for a specific product.
Materials selection for NFC goods is a relatively recent field of study. Because of the vast
number of individual constituent materials in NFC, the work of selecting the best NFC
for a specific product is regarded as challenging and time-intensive (Marques T, Esteves
JL, Viana J, Loureiro N, Arteiro A. Design for sustainability with composite systems. In:
15th international conference on experimental mechanics (ICEM15); 2012:1–2 pp).
Whereas conceptual design is a crucial activity in the design process in modern design,
it is continually highlighted that improper conceptual design can lead to extensive
rework and problems after the product is produced. According to (Pugh S. Total design:
integrated methods for successful product engineering. Wokingham, England:
Addison-Wesley Publishing; 1991), conceptual design activity is creating and assessing
design solutions to meet the PDS. Because many design features are distinct in composites, and the tailor-made nature of composites has caused the design approach to be

*Corresponding author: Salit Mohd Sapuan, Laboratory of Biocomposite Technology, Institute of
Tropical Forestry and Forest Products, Universiti Putra Malaysia, 43400 UPM Serdang, Selangor,
Malaysia; and Department of Mechanical and Manufacturing, Faculty of Engineering, Universiti Putra
Malaysia, 43400 UPM Serdang, Selangor, Malaysia, E-mail: sapuan@upm.edu.my
Nurul Ain Maidin, Department of Mechanical and Manufacturing, Faculty of Engineering, Universiti
Putra Malaysia, 43400 UPM Serdang, Selangor, Malaysia; and Department of Manufacturing
Engineering Technology, Faculty of Mechanical and Manufacturing Engineering Technology, Universiti
Teknikal Malaysia Melaka, 76100 Durian Tunggal, Melaka, Malaysia
Mastura Mohammad Taha, Department of Manufacturing Engineering Technology, Faculty of
Mechanical and Manufacturing Engineering Technology, Universiti Teknikal Malaysia Melaka, 76100
Durian Tunggal, Melaka, Malaysia
Mohd Yusoff Mohamed Zuhri, Department of Mechanical and Manufacturing, Faculty of Engineering,
Universiti Putra Malaysia, 43400 UPM Serdang, Selangor, Malaysia

As per De Gruyter's policy this article has previously been published in the journal Physical Sciences Reviews. Please cite as:
N. A. Maidin, S. M. Sapuan, M. Mohammad Taha and M. Y. M. Zuhri "Material selection and conceptual design in natural
fibre composites" *Physical Sciences Reviews* [Online] 2022. DOI: 10.1515/psr-2022-0073 | https://doi.org/10.1515/
9783110769227-002

different, conceptual design with NFC is typically different from metals. Designing with NFC is similar to designing with a traditional composite product in terms of concept. This activity entails numerous processes, including creating a design brief, information collecting, market research, and product design specifications (PDS).

Keywords: conceptual design, design process, material selection, natural fibres, natural fibre composites.

2.1 Natural fibres composites

As science and technology progress, manufacturing industries change toward a more environmentally friendly economic output. Polymer composites made from natural fibres have aroused industrial players' and researchers' interest because they are "greener" and contribute to sustainable practice. Some businesses have turned to sustainable technology to strike a better balance between environmental protection and social and economic concerns [1]. Figure 2.1 shows that natural fibres can be categorised into animal, mineral, and plant fibres. Plant fibres are the most commonly accepted and carefully investigated for both industry and science. Short growing periods, renewability, and more excellent supply contribute to this [2]. It's possible to extract vegetable fibres from bast, leaf, seed, fruit, wood, stalk, and grass/reed. Vegetable fibres are made up of cellulose and hemicellulose.

Many advantages of natural fibres include their availability, low cost, low density, appropriate modulus-weight ratio, good acoustic damping and low manufacturing energy consumption, and a minimal carbon footprint [7]. According to some writers, these natural fibres are less expensive and use less energy to generate than standard reinforcing fibres like glass and carbon [8]. On the other hand, natural fibres have limitations because of the variability in their qualities and quality. In terms of physical and mechanical variability, these fibres are more prone to moisture absorption, less durable, less intense, and easier to process [9–13]. Different plant species, growth circumstances, and fibre extraction techniques are responsible for the wide range of qualities found. Each form of cellulose is affected by the fibre cell's shape and the

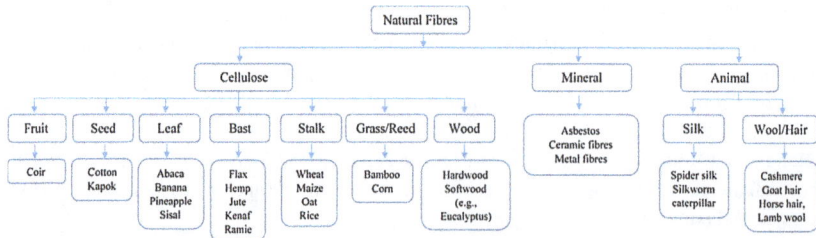

Figure 2.1: Natural fibres classification [3–6].

degree of polymerisation [14]. Hemicelluloses and lignin form intimate associations with linear cellulosic macromolecules linked by hydrogen bonds to give fibre its rigidity. In addition to holding fibres together, cellulose in the cell wall of the fibre is also held together by this substance.

Bio-based composites can be categorised as either environmentally friendly or environmentally friendly, depending on the ingredients used in their construction [15]. Carbon dioxide emissions can be reduced by using renewable resources in the composition of green composites, thus minimising the need for petroleum-based products. One of the fibres or components of the matrix is not generated from renewable resources, even though it is environmentally friendly in part [16, 17]. In that they absorb carbon dioxide and release oxygen, natural fibres are not only biodegradable but also ecologically benign. In comparison to synthetic fibres, they're a better value. Initially, these plant-based products were employed for residential reasons [18]. They are now being used in various industries because of their superior mechanical, physical, and chemical qualities [18].

Whether the polymers used in the composite are petroleum-based or biobased, natural fibre composites can be classed as biocomposites or complete green composites. The biodegradability of these biocomposites is dependent on the polymeric matrix utilised and comprises both natural fibre-based composites and polymer composites derived from petrol-based polymers. For example, PP-based nonbiodegradable composites and PBAT-based biodegradable composites. The biodegradability of whole green composites depends on the polymer matrix used, which is dependent on the natural fibres used in the composite and how they are incorporated into the matrix. Composites based on natural fibre and PLA, such as natural fibre-bio-nylon composites [19].

It's a promising material type, and it's currently the focus of a lot of research. Because of their ability to replace synthetic fibre-reinforced polymers at a reduced cost

Table 2.1: Advantages and disadvantages of NFCs [1, 9, 20–23].

Advantages	Disadvantages
– Fibres are ably produced at a lower cost than synthetic fibres	– Lower strength, particularly impact strength, when compared to synthetic fibre composites
– Fibres are a renewable resource, requiring little energy to produce and absorb CO2 while returning oxygen to the environment	– Lower durability than synthetic fibre composites, but can be significantly improved with treatment
– Low density combined with high specific strength and stiffness	– High moisture absorption, resulting in swelling
– Less abrasive damage to processing equipment than synthetic fibre composites	– Lower processing temperatures, limiting matrix options
– Low-risk manufacturing processes	– Greater variability of properties
– Low toxic fume emission when subjected to heat and during end-of-life incineration	

and improve environmental sustainability, NFC technology is becoming increasingly popular. Table 2.1 lists the positives and negatives of each.

2.2 Material selection

The importance of material selection in the design process has been well-documented for more than two decades [24]. Choosing the wrong materials can lead to the destruction or failure of an assembly and reduce its performance significantly. Multi-criteria decision making (MCDM) strategies have been popular in this industry because it has been shown that the least price may not be the most encouraging route to achieving the best material. An established set of features or attributes can be found in every differentiated substance. Designers are usually looking for a certain blend of these characteristics or a property profile for their products. That's why it's so important to know the unique property profile of a given substance. The physical qualities, such as density, thermal and electrical conductivity, modulus, toughness, strength, and so on, are all the same [25].

Theoretical and experimental properties of engineering materials are never the same [26–28]. It's common for theoretical properties to be more important than those tested in the labs. Inhomogeneity, internal fissures and imperfections, and manufacturing problems are some of the many causes of this disparity. As a result, the useful life of the mechanical component will be reduced. Component failure is the most difficult problem to avoid when it comes to designing. Among the most common causes of member failure include design defects, manufacturing flaws, poor material selection, components overload, and insufficient maintenance. Engineered parts need proper planning and design to avoid unplanned failures. On the other hand, engineers should take mechanical failure into account when developing a product because it has a considerable impact on the product's lifespan.

As a result, selecting appropriate materials and implementing a good design process will have a major impact on and improve service performance and engineering design. As shown in Figure 2.2, material qualities directly affect the end product's

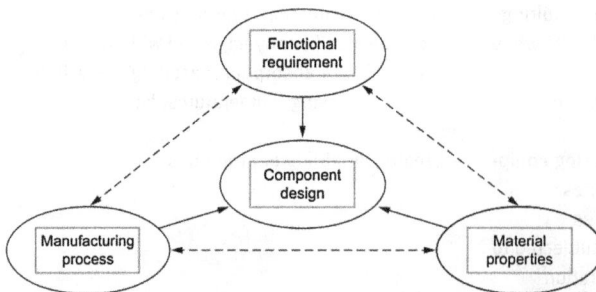

Figure 2.2: The integration of factors that affect the design of a component [28, 29].

behaviour and have a direct and indirect relationship to the shape of the part and the manufacturing process. The component's design is at the heart of three interconnected or cooperating aspects. The selection of materials is based on the interplay of the product's functional needs with the material's qualities and the manufacturing method. In other words, engineers should begin by determining the mechanical and environmental requirements. A product's success, competitiveness, and reliability are all dependent on the selection and design of the materials at the advanced level.

Natural resource use has been promoted due to the recent surge in environmental concern and public awareness, as well as government policies aimed at fostering long-term sustainability. These natural fibre-reinforced composites have developed as a viable alternative material in modern industries because of their versatility [30]. Polymer-based matrices in natural fibre composites use natural fibres as fillers or reinforcements (jute, sisal, hemp, oil palm, date palm, kenaf, rami, and flax). This would lessen the amount of trash disposal concerns and improve environmental pollution control. Traditional glass/carbon composites are being replaced with natural fibre composites because they are more environmentally benign. Disposable accessories, construction, packing, furniture, insulation, and the automobile industry [31] are just a few of the many uses for these products.

On the other hand, materials play an important role in engineering design and applications. As a result, the final product's attributes would be greatly influenced. As a result, engineers, designers, and materials scientists are often the ones responsible for the task of material selection. To select the best materials from a collection of probable candidates, evaluation criteria must be established after a design has been finished. Complex designs make material selection more difficult because there are many functional and economic considerations to weigh before deciding on a final choice. It is more efficient to produce effective, sustainable products if appropriate materials are picked utilising systematic procedures and techniques [32–34]. Modern companies place a high value on the compatibility of materials with their intended use, and the benefits that recycling and protecting the environment may provide. Furthermore, inventing new materials with desirable differentiating properties would substantially extend new design options.

Additionally, various factors and restrictions can impact the selection of a particular material for a given application. Because of this, they are deciding which material to use in a given application is a complex, unpredictable decision. Based on the findings of multiple studies with divergent conclusions [34–36]. Despite the wide selection of materials available to design engineers, they face a considerable problem in this situation. Therefore, using the best materials to provide the needed performance while also taking environmental concerns and sustainability into consideration is critical. As a result, engineering's material selection procedure has become an essential component [54].

2.3 Conceptual design

It is vital to note that faulty conceptual designs can lead to many rework and problems once the product has been created [37] in the current design. According to Pugh [38], conceptual design activity is an activity that develops design solutions and analyses them to meet the PDS requirements. Composites include various design aspects that aren't present in metals, which necessitates a different conceptual approach in NFC. Composites have a unique design approach because of their customizability. In many ways, working with NFC is like working with a typical composite material. To complete this task, you must first create a design brief, gather data and conduct market research before producing product design requirements (PDS). Pre-conceptual design is the term for this stage [39].

Sapuan et al. [40] used Pugh's approach to conceptualise NFC goods. They conducted a case study on designing an NFC laptop case with an external cooling fan as part of the research. For the most part, the Theory of Inventive Problem Solving was employed (TRIZ). Various design tools and methodologies have been used in contemporary NFC conceptual design work. Analysis Hierarchy Process (AHP), Function Analysis Diagram, Failure Modes and Effects Analysis (FMEA), TRIZ, and Morphological Chart are existing tools used in the conceptual design of natural fibre-reinforced composites.

Mastura [41] claims to have employed a variety of methodologies to theoretically create an NFC product, especially an anti-roll bar for automobiles. Using a business strategy known as blue ocean strategy (BOS) as a conceptual design tool for NFC was described by her. Her study focused on constructing an automotive anti-roll bar built of thermoplastic composites, including glass and natural fibres. She used BOS techniques such as a four-action framework and a strategy canvas (value curve) to explore several designs for vehicle anti-roll bars made of natural fibre-glass reinforced polymer hybrid composites.

However, Mansor et al. [42] described the conceptual design development of a green biocomposites product using a bicycle frame as a case study [42]. Designing a new bicycle frame using NFC instead of steel was done using the TRIZ method, a problem-solving technique. The TRIZ 40 engineering contradiction tool was employed to represent the challenge, followed by the contradiction matrix and 40 innovative principle tools. A novel design technique can alleviate the present NFC material restriction, which is low strength, through geometrical design solutions such as ribs and triangulation. Mansor and colleagues [43] also exhibit the integrated TRIZ and Morphological chart method in creating a new conceptual design of an automotive rear spoiler component utilising kenaf composites in another report. Kenaf composites had a structural strength constraint that necessitated the application of TRIZ techniques, including engineering contradiction, contradiction matrix, and 40 innovative principles. New concepts for kenaf fibre polymer composite vehicle parking brake lever components were generated using an integrated TRIZ-Morphological chart approach [44]. There was also a similar approach taken by Mahmood, Yusof and Shaharuzaman [45–47] in conducting NFC conceptual design studies.

Conceptual design for takeout food containers composed of natural fibre-reinforced biopolymer composites was described by Salwa et al. [48]. Takeout food containers were studied using the Kano model, and the results were included in QFDE. To produce conceptual designs, the morphological chart (MC) was used to refine the design attributes with the highest weight of consumer and environment voices (VOCE). AHP was utilised to choose the best idea design as a final step. The concept design that received the largest votes (83.3%) was selected as the final design.

2.3.1 Case study on conceptual design development using brainstorming method

Conceptual design is used in NFC product development to generate the product's first shape or form. Through conceptual design, customers can get a better sense of a product's final form. Several conceptual designs are often created based on the supplied product design parameters. The gallery approach, TRIZ, brainstorming, SWOT analysis, and organised exploitation of proven ideas from experience are only a few of the techniques designers have used to produce conceptual designs for goods, according to Sapuan et al. [49–52].

For conceptual design, brainstorming is one of the most fundamental strategies. By conducting brainstorming sessions with other people, an instructor can help guide participants through an organised and systematic process [53]. To get the most out of the brainstorming process, it's important to link the generated ideas to the project's goals, illustrate the ideas with sketches, pick out the most promising ones, develop them further, and make connections between the developed ideas (or concepts) to generate more innovative ones (or concepts).

The brainstorming method was used to develop design thoughts for this project [1]. Group discussions have taken place at the Advanced Engineering Materials and Composites Research Centre (AEMC), Universiti Putra Malaysia's Mechanical and Manufacturing Engineering Department. Each idea design was described and listed based on the initial design's discussion outputs and PDS documents. Roselle fibre-reinforced polymer composite smartphone holders were then designed in five different ways, as shown in Figure 2.3.

Therefore, it is clear that correct NFC product design and manufacturing techniques are needed to enhance the products' qualities and their materials to optimise strength and usefulness. Design procedures and techniques such as brainstorming, the voice of the customer (VOCs), and morphological charts are vital for optimising the power and effectiveness of natural fibre composite products. These methods could help users pinpoint their issues and narrow down the scope of a solution. As a result, the

~ Concept inspired by the bottle cap.
~ Composes of a main pad and two stand legs.
~ Has a good artistic design.

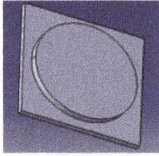

~ Comprises a bottom rectangle plate and circle shape pad on the top.
~ Simple and minimalist design.
~ High stability due to a wider baseline.

~ Concept inspired by a square coffee table.
~ Composed of four legs, one ring to hold the mug, and one square pad.
~ Complex shape with artistic value.

~ Simple and minimalist design.
~ Easy for storage.
~ Easy to fabricate and manufacture.
~ High stability
~ High strength due to wider baseline.
~ Concept inspired by a leaf.

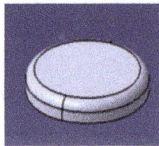

~ Simple and minimalist design.
~ Circular and isometric design, hence easy to fabricate.
~ Easy for storage.
~ Easy to fabricate and manufacture.
~ High stability

Figure 2.3: Proposed conceptual designs of a roselle fibre biocomposite mug pad [1, 52].

product's design and its intended use must be considered while determining the best manufacturing method.

References

1. Azman MA, Asyraf MRM, Khalina A, Petrů M, Ruzaidi CM, Sapuan SM, et al. Natural fiber reinforced composite material for product design: a short review. Polymers 2021;13:1–24.
2. Cicala G, Cristaldi G, Recca G, Latteri A. Composites based on natural fibre fabrics. In: Dubrovski P, editor. Woven fabric engineering. London, UK: InTech; 2010.
3. Sanjay MR, Arpitha GR, Yogesha B. Study on mechanical properties of natural—glass fibre reinforced polymer hybrid composites: a review. Mater Today Proc 2015;2:2959–67.
4. Bharath KN, Basavarajappa S. Applications of biocomposite materials based on natural fibers from renewable resources: a review. Sci Eng Compos Mater 2016;23:123–33.
5. Aditya PH, Kishore KS, Prasad DVVK. Characterization of natural fiber reinforced composites. Int J Eng Appl Sci 2017;4:26–32.

6. Peças P, Carvalho H, Salman H, Leite M. Natural fibre composites and their applications: a review. J Compos Sci 2018;2:1–20.
7. Mohanty A, Misra M, Drzal L, Selke S, Harte B, Hinrichsen G. Natural fibers, biopolymers, and biocomposites, Mohanty A, Misra M, Drzal L, editors. Boca Raton, FL, USA: CRC Press; 2005.
8. Huda MS, Drzal LT, Ray D, Mohanty AK, Mishra M. Natural-fiber composites in the automotive sector. In: Properties and performance of natural-fibre composites. Oxford, UK: Woodhead Publishing; 2008.
9. Pickering KL, Efendy MGA, Le TM. A review of recent developments in natural fibre composites and their mechanical performance. Compos Appl Sci Manuf 2016;83:98–112.
10. Rohan T, Tushar B, Mahesha GT. Review of natural fiber composites. IOP Conf Ser Mater Sci Eng 2018;314:012020.
11. Cheung H, Ho M, Lau K, Cardona F, Hui D. Natural fibre-reinforced composites for bioengineering and environmental engineering applications. Compos B Eng 2009;40:655–63.
12. Faruk O, Sain M. Biofiber reinforcements in composite materials. Amsterdam, Netherlands: Elsevier; 2014.
13. Carvalho H, Raposo A, Ribeiro I, Kaufmann J, Götze U, Peças P, et al. Application of life cycle engineering approach to assess the pertinence of using natural fibers in composites—the rocker case study. Procedia CIRP 2016;48:364–9.
14. Ho M, Wang H, Lee JH, Ho C, Lau K, Leng J, et al. Critical factors on manufacturing processes of natural fibre composites. Compos B Eng 2012;43:3549–62.
15. Chen H. Biotechnology of lignocellulose. Dordrecht, The Netherlands: Springer; 2014.
16. Mitra BC. Environment friendly composite materials. Defence Sci J 2014;64:244–61.
17. Dicker MPM, Duckworth PF, Baker AB, Francois G, Hazzard MK, Weaver PM. Green composites: a review of material attributes and complementary applications. Compos Part A Appl Sci Manuf 2014;56:2809.
18. Ekundayo G. Reviewing the development of natural fiber polymer composite: a case study of sisal and jute. Am J Mech Eng 2019;3:1.
19. AL-Oqla FM, Salit MS. Natural fiber composites. In: Materials selection for natural fiber composites; 2017:23–48 pp.
20. Cao Y, Wu Y. Evaluation of statistical strength of bamboo fiber and mechanical properties of fiber reinforced green composites. J Cent South Univ Technol 2008;15:564–7.
21. Lee BH, Kim HJ, Yu WR. Fabrication of long and discontinuous natural fiber reinforced polypropylene biocomposites and their mechanical properties. Fibers Polym 2009;10:83–90.
22. Li X, Tabil LG, Panigrahi S. Chemical treatments of natural fiber for use in natural fiber-reinforced composites: a review. J Polym Environ 2007;15:25–33.
23. Mehta G, Mohanty AK, Thayer K, Misra M, Drzal LT. Novel biocomposites sheet molding compounds for low cost housing panel applications. J Polym Environ 2005;13:169–75.
24. Sandström R. An approach to systematic materials selection. Mater Des 1985;6:328–38.
25. Al-Oqla FM, Salit MS. Materials selection for natural fiber composites. In: Materials selection for natural fiber composites; 2017.
26. AL-Oqla FM, Almagableh A, Omari MA. Design and fabrication of green biocomposites. Green biocomposites. Cham, Switzerland: Springer; 2017.
27. Edwards K. Selecting materials for optimum use in engineering components. Mater Des 2005;26:469–73.
28. Sapuan SM, Haniffah W, AL-Oqla FM. Effects of reinforcing elements on the performance of laser transmission welding process in polymer composites: a systematic review. Int J Perform Eng 2016;12:553.
29. Almagableh A, AL-Oqla FM, Omari MA. Predicting the effect of nanostructural parameters on the elastic properties of carbon nanotube-polymeric based composites. Int J Perform Eng 2017;13:73.

30. Barbero EJ. Introduction to composite materials design. Boca Raton, FL, USA: CRC Press; 2010.
31. AL-Oqla FM, Sapuan SM, Jawaid M. Integrated mechanical-economic Environmental quality of performance for natural fibers for polymeric-based composite materials. J Nat Fibers 2016;13:651–9.
32. AL-Oqla FM, Sapuan SM, Ishak MR, Aziz NA. Combined multi-criteria evaluation stage technique as an agro waste evaluation indicator for polymeric composites: date palm fibers as a case study. Bioresources 2014;9:4608–21.
33. AL-Oqla FM, Sapuan SM, Ishak MR, Nuraini A. A decision-making model for selecting the most appropriate natural fiber—polypropylene-based composites for automotive applications. J Compos Mater 2015. https://doi.org/10.1177/0021998315577233.
34. AL-Oqla FM, Sapuan SM, Ishak MR, Nuraini A. Predicting the potential of agro waste fibers for sustainable automotive industry using a decision making model. Comput Electron Agric 2015;113: 116–27.
35. Al-Widyan MI, AL-Oqla FM. Utilization of supplementary energy sources for cooling in hot arid regions via decision-making model. Int J Eng Res Afr 2011;1:1610–22.
36. Saaty TL. The modern science of multicriteria decision making and its practical applications: the AHP/ANP approach. Oper Res 2013;61:1101–18.
37. Mansor MR, Sapuan SM. Concurrent conceptual design and materials selection of natural fiber composite products; 2018.
38. Pugh S. Total design: integrated methods for successful product engineering, 1st ed. Wokingham, England: Addison-Wesley Publishing; 1991.
39. Sapuan SM. Composite materials: concurrent engineering approach. Oxford: Butterworth-Heinemann; 2017.
40. Sapuan SM, Mansor MR. Design of natural fiber-reinforced composite structures. In: Campilho RDSG, editor Natural fibre composites: overview and recent developments. Boca Raton: CRC Press; 2016:255–78 pp.
41. Mastura MT. Concurrent conceptual design and materials and manufacturing process selection of hybrid natural/glass fiber composite for automotive anti roll bar [Ph.D. thesis]: Universiti Putra Malaysia; 2017.
42. Mansor MR, Sapuan SM, Salim MA, Akop MZ, Musthafah MM, Shaharuzaman MA. Concurrent design of green composite products. In: Verma D, Jain S, Zhang X, Gope PC, editors. Green approaches to biocomposite materials science and engineering. Hershey, USA: IGI Global; 2016: 48–75 pp.
43. Mansor MR, Sapuan SM, Hambali A, Zainudin ES, Nuraini AA. Conceptual design of kenaf polymer composites automotive spoiler using TRIZ and morphology chart methods. Appl Mech Mater 2015; 761:63–7.
44. Mansor MR, Sapuan SM, Zainudin ES, Nuraini AA, Hambali A. Conceptual design of kenaf fiber polymer composite automotive parking brake lever using integrated TRIZ—morphological chart—analytic hierarchy process method. Mater Des 2014;54:473–82.
45. Mahmood A, Sapuan SM, Karmegam K, Abu AS. Development of kenaf fibre-reinforced polymer composite Polytechnic chairs. In: Proceedings of the 5th postgraduate seminar on natural fiber composites. Serdang, Selangor, Malaysia; 2016:38–42 pp.
46. Yusof NSB, Sapuan SM, Sultan MTH, Jawaid M, Maleque MA. Development of automotive crash box: a review. In: Proceedings of the 5th postgraduate seminar on natural fiber composites. Serdang, Selangor, Malaysia; 2016:27–29 pp.
47. Shaharuzaman MA, Sapuan SM, Mansor MR. Composite side door impact beam: a review. In: Proceedings of the 5th postgraduate seminar on natural fiber composites. Serdang, Selangor, Malaysia; 2016:70–73 pp.
48. Salwa HN, Sapuan SM, Mastura MT, Zuhri MYM. Conceptual design and selection of natural fibre reinforced biopolymer composite (NFBC) takeout food container. J Renew Mater 2021;9:803–27.

49. Sapuan SM, Ilyas RA, Asyraf MRM, Suhrisman A, Afiq TMN, Atikah MSN, et al. Application of design for sustainability to develop smartphone holder using roselle fiber-reinforced polymer composites. In: Sapuan SM, Razali N, Radzi AM, Ilyas RA, editors. Roselle: production, processing, products and biocomposites. Cambridge, MA, USA: Elsevier Academic Press; 2021:1–300 pp.
50. Sapuan S. Concurrent engineering in natural fibre composite product development. Appl Mech Mater 2015;761:59–62.
51. Sapuan SMA. Conceptual design of the concurrent engineering design system for polymeric-based composite automotive pedals. Am J Appl Sci 2005;2:514–25.
52. Ilyas RA, Asyraf MRM, Sapuan SM, Afiq TMN, Suhrisman A, Atikah MSN, et al. Development of roselle fiber reinforced polymer biocomposites mug pad using hybrid design for sustainability and Pugh method. In: Sapuan SM, Razali N, Radzi AM, Ilyas RA, editors. Roselle: production, processing, products and biocomposites. Cambridge, MA, USA: Elsevier Academic Press; 2021: 1–300 pp.
53. Sutton RI, Hargadon A. Brainstorming groups in context: effectiveness in a product design firm. Adm Sci Q 1996;41:685–718.
54. Marques T, Esteves JL, Viana J, Loureiro N, Arteiro A. Design for sustainability with composite systems. In: 15th international conference on experimental mechanics (ICEM15); 2012:1–2 pp.

Aizat Ghani*, Zaidon Ashaari, Seng Hua Lee, Syeed Saifulazry,
Nabilah Ahamad, Fatimah A'tiyah and Chuan Li Lee

3 Amine compounds post-treatment on formaldehyde emission and properties of urea formaldehyde bonded particleboard

Abstract: Post-treatment of particleboards using three types of amine compounds namely ethylamine, methylamine and propylamine, were produced with 8% urea formaldehyde (UF) resin content and were hot-pressed for 270 s at 180 °C was carried out. The amine compounds were applied over the surface of the particleboard surface at rate of 40–60 g/m², and the formaldehyde emission value from the board was significantly reduced. In general, the findings of the post-treatment application showed that formaldehyde emissions were successfully decreased from 0.70 mg/L to 0.59–0.33 mg/L without having a substantial impact on the particleboard's other qualities.

Keywords: amine compound; formaldehyde emission; post-treatment; particleboard; urea-formaldehyde.

3.1 Introduction

Particleboard production continues to rise globally as one of the most important forest products [1]. Because of its low cost and high reactivity, urea formaldehyde (UF) is still widely used as binding agent in the particleboard production manufacturing industry. Particleboard bonded with UF resin, on the other hand, has been linked to excessive formaldehyde emissions. Which are known to be detrimental to user's health [2].

Because of the problems produced by formaldehyde emissions, many researchers and particleboard manufacturers have worked hard to find ways to reduce formaldehyde emission through various methods in order to meet the worldwide low-emission

*Corresponding author: Aizat Ghani, Institute of Tropical Forestry and Forest Product, Universiti Putra Malaysia, 43400 UPM, Serdang, Selangor, Malaysia, E-mail: aizatabdghani@gmail.com, muhammad.aizat@upm.edu.my

Zaidon Ashaari and Nabilah Ahamad, Department of Wood and Fiber Industries, Universiti Putra Malaysia, 43400 UPM, Serdang, Selangor, Malaysia

Seng Hua Lee, Syeed Saifulazry and Chuan Li Lee, Institute of Tropical Forestry and Forest Product, Universiti Putra Malaysia, 43400 UPM, Serdang, Selangor, Malaysia

Fatimah A'tiyah, Faculty of Biotechnology and Biomolecular Sciences Universiti Putra Malaysia, 43400 UPM, Serdang, Selangor, Malaysia

As per De Gruyter's policy this article has previously been published in the journal Physical Sciences Reviews. Please cite as: A. Ghani, Z. Ashaari, S. H. Lee, S. Saifulazry, N. Ahamad, F. A'tiyah and C. L. Lee "Amine compounds post-treatment on formaldehyde emission and properties of urea formaldehyde bonded particleboard" *Physical Sciences Reviews* [Online] 2022. DOI: 10.1515/psr-2022-0075 | https://doi.org/10.1515/9783110769227-003

market, such as Japan. To achieve low emission levels, several methods had been conducted, and the common way is reducing the F/U (formaldehyde/urea) molar ratio. Another efficient and practical method is to incorporate a formaldehyde scavenger directly into the UF resin.

According to Zaidon et al. [3] formaldehyde scavenger can be applied it two ways; incorporating directly into the resin during the resin preparation also can be denoted as add-in and can be applied after the conditioning of the wood product production (denoted as post-treatment). Post-treatment usually applied the formaldehyde scavenger on the particleboard surface, and this might be influencing the surface characteristics of the particleboard panels.

Lum et al. [4] points out the advantages of using post-treatment formaldehyde scavenger where this treatment provides a much more acceptable result and additionally, moreover this application is more effective than add-in formaldehyde scavenger. Furthermore, the application of formaldehyde scavenger post-treatment has been proven can reduce the formaldehyde emission without adversely affect the mechanical and physical properties of the wood product [3, 4]. Ding et al. [5] also mention in this study that, formaldehyde emission of plywood made from poplar and bonded by UF resin reduce from 57.6 to 74.4% after applying sealing treatment (post-treatment).

There are several chemical compounds that could be used as formaldehyde scavenger either post-treatment or add-in application. A variety of ammonium salts, ammonium and alkali metals salts with sulphur-contains anions, urea and another compound with -NH functionality were suggested by [6]. In addition, Zaidon et al. [3] used urea and ammonium carbonate on phenolic resin treated Sesenduk, while Lum et al. [4] used a solution of amino compound on the particleboard panel's surface and Ding et al. [5] treated the plywood surface using polyethylene wax. Nevertheless, Boran et al. [7] employs amine compound as a formaldehyde scavenger directly in the UF resin for the medium density fibreboard production.

Apart from the numerous benefits derived from the usage of post-treatment and various types of formaldehyde scavengers to limit formaldehyde emission, a study was undertaken to determine the influence of post-treatment amine compounds on particleboard physical and mechanical properties as the effectiveness of the spread rate quantity of the formaldehyde scavenger utilised.

3.1.1 Objectives

The objectives of this study were;
1. To investigate the effect of amine compound post-treatment on formaldehyde emission of UF-bonded particleboard.
2. To assess both physical and mechanical strength properties of post-treated particleboard as a function of amine compound and spread rate amount.

3.2 Materials and methods

3.2.1 Preparation of materials

Rubberwood particles used in this study were supplied from particleboard plant manufacturing located in Gemas, namely HeveaBoard Berhad. Urea formaldehyde (UF) resin type E1 was used as binding agent and obtained from Aica Chemicals (M) Sdn. Bhd from Senawang. Ammonium sulphate (NH_2SO_4) was used as hardener. Three types of amine compound used as formaldehyde scavengers were methylamine, ethylamine and propylamine.

3.2.2 Particleboard fabrication

A homogeneous (single layer) UF-bonded particleboard with a targeted density of 650 kg/m^3 and dimension 340 × 340 × 12 mm in thickness was fabricated. 1% of ammonium sulphate as hardener based on resin solid weight then mixed with 8% rubberwood particles were used to make UF resin based on their dry weight. A 0.5% of wax based on dry particle weight was also mixed together into the UF resin mixture as a water repellent and then was sprayed evenly onto the rubberwood particles while blending in the blender for 3 min. After blending, the resinated particles called furnish was then formed in a former, pre-pressed and subsequently pressed in a hot-pressed machine maintained at 180 °C for 270 s with pressure of 4 MPa. The produced particleboard was conditioned in a conditioning room until constant weight was achieved prior to properties assessment.

3.2.3 Post treatment with amine compound

To see how effective amine compound post-treatments on the particleboard properties, post-treatment with amine compound was applied to the particleboard using brushed techniques as been shown in Figure 3.1 below. 40, 50 and 60 g/m^2 of amount of spread

Figure 3.1: Particleboard samples were subjected to a post-treatment technique.

rate of methylamine, ethylamine and propylamine were weighed respectively. The particleboards surface was brushed with aqueous amine compound and were stacked and stored. The particleboards sample were namely following the type of amine compound and amount of spread rate used (Figure 3.2).

3.2.4 Properties evaluation of particleboard

3.2.4.1 Formaldehyde emission test by desiccator method

Formaldehyde emission of particleboards was conducted using desiccator method in accordance with Japanese Industrial Standard (JIS) A 1460:2001 [8]. A total nine pieces particleboard samples (50 mm width × 150 mm length × 12 mm thickness) with approximately 1800 cm^2 total surface areas were placed in desiccator. Then, crystallizing glass dish was filled with 300 mL of distilled water were placed in the bottom of the desiccator right below the particleboard samples and kept for 24 h. After 24 h, the water in the crystallizing dish was collected and the concentration of formaldehyde in the solution was determined by acetylacetone molecular absorption spectrometry.

3.2.4.2 Evaluation of physical properties

The thickness swelling (TS) and water absorption (WA) test were performed according to the technique outline in Japanese Industrial Standard (JIS) A 5908:2003 [9]. Five

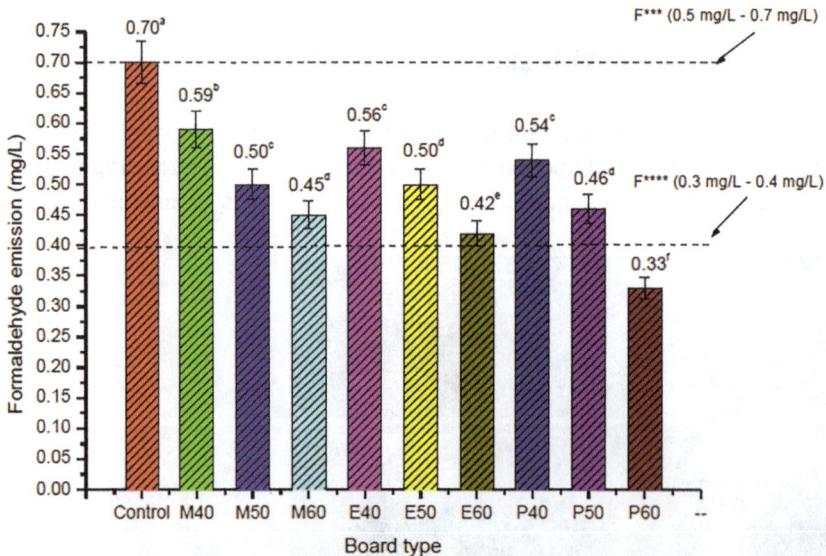

Figure 3.2: Formaldehyde emission of post-treated UF-bonded particleboard. *means followed by the same letter is not significant at $p \leq 0.05$; **M = methylamine, E = ethylamine, p = propylamine; 40: 40 g/m^2, 50: 50 g/m^2; 60: g/m^2.

particleboard samples nominal sample size of 50 mm wide × 50 mm length × 12 mm thick were prepared. The initial dimension including initial thickness and initial weight were measured first before submerging in water for 24 h. The dimensions of the particleboard sample were re-measured after soaking for 24 h to calculate the TS and WA. The percentage of TS and WA were calculated using equations (3.1) and (3.2) below.

$$TS\,(\%) = (T_2 - T_1)/T_1 \tag{3.1}$$

$$WA\,(\%) = (W_2 - W_1)/W_1 \tag{3.2}$$

where T_1 is the initial particleboard sample thickness before immersion in water (mm), T_2 is the thickness of the particleboard sample after immersion in water (mm), W_1 is the initial weight of the particleboard sample before immersion in water (g) and W_2 is the weight of the particleboard sample after immersion in water (g).

3.2.4.3 Evaluation of mechanical properties

The mechanical properties such as modulus of rupture (MOR), modulus of elasticity (MOE), and internal bonding (IB) were performed according to the procedure specified in JIS A 5908:2003 [9]. For bending strength (MOR and MOE) size of the sample was prepared in 230 mm in length × 50 mm in width × 12 mm in thickness, and in dimension. The bending strength was tested using universal testing machine (INSTRON 50 kN) by applying loading approximately 10 mm/min. The centre loading was applied at a span length of 150 mm at mean deformation speed from the surface of the particleboard sample, and the maximum load was measured. For IB determination, the test pieces sample with size of 50 mm in length × 50 mm in width × 12 mm in thickness was adhered to aluminium block using hot melt adhesive and attached to universal testing machine (INSTRON 50 kN). A persistent load of 5 mm/min was applied and the maximum breaking load was recorded. The bending strength (MOR, MOE) and IB were calculated based on the equations (3.3), (3.4) and (3.5), respectively.
    Bending strength;

$$MOR\,(N/mm^2) = P_m L / bd^2 \tag{3.3}$$

$$MOE\,(N/mm^2) = P_l L^3 / 4bd^3 \sigma \tag{3.4}$$

where P_m is the maximum breaking load (N), P_l is the load at proportional limit (N), L is the length span of the particleboard sample (mm), b is the width of particleboard sample (mm), d is the thickness of the particleboard sample (mm), and σ is the deflection at mid-span at limit of proportionally (mm).
    Internal bonding;

$$IB\,(N/mm^2) = P'/2bl \tag{3.5}$$

where P' is the maximum load at the time failing force (N), b is the width of the particleboard sample and l is the length of the particleboard sample.

3.2.5 Statistical analysis

The data then were evaluated statistically to verify any significance difference of the variables studied using Statistical Packages for the Social Sciences (SPSS). The collected data were analysed using one-way analysis of variance (ANOVA). Then, to separate the means, Tukey's HSD test was performed.

3.3 Results and discussion

3.3.1 Formaldehyde emission of particleboard

Figure 3.1 shows the formaldehyde emission values of post-treated UF-bonded particleboard with various amine spread rates. Statistically from the ANOVA, post treatment significantly affects the rate of formaldehyde emission from the particleboard. Apparently, formaldehyde emission of post-treated panels had decreased from 0.59 mg/L to 0.33 mg/L and about 15.71–52.86% from the control panel particleboard. The formaldehyde emission of the control particleboard using E1 UF resin was 0.7 mg/ L, which is in F*** class according to JIS standard. The formaldehyde emission was successfully lowered by using an amine compound as a post-treatment. According to the standard, all post-treated particleboard should emit at least F*** (0.5–0.7 mg/L) emission. As the amount spread rate of methylamine, ethylamine and propylamine increased from 40 to 60 g/m^2, there was a noticeable reduction in formaldehyde emission.

Regardless type of amine, the highest reduction was examined from 60 g/m^2 board type about 35–52%. Therefore, the combination between propylamine and 60 g/m^2 resulted the lowest formaldehyde emission and met the F**** (0.3–0.5 mg/L) emission level. This value was approximately equivalent to European E0 class emission level [10]. As indicated by the study, particleboard post-treated with 60 g/m^2 spread rate have the lowest rate of formaldehyde emission. This demonstrates that a higher spread rate will significantly reduce formaldehyde emissions.

However, different amine compound generated different amount of formaldehyde emission due to the different chemical structure of the amine compound. As formaldehyde scavenger, propylamine portrays higher effectiveness compared to methylamine and ethylamine due to its higher reactivity. According to Reusch [11], differences in C–H bond dissociation energy ascribed to the differences in amine compound reactivity as for example, the dissociation energy for methylamine is 103 kcal/mol, and for ethylamine is 98 kcal/mol while propylamine dissociation energy is 95 kcal/mol.

Because of propylamine has the weakest link, it can break readily and bind with a greater amount of free formaldehyde generated from the particleboard surface than the other ethylamine and methylamine, making it the most reactive chemical.

According to Ghani et al. [12] as formaldehyde scavenger, the amines compound is able to react with the formaldehyde to convert aldehyde (formaldehyde) to imine. It is believed, when amine reacts with free formaldehyde released from particleboard panel, the mechanism of addition-elimination was occurred. The probable theory for reducing formaldehyde emission is depicted in Figure 3.3.

When the surface area of the particleboard was post-treated or coated with the amine compound, it inhibits thus act like a barrier to formaldehyde from release to environment. As seen in Figure 3.4, the formaldehyde emitted from the particleboard panel is contained, resulting in a reduction in subsequent formaldehyde emissions. So, higher amount of amine compound spread rate, consequently, will provides more formaldehyde scavenger to react with the free formaldehyde emits from the particleboard and further decrease the formaldehyde emission.

Ghani et al. [12] also mentioned again in his study, the reaction of free formaldehyde with methylamine, ethylamine and propylamine were liberated N-methylmethanimine, N-ethylmethanimine and N-propylmethanimine, respectively and water as a byproduct for all amine and aldehyde reaction. As a result, this amines compound has the competency to reduce the formaldehyde emission with satisfactory result.

This finding is consistent with previous studies that looked at the effectiveness of utilizing an amine compound as a formaldehyde scavenger in lowering formaldehyde emission. From all the works done, propylamine shows the best ability in decreasing the formaldehyde emission from the medium density fibreboard (MDF) and particleboard panels, compared to other amines used (Boran et al. 2011) [12, 13]. Even though

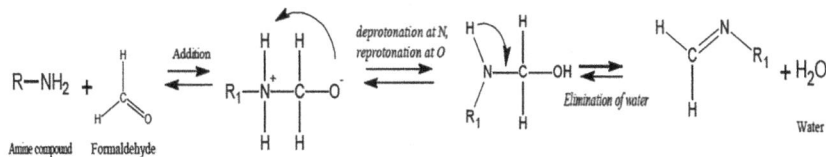

Figure 3.3: Addition and elimination mechanism of free formaldehyde and amine Source: Ghani et al. [12].

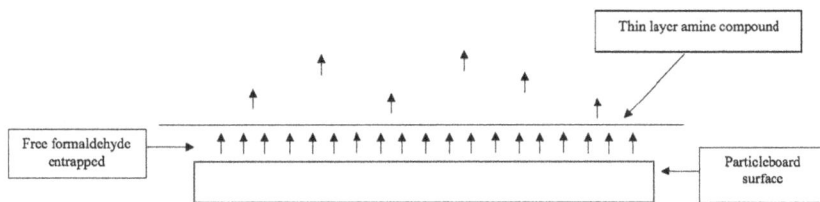

Figure 3.4: An illustration of post-treatment mechanism on the particleboard surface.

the incorporation of amine compound into the UF resin was really efficient in decreasing formaldehyde emission, however this add-in method impairs the physical and mechanical strength of the wood products.

This result is consistent with the previous research of Zaidon et al. [3] where the formaldehyde emission from Sesenduk *compreg* phenolic resin was greatly reduce about 75% after post-treated with ammonium carbonate. Furthermore, increasing the dose of ammonium carbonate reduced the formaldehyde emission of the treated Sesenduk wood by roughly 35% without affecting the wood's qualities except for the water uptake. In accordance with the present results, Ding et al. [5] also discovered that by applying polyethylene wax to coat the surface and edges of plywood efficiently reduces formaldehyde emission. Another study by Lum et al. [4] also discovered the post treatment formaldehyde catcher containing amino compound has magnificently lowered the formaldehyde emission from particleboard about 59–71% while increasing the formaldehyde catcher from amount 40–60 g/m^2 with positive result plus insignificant effect on the particleboard properties. According to Nemli et al. [14] coating particleboard surfaces reduced formaldehyde emission, depending on the type of coating applied. In practice, McVey [15] stated that any post-treatment, surface, or coating treatment will likely result in a reduction in formaldehyde emission.

3.3.2 Physical and mechanical properties of the post-treated particleboard

The particleboard's density and moisture content ranged from 663 to 685 kg/m^3 and 6.08–7.93%, respectively. All the post-treated particleboard panels have achieved the targeted density of 650 kg/m^3. The thickness swelling (TS) and water absorption (WA) of amine compound post-treated particleboard are listed in Table 3.1. As can be seen

Table 3.1: Physical properties values for control and post-treated particleboard ($n = 5$).

Type of board	Thickness swelling (%)	Water absorption (%)
M40	33.19[b,c] (4.44)	49.27[b] (0.82)
M50	34.67[b,c] (2.21)	49.56[b] (1.53)
M60	36.32[b,c] (3.88)	49.85[b] (1.11)
E40	36.40[b,c] (2.36)	50.23[a,b] (1.30)
E50	36.48[b,c] (4.46)	50.47[a,b] (1.31)
E60	36.81[b,c] (4.11)	50.67[a,b] (0.34)
P40	38.38[b,c] (7.62)	51.30[a,b] (1.36)
P50	39.67[a] (3.97)	51.74[a,b] (1.44)
P60	42.18[a] (2.19)	52.37[a] (0.93)
Control	28.01[c] (2.9)	46.07 (1.5)

[a]number in the parentheses are standard deviation; Means followed by the same letter is not significant at $p \leq 0.05$
[b]M = methylamine, E = ethylamine, P = propylamine; 40: 40 g/m^2, 50: 50 g/m^2; 60: g/m^2.

from the Table 3.1, TS and WA values of the post-treated particleboard panels were relatively higher than untreated (control) particleboard where the values were 28.01 and 46.07%, respectively. As from the analysis of variance (ANOVA), the values of WA and TS of amine compound post treated particleboard were increased significantly within the range of 33.19–42.18% and 49.27–52.37%, respectively. Regardless type of amine compound used, both TS and WA values increased with the rising of spread rate amount. In general, all post-treated particleboard panels TS values are not able to satisfy the minimum requirement of TS which is 12% as allowed by the JIS A 5980 standard.

The increment of TS and WA values from the post-treated particleboard panels might be influence by the reaction between amine and the free formaldehyde release from the particleboard surface.

As can be seen from the Figure 3.2, general reaction of the amine and formaldehyde will produce water as a by-product. Then, this by-product will be remained on the particleboard surface then make the particleboard surface in the hydrophilic condition thus attracted water to enter inside the particleboard consequently increase the amount of TS and WA as well. Furthermore, applying an amine compound to the surface of particleboard can reduce surface tension and increase wettability, making the particleboard more prone to water absorption. The result obtained agrees with previous studies carried out by Lum et al. [4] discovered that post-treated particleboard with amino-compound formaldehyde catcher generally had higher TS and WA although it not significant. Zaidon et al. [3] also revealed same trend on ammonium carbonate post-treated Sesenduk wood treated with low molecular weight phenol formaldehyde (LmwPF) resin and found positive correlation between TS and WA with ammonium carbonate dosage, where the values increased as the dosage of ammonium carbonate dosage increased.

Table 3.2 lists the values mechanical properties for post-treated and control particleboard panels produced. Modulus of rupture (MOR), modulus of elasticity (MOE) and

Table 3.2: Mechanical properties of control and post-treated UF-bonded particleboard ($n = 5$).

Type of board	Modulus of rupture (N/mm^2)	Modulus of Elasticity (N/mm^2)	Internal bonding (N/mm^2)
M40	15.17[a] (0.97)	1724[a] (259)	1.29[a] (0.22)
M50	15.48[a] (3.89)	1914[a] (337)	1.24[a] (0.17)
M60	14.63[a] (0.73)	1650[a] (205)	1.30[a] (0.18)
E40	14.80[a] (1.56)	2020[a] (277)	1.34[a] (0.21)
E50	14.94[a] (1.22)	2224[a] (466)	1.41[a] (0.22)
E60	15.03[a] (2.06)	1838[a] (331)	1.42[a] (0.20)
P40	14.87[a] (2.06)	2231[a] (385)	1.54[a] (0.24)
P50	14.68[a] (3.85)	1980[a] (387)	1.44[a] (0.41)
P60	15.65[a] (0.76)	2126[a] (163)	1.24[a] (0.13)
Control	15.13[a] (1.9)	1772[a] (231)	1.40[a] (0.31)

[a]Number in the parentheses are standard deviation; Means followed by the same letter is not significant at p ≤ 0.05.
[b]M = methylamine, E = ethylamine, P = propylamine; 40: 40 g/m^2, 50: 50 g/m^2; 60: g/m^2.

internal bonding (IB) for control particleboard were 15.13, 1772 and 1.40 N/mm^2, respectively. Since the process of post-treatment on the particleboard surface does not involving the UF resin curing system during the particleboard production therefore, this post-treatment seems does not influence the mechanical properties.

Statistically, it was proven by analysis of variance where it indicates that, amine compound post-treatment does not affect all the mechanical properties of the particleboard. From the Table 3.2, one can see that, all the post-treated particleboards produce is able to meet minimum requirement for MOR and IB, whereas specified in JIS A 5908, the minimum MOR and IB requirement of particleboard is 13 and 0.2 N/mm^2, respectively. Hence, according to the result obtained all the post-treated particleboard panels were higher than then minimum requirement. This finding corroborates the ideas of Lum et al. [4] and Zaidon et al. [3] suggested that the use of post-treatment that was only applied to the surface of the particleboard had no effect on the resin's curing behaviour and had no effect on the mechanical qualities of the wood product. Similar examples were also reported by Nemli et al. [14] in which the surface treatment had no little effect on the bending strength of post-treated particleboard after several coating treatments, implying that wood particles would require additional treatment to improve dimensional stability.

3.4 Conclusions

Formaldehyde emission of the post-treated with amine compound were reduced significantly. Regardless type of amine used; this post-treatment was found further decreased the formaldehyde emission of particleboard when spread rate of the amine compound increases from 40 to 60 g/m^2. The result showed that, particleboard post-treated with propylamine and 60 g/m^2 spread rate have the lowest formaldehyde emission thus achieved F**** emission level. In terms of mechanical properties such as bending and internal bonding strength, there is no significant different was observed between the strength in the post-treated particleboards. Yet, all the particleboard panels have reached the minimum requirement in JIS A 5908 standard. In addition, as for dimensional stability, the TS and WA found were influences by the application of the post-treatment. Irrespective of the spread rate and amine compound used, the application of the formaldehyde scavenger on the surface of particleboard panel may alter its surface characteristics and slightly increased the particleboard water uptake and the degree of the swelling. Nonetheless, all post-treated particleboard produced still unable to comply with the JIS A 5908 standard. In this study, the result has proven that amine compound has a great potential as formaldehyde scavenger and the same time did not adversely affect the particleboard mechanical properties. However, in order to improve the dimensional stability of the post-treated particleboard, additional treatment such as higher dosage of wax or addition of melamine can use, or improvement of the processing parameters would be necessary to achieve ideal thickness swelling with a lowest formaldehyde emission.

References

1. Guler C, Sahin HI, Yeniay S. The potential for using corn stalks as a raw material for production particleboard with industrial wood chips. Wood Res 2016;61:299–306.
2. Moubarik Amine, Mansouri Hamid Reza, Pizzi Antonio, Charrier Fatima, Allal Ahmed, Charrier Bertrand. Corn flour-mimosa tannin-based adhesives without formaldehyde for interior particleboard production. Wood Sci Technol 2013;47:675–83.
3. Zaidon A, Hua LS, bin Radzali MNE. Effects of ammonium carbonate post treatment on phenolic resin treated sesenduk (Endospermum diadenum) wood. Sains Malays 2015;44:987–94.
4. Lum WC, Lee SH, H'ng PS. Effects of formaldehyde catcher on some properties of particleboard with different ratio of surface to core layer. Asian J Appl Sci 2013;6:1–8.
5. Ding W, Li W, Gao Q, Han C, Zhang S, Li J. The effects of sealing treatment and wood species on formaldehyde emission of plywood. Bioresources 2013;8:2568–82.
6. Myers G. E. Advances in methods to reduce formaldehyde emission. In: Composite board products for furniture and cabinet-innovations in manufacture and utilization. Madison: Forest Products Society; 1989.
7. Boran S., Usta M., Gümüşkaya E. Decreasing formaldehyde emission from medium density fiberboard panels produced by adding different amine compounds to urea formaldehyde resin. International Journal of Adhesion and Adhesives 2011;31:674–8.
8. Japanese Industrial Standard (JIS) A 1460. Building boards determination of formaldehyde emission-desiccator methods. Tokyo;: Japan Standard Association (JSA); 2001.
9. Japanese Industrial Standard (JIS) A 5908. Particleboard. Tokyo: Japan Standard Association (JSA); 2003.
10. Athanassiadou E, Tsiantzi S, Markessini C. Towards composites with formaldehyde emission at natural wood levels. In: Proceedings of the 2nd conference of cost action E; vol 49; 2007. 28–9 pp.
11. Reusch W. Reaction of alkenes, virtual textbook of organic chemistry. Michigan: Department of Chemistry, Michigan State University; 1999. Available from https://www2.chemistry.msu.edu/faculty/reusch/VirtTxJml/funcrx1.htm [Accessed 26 Apr 2018].
12. Ghani A, Ashaari Z, Bawon P, Lee SH. Reducing formaldehyde emission of urea formaldehyde-bonded particleboard by addition of amines as formaldehyde scavenger. Build Environ 2018.
13. Ghani A, Bawon P, Ashaari Z, Wahab MW, Hua LS, Chen LW. Addition of propylamine as formaldehyde scavenger for urea formaldehyde-bonded particleboard. Wood Res 2017;62: 329–34.
14. Nemli G, Örs Y, Kalaycıoğlu H. The choosing of suitable decorative surface coating material types for interior end use applications of particleboard. Construct Build Mater 2005;19:307–12.
15. McVey DT. Great strides forward: formaldehyde emission from the production standpoint [Particleboard]. In: Proceedings of the Washington State University international symposium on particleboard; 1982.

Abdan Khalina*, Ching hao Lee and Aisyah Humaira

4 Manufacturing defects of woven natural fibre thermoset composites

Abstract: Thermoset polymer are components with high strength, chemical inert and thermally stable, due to its high degree of cross-linking. Natural fibre composite is providing a winning solution for extraordinary performances yet biodegradable. Woven form fibre even found better in specific energy absorption and stronger in strength. Fabricating woven thermoset composites may be done in a variety of ways. However, processing errors or manufacturing defects often occur by many factors, especially thermoset composites with natural fibre reinforcement. It is nearly impossible to achieves in detect-free when in lab scale production. Hence, it is important to study and understand the factors that causing the defects. Processing parameters, compatibility of matrix/fibre combination, yarn production and woven waiving skills may be the reasons of composite's defects. In this chapter, several fabrication methods for woven thermoset composite were introduced. Some major defects on manufacturing the thermoset composites were highlighted. Some future perception of the woven natural fibre thermoset composite also have been discussed. This chapter set as a guidance to avoid or minimizes manufacturing defects upon thermoset composite processing.

Keywords: defects; manufacturing process; natural fibre composite; thermoset composites; woven fibre.

4.1 Introduction

Natural fibre composite is gaining much attention to the researcher around the globe. Woven natural fibre composite offers better strength than short reinforcement natural filler in composites systems [1]. The woven natural fibre may be derived from kenaf,

*Corresponding author: Abdan Khalina, Laboratory of Biocomposite Technology, Institute of Tropical Forestry and Forest Products (INTROP), Universiti Putra Malaysia, 43400 Serdang, Selangor, Malaysia; and Department of Biological and Agricultural Engineering, Faculty of Engineering, Universiti Putra Malaysia, 43400 Serdang, Selangor, Malaysia, E-mail: khalina@upm.edu.my
Ching hao Lee, Laboratory of Biocomposite Technology, Institute of Tropical Forestry and Forest Products (INTROP), Universiti Putra Malaysia, 43400 Serdang, Selangor, Malaysia; and Department of Mechanical Engineering, School of Computer Science and Engineering, Taylor's University, Subang Jaya, Malaysia
Aisyah Humaira, Department of Mechanical and Manufacturing Engineering, Faculty of Engineering, Universiti Putra Malaysia, 43400 Serdang, Selangor, Malaysia; and Universiti Pertahanan Nasional Malaysia, Sungai Besi, Malaysia

As per De Gruyter's policy this article has previously been published in the journal Physical Sciences Reviews. Please cite as: A. Khalina, C. h. Lee and A. Humaira "Manufacturing defects of woven natural fibre thermoset composites" *Physical Sciences Reviews* [Online] 2022. DOI: 10.1515/psr-2022-0077 | https://doi.org/10.1515/9783110769227-004

jute, ramie and flax. The woven fabric often applied in thermoset polymer composite system. The most extensive matrix used are epoxy and polyester. The epoxy matrix is more reliable for structural application and with the addition of woven natural fibre, reinforcement mechanism become more strengthen. On the other hand, polyester also gaining more popular among the composite manufacturer, beside of its lower cost and acceptable properties.

Unfortunately, neither woven nor natural fibres are perfect, and they have notable drawbacks. The natural fibre's natural structure permits a very high moisture absorption which made poor interfacial conditions with the matrix. Water absorption properties is very crucial for natural fibre composites for many applications. Al-Maharma and Al-Huniti (2019) [2] have reviewed the effectiveness of moisture absorption treatments for natural fibre composites [3], in resulted around 47 and 45% reduction of tensile strength and flexural strength, respectively. The moisture absorption also increased composite thickness, thereby magnifying the effects of fibre debonding, delamination, and affection on mechanical properties too [4]. In contrary, glass fibre reinforced polymer composite showed only 28% of tensile strength deterioration, after eight years of salt water immersion [5]. This outcome tells that the natural fibres substitution still needs to solve tons of challenging. Furthermore, combination between the natural fibre and matrix are considered a challenge because of the different nature of both fibres and matrix. It altered stress transfer mechanism based on interface of the produced composites. The performance of composites are highly depending on the interfacial conditions. There are four interface mechanisms known as interdiffusion, electrostatic adhesion, chemical adhesion, and mechanical interlocking [6]. Furthermore, the gap between warp and weft acts as a mechanical interlock among the polymer matrix provides additional strong fibre/matrix bonding conditions [7]. Therefore, woven natural fibre thermoset composite always show explicit premier strength.

The hydrophilic nature of natural fibre is always incompatible with hydrophobic polymer matrix [8]. Although treatments could enhance the interfacial strength [9]. Deviation from the optimized treatment's parameters may ended with opposed performance [10]. Unlike synthetic fibres, good and consistent bonding conditions require no treatment and minimizing possibility of defects on the ended products [11]. After reviewing all the drawbacks of the natural fibres, it is still popular to be considered in current research and development due to its biodegradable and natural raw materials criteria, which synthetic fibres will never compete with.

Thermoset plastics shall strengthen when cured but cannot be reshape after formed. This is due to its high degree of cross-linking bonding inside the polymer, making it difficult to melt for remould. However, it gives high thermally stable, strength and geometry integrity. The thermoset plastic components are used extensively in a wide range of industries, in the automotive, appliance, electrical, lighting, and energy markets due to its excellent chemical and thermal stability along with superior strength, hardness, and moldability [12]. It has lower relative cost to the metal but comparable performance. Furthermore, insertion of woven natural fibre may again enhance composite's strength, inhibits certain degree of biodegradability.

Woven composites generally consist of a fabric element, made up from woven materials where vertical yarn (warp) and horizontal yarn (weft) are interlaced to form a fabric in a regular pattern or weave style. The woven structure of a composite determines the orientation of the fibres, which are responsible for technical properties of the composite specifically mechanical strength. The woven material could be obtained in various pattern either plain, satin and twill pattern. The mechanical interlocking of the fibres ensures the fabric's integrity. In consideration of woven composites, the physical and mechanical characteristics of woven composites were discussed critically in many works [13–16].

However, processing errors or manufacturing defects often occur by many factors, especially thermoset composites with natural fibre reinforcement. It is very challenging to produce zero defect even not mass producing. Yet minimizing defects and maintaining product's consistency are achievable. Processing parameters always the important criteria to control defect yield [17]. Synchronization between manufacturing period, temperature and pressure must investigate before conducting any experimental works. Besides that, the natural incompatible hydrophilicity between thermoset resin and natural fibre woven, resulted in poor interfacial bonding conditions. Voids formation shall always be detected in the natural fibre thermoset composites, which deteriorating composite's performances. Besides, that, the yarn production and waiving pattern may also cause some damages on the woven itself, leading to defects yield during composite processing.

In this chapter, advantages of woven composite have been discussed. Several fabrication methods for woven thermoset composites were also introduced, which focused on closed-molding technique that producing less waste and more complex product. Afterward, some major defects on manufacturing the thermoset composites were highlighted on processing parameters. Lastly, future direction of the woven natural fibre thermoset composite was mentioned at the end of the chapter. This chapter set as a guidance for audiences to avoid or minimizes manufacturing defects upon thermoset composite processing.

4.2 Advantageous of woven composites

Woven composites are widely employed in a variety of advanced materials, including composite component, biomedical, army, and protective helmets due to their excellent mechanical strengths [18–23]. The most significant advantages that woven composite can offers is excellent energy absorption, attributed from the fabric structure of the woven materials. Ghoushji et al. utilize ramie fibre in the plain fabric form to produced square composite tubes [24]. They found that higher energy-absorbing capability of woven ramie in the composite tubes compared to other tubes due to its fabric structure and low density of ramie that makes it better in specific energy absorption (SEA) value than the synthetic fibres in the composite structures. The fabric weave offers penetration resistance in addition to the high strength yarn utilized for energy absorber structures and impact applications.

Woven composite has been discovered own better mechanical properties and that properties were influenced by many factors such as yarn properties, fabric counts and weave patterns. Higher mechanical properties specifically tensile strength and tensile modulus are related to differences in the load-distribution properties of the yarns in the longitudinal and transverse directions of the fabric, resulting in higher stress uptake capacity [25]. During the tensile loading, the crimped yarns tend to straighten and causes transverse loads at the warp-weft yarn overlap area or yarn interlacement. This lowers long-term fatigue and creep rupture performance by reducing the translation of fibre strength to fabric strength [26]. A part of this, manner of stacking sequences do influences the composite strength, showing high flexibility on custom-made on specific application to achieves best performances [27].

The existence of yarn interlacement in the woven fabric also attributed to the uniform load distribution, thus the fractures ran transversally in all directions, perpendicular to the loading direction, due to isotropic woven packing. As a consequence, it can resist higher tensions that keep each other together, allowing stress to gradually pass to the next yarn and resulting in less structural slippage. Furthermore, according to Liu and Hughes, the inclusion of woven material enhances the fracture toughness of flax composites [28].

4.3 Fabrication method of woven thermoset composites

Fabricating woven thermoset composite components may be done in a variety of ways. In general, thermoset polymer composite production is classified into open- and closed-molding process, which their differences were listed in Table 4.1. Those methods provide certain advantages with some limitation. Hence, selection of processing method mainly depends on the skilled worker availability, product's profile and production conditions. Several parameters such as raw materials, design, and applications shall decide the optimum fabrication method. In regard to this matter, researchers aim to diversify the applicability of woven composite in various fabrication methods, including hand lay-up, wet laminate, resin transfer, and vacuum bagging, as discussed in the subsections that follow.

4.3.1 Hand lay-up and wet laminate

Hand layup is the most simple and basic woven composite production process, which generally comprises of several woven materials, placing manually by hand onto a plate or mould to make a laminating stack. There are two ways to apply resin after the layup

Table 4.1: Comparison between open molding and closed molding [29].

Open molding	Closed molding
Top layer of the laminates and matrix are exposed to the atmosphere, resulting in uncontrolled surface condition	Both side of the specimen are inside the mold
Tooling fabrication process is relatively simple and low cost, rapid product development cycle is possible to be implemented	Process is normally automated and required special equipment and capable of producing large part and volume
Wide part size potential, tooling fabrication process is relatively simple and low cost	Allows for more complex part geometries
Secondary finishing processes needed as only one side of the finished part will have a good surface finish	Less waste produced
Example of open molding 1. Hand lay-up 2. Filament winding 3. Spray-up	Example of closed molding 1. Resin transfer molding 2. Compression molding 3. Vacuum infusion 4. Vacuum bag

is finished; (1) resin is applied to the dry fabric using a brush followed by hand roller to ensure uniform resin distribution, or (2) fabric pre-impregnated with resin. Each layer is covered with resin and compacted when it is put in a wet layup variant, and the composite is allowed to cure under standard atmospheric conditions for curing stage. There are a variety of curing options including initiated using catalyst or hardener (Figure 4.1a), pressure (Figure 4.1b) or heat. This method enables for the production of a broad variety of applications, allows production of large parts and low-cost tooling. However, there are several drawbacks, including a low reinforcement volume,

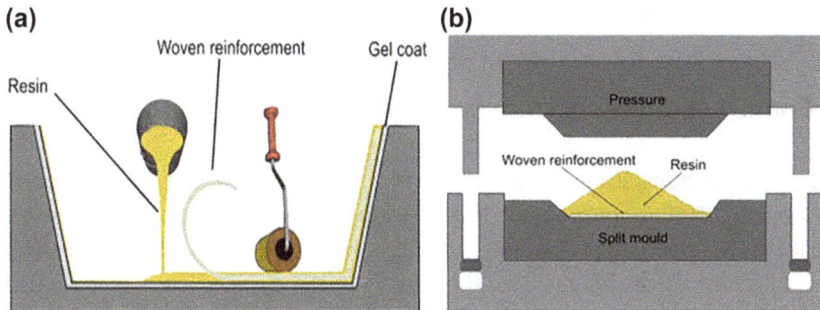

Figure 4.1: Schematic representation of curing in hand lay-up process, (a) self-curing using hardener and pressure and (b) autoclaved under pressure at steam atmosphere or using heat. Source: Figure reproduced with copyright permission from Raji et al. (2019) with permission from Elsevier [30].

PREFORM LAY-UP   CLOSING MOLD   INJECTION PHASE   CURING PHASE

PREFORM

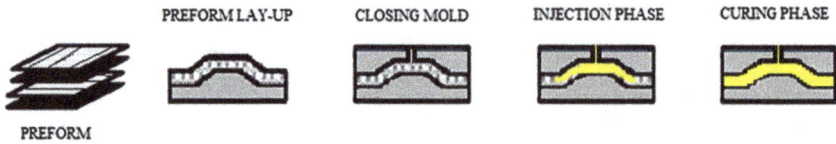

Figure 4.2: Basic steps in the resin transfer moulding (RTM) method. Source: Figure reproduced with copyright permission from Laurenzi et al. with permission from Intechopen [31].

nonuniform quality of the reinforcement material, uneven matrix distribution as well as unfriendly to the environment since it is an open molding process.

4.3.2 Resin transfer moulding (RTM)

One of the most essential processes for making woven thermoset composites is resin transfer moulding (RTM), where a closed mould is used to impregnate resin to the fabric lamination sequence or preform. Figure 4.2 shows basic steps during the RTM process. The preform is placed into the mould, then the mold is closed and the preform was compressed as well. The resin is injected under pressure to ensure that the preform is completely impregnated, thus avoid rough and irregular surface to remove trapped air and speeds up the RTM process. When the resin has filled the mould, the gates are clamped and the preform is impregnated then cure phase is taking place. Finally, the mold is opened, and composite is demolded. This method is able to produce near-net-shape complicated parts with tight dimensional tolerances, and excellent smooth surface quality on both sides. It produces a viable alternative to the prepreg approach, allowing for superior finish quality and fibre direction control. Additionally, it also eliminates voids, helps in improving composite mechanical strength. The wetting of the fabric was depended by weave design, fabric permeability, fabric crimps, resin viscosity, and operating temperature. To produce a high-quality woven composite, careful process design is required upon the RTM processing.

4.3.3 Compression resin transfer moulding (CRTM)

Compression resin transfer moulding (CRTM) is a method of compressing on regular RTM method. In CRTM, in contrast with RTM, the top mould platen is lowered to leave a small space between mould platen walls and fibre preform. Resin is supplied into this space via injection valves following to the closure of the upper plate. The resinated preform is then undergoes compression to obtain proper volume fraction and left for curing phase (Figure 4.3). The composite is ready for demolding now. This method offers rapid component manufacturing, cheap labor costs, food grade surface finishing on both sides, but required strong understanding on process parameters [32].

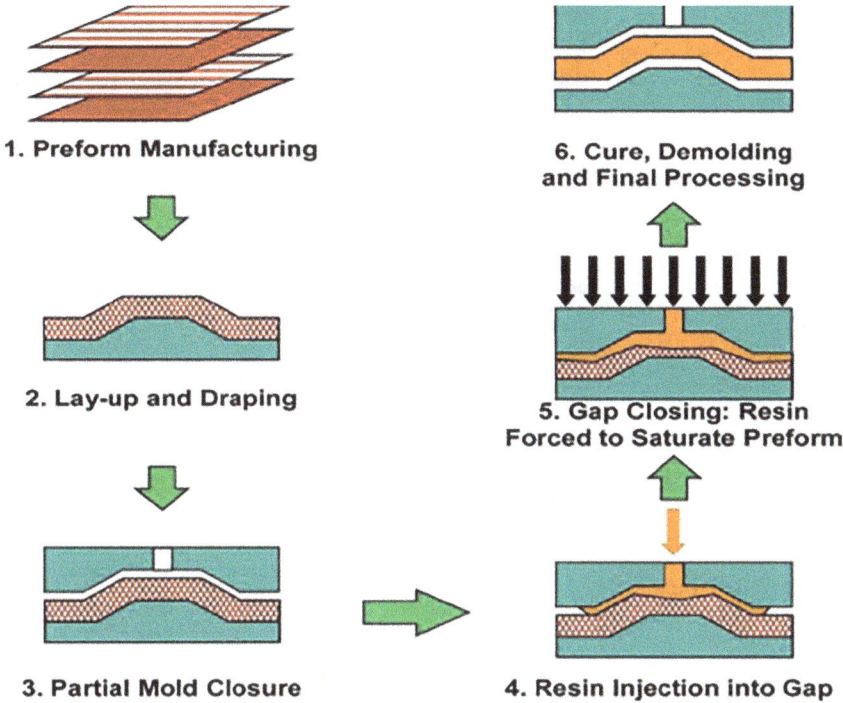

1. Preform Manufacturing

2. Lay-up and Draping

3. Partial Mold Closure

4. Resin Injection into Gap

5. Gap Closing: Resin Forced to Saturate Preform

6. Cure, Demolding and Final Processing

Figure 4.3: Steps in the compression resin transfer moulding (CRTM) method. Source: Figure reproduced with copyright permission from Simacek and Advani with permission from *Revue Européenne des Eléments*, [33].

4.3.4 Vacuum assisted resin transfer molding (VARTM)

In the vacuum assisted resin transfer molding (VARTM) method, a vacuum is applied before resin to enhance resin flow and minimizes void formation during composite fabrication. This method also known as vacuum infusion process (VIP). The preform is manual lay-up onto mold and a vacuum bag is cover entirely on top to forms a vacuum-tight sealing (Figure 4.4). The vacuum applied as a driving force for resin penetration and impregnation through prepreg and after achieving a complete vacuum, resin is practically sucked into the laminate using properly arranged tubing. This method is advantageous for gigantic product fabrication such as wind turbine blades, boat hulls, and constructions material. The VARTM fabrication method is also capable of producing high-quality products at low costs, as well as complicated composite components with high production rates. Good resin distribution can be achieved as the resin flows are in both plane and transverse directions of the preform. However, it only produce premier finishing from one surface, which limiting its application that requires two sided surface finishes.

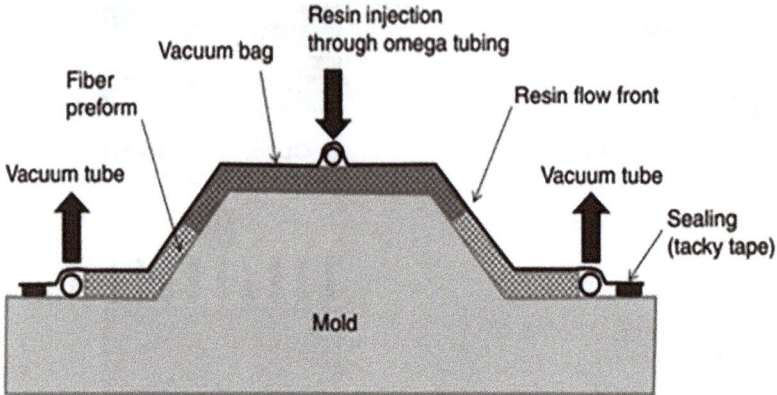

Figure 4.4: Schematic diagram of vacuum assisted resin transfer moulding (VARTM) method. Source: Figure reproduced with copyright permission from Advani and Sozer with permission from CRC press [34].

4.3.5 Vacuum bagging process

Vacuum bagging process is similar to the hand lay-up processing, where in this method, a vacuum bag is fastened over the mould and vacuum is applied as shown in Figure 4.5. Hence it is a closed-mold method and producing high-quality composite materials. The advantages of this method are the ability to produced large, complicated composite products and suitable for thicker components with excellent quality. In term of mould, this method offers flexible mould tooling design and selection of mould materials. However, the tools such as vacuum bag, flow media, peel ply, sealing tape,

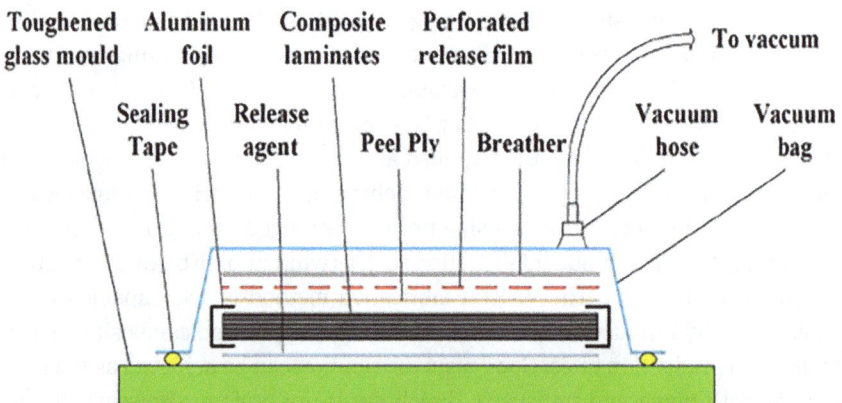

Figure 4.5: Vacuum bagging system in the composite production. Source: Figure reproduced with copyright permission from Hang et al. with permission from *Proceedings of the Institution of Mechanical Engineers, Part B: Journal of Engineering Manufacture* [35].

and resin tubing may not be reused, creating high amount of wastages thereby increasing production cost. The possibility of air leakage is highly depending on the worker's competence and experience.

4.4 Epoxy composite failure and manufacturing

Epoxies have a low viscosity before curing, which helps with fibre saturation and wetting. This reduces air gaps and improves adherence of polymer matrix to reinforcing fibre. These components work together to determine the total strength of the epoxy composite. The low viscosity also allows multiple processing methods as mentioned above including resin transfer moulding (RTM, or infusion), filament winding, pultrusion, and prepreg for composite production. The epoxy polymers could cure to a high level of chemical resistance, allowing composite products work under hazardous conditions. Besides, the epoxy polymers can also be easily modified to meet the physical requirements of service, by adding reinforcement fibre fillers, to improve toughness against impact and obtain flexibility when needed, despite their reputation as hard and stiff materials. Hence, the epoxy composites often tagged with high performances and great flexibility. For example, aircraft components made by the epoxy composites aimed to reduce weight but maintaining strength as metal components. Intense cross-linking found in the epoxy polymer capable to withstand high loading before failure. Besides, good mechanical and thermal properties with electrical inert is claimed to be excellent for electronic packaging [36, 37]. However, defects of the epoxy composites may find and due to many factors, such as processing parameters, material preparation and/or complexity of product's geometry.

Processing parameters is one of the most crucial factors affecting performances of final epoxy composite product. Under- or over-optimum parameters may cause partially cured or defects. Pressure helps in directing and pushing the resin into the mold. There shall be only a small pressure range for resin and mold, in order to avoid defect occurred, showed in Figure 4.6. Besides, epoxy thermoset resin requires hardener to solidify or create strong cross-linkages. Hence, the hardener's type and quantity does play a vital role. ANOVA modelling shows 12.28% of strength influences by epoxy-hardener ratio [38]. On the other hand, increment of curing temperature had shorten the curing time with evident of disappearing of amine, imide and epoxide groups in earlier time [39]. However, it may damage the natural fibres, which low in thermal resistance. Thermal decomposition temperature of fibre's constituents ranging from 180 to 400 °C. High working temperature will 'burn' the fibre and losing its load bearing capability [40].

The time period to process a product with good quantity and quality is one of the important parameters considered in the manufacturing industry. For a manufacturer that produces a product requires a fast and quick process with the quality of composite products at a good level. However, speed of the composite manufacturing process can

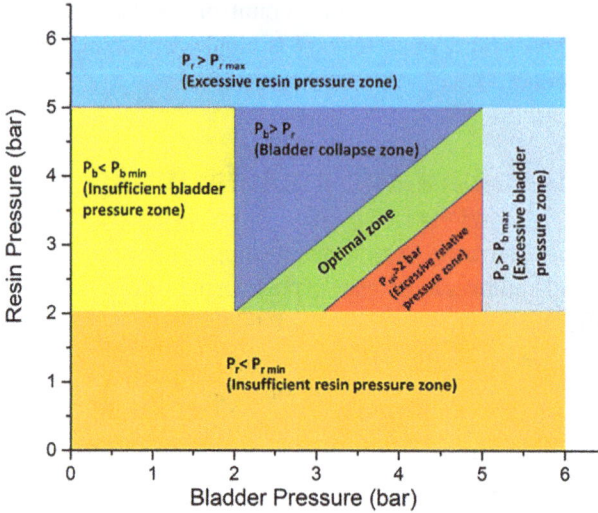

Figure 4.6: Manufacturing optimization zone diagram at different pressure boundary conditions during B-RTM process [41].

affect changes in the micro level of the structure of the material in matrix or reinforcement. Therefore, the rate of change of a material to the time factor must be examined to avoid product defects such as voids and delamination. A composite production process should be optimized to avoid major manufacturing defects such as voids, penetrations and areas of poor or rich resin in the composite. Unoptimized processing parameters may led to delamination of natural fibre and matrix. Thin-walled laminated fibre epoxy composites show the greatest affection from this defect. During the compression process, the delaminated epoxy composite would have low capability to carry compressive loads.

It is also reported that debonding at fibres-matrix interface may be generated due to low adhesion polymer resin to natural fibres [42]. All the defects mentioned above depend on microstructure and thermomechanical manufacturing history and contribute to decrease mechanical properties of the composite material and consequently composite structures. In order to manufacture homogenous natural fibre/thermoplastic matrix composites with low content of voids and high mechanical properties, answering industrial and financial needs, several manufacturing conditions must be met: These ones can be found in the literature and for example in the following references for hot press manufacturing [43] and in [44] for rapid manufacturing process. The insertion of natural fibres has created numerous of defects although it is the trend of future for environmental fighting. The flow of epoxy resins under the natural fibre woven appearances required longer time for complete wetting. This is due to incompatibility of hydrophilicity between hydrophobic epoxy and hydrophilic natural fibres. The heterogeneity in thermodynamic and rheological on both fibre and epoxy matrix, would create voids during the process. Besides that, improper resin injection may cause turbulence flow and creating air bubbles, especially for fast flowing resin. Hence, it requires longer period to flow and cover the whole specimen, making homogeneous wetting more difficult. Consequently, hardening of epoxy before completion of resin wetting, creating the void defects in the composite. Besides, gasses produced by chemical reactions upon curing may also trapped as the voids. This defect eventually deteriorated epoxy composites' performances.

The non-uniform permeability of the fibre preforms which in turn causes the resin velocity to vary from point to point at a micro scale. The capillary pressure, which also prevails at this length scale, exacerbates the spatial variation of resin velocity. The resulting microscopic perturbations in the resin flow front allow voids to form. Thought fibre wetting could be improved by applying pressure and/or fibre treatment. Yet, over-treated fibre observed damages on fibre surface [45]. This defect may be due to high interlaminar stresses of inserted woven failed to transport the loads.

The heterogenous profile of natural fibre from batch-to-batch and even in every single fibre was one of the major factors causing uneven load transfer mechanism in the epoxy composite. Each fibre reported porosity content ranging at 7–12%, was shaped by external surrounding conditions and internal micromolecular arrangement. Fibre extraction, retting methods and/or parameters was then may further damage the natural fibre. Fibre surface flaw, surface peeling, crack branching, lumen crack departure, intra-lumen damage, inter-lumen damage, inter-fibre cracking and branching and interfacial cracking, shown in Figure 4.7, were the reasons of high variability in tensile values. The defects may be even more complexity when the natural fibre applied as reinforcements in epoxy composites. The load transfer between fibre and matrix and composite's failure mechanism shall depends on boarder consideration such as fibre conditions, filler ratio, geometrical arrangement and most importantly, interfacial conditions between fibre and matrix. Surprisingly, defects were found upon yarn spinning production had influences the quality of fibre epoxy composite. A number of fibres were twisted into a continuous strand, had produced radial forces and caused some fibre breakages and slippages on the yarn (Figure 4.8). The load capabilities of defected yarns inserted epoxy composite was significantly reduced. Besides, weaving pattern may caused more defects which are voids in matrix pockets, tow-matrix separation, warp-fill interface separation and matrix crack in fill tow, which explicated in Figure 4.9. The major defect among all is fibre tow-matrix pocket interface separation, which would result in a sharp reduction on stress-strain curves. Modelling analysis also showed significant defects observed between the ply of weave textiles, when more than one ply is stacking [46].

When the geometry of mold is too complicated for the natural fibre epoxy composite products, there might found some defects on the edges due to the fibre filler unreachable. These edges, especially sharp internal corners, may act as stress concentration spots, leading to product's defects or failure. Besides, thicker wall thickness may prone to cracking, warpage and breaking defects during processing, which caused

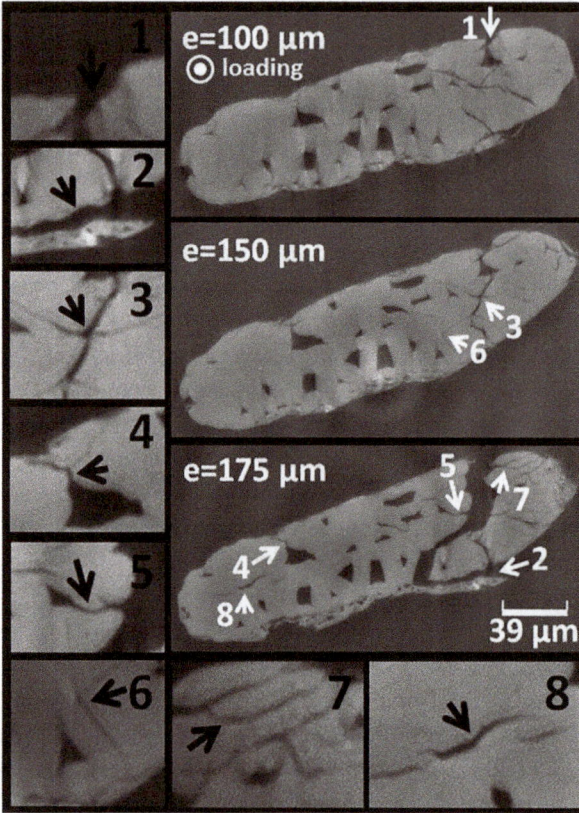

Figure 4.7: Damage mechanism in natural fibre, 1: µ-crack departure from surface flaw (crack depth 3 µm), 2: surface peeling, 3: crack branching, 4: lumen crack departure, 5: intra-lumen damage, 6: inter-lumen damage, 7: inter-fiber cracking and branching, 8: interfacial cracking [47].

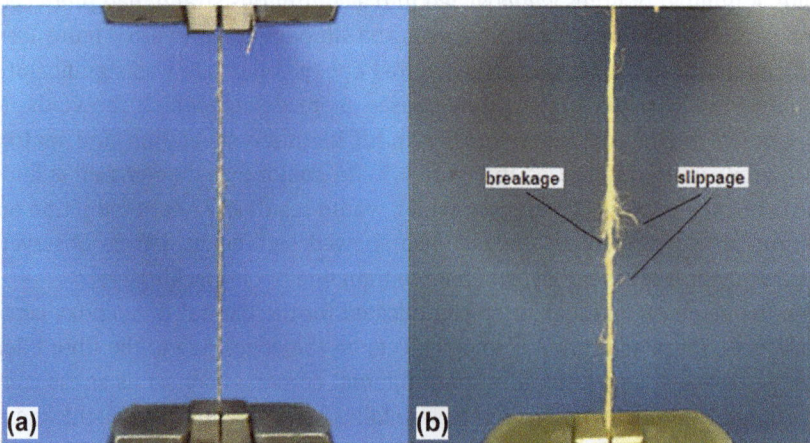

Figure 4.8: A single-strand natural yarn fibre in (a) tensile tension and, (b) early failure due to yarn's defects [48].

by higher epoxy shrinkage rate for thick product [49]. Hence, designing a mold for epoxy composite products should done in extra care, to avoid potential defects. The mastery of thermoset composite processing by skilled labor have directly relevant to the severity of defects shall be found in the final products. Besides, carefully inspection on quality of woven, natural fibre will reduce the unnecessary defects on woven thermoset composite products. Hence, minimizing the defects required closed cooperation from upstream to downstream.

4.5 Future direction toward sustainable and green materials

According to the COP27: UN climate change conference 2022, majority of countries had agreed to work together to accelerate global efforts to confront climate crisis [50]. The final limit global warming of 1.5 °C might breach earlier than what scientists expected. Hence more aggressive actions must be taken to avoid the most devastating impacts of the climate change to the humanity and environmental.

Promotion of sustainable and green materials is a promising method upon dealing with climate change issue and this is well-fitted into the sustainable development goals (SDG), responsible consumption and production (SDG12) and Industry, Innovative & Infrastructure (SDG9) [51]. Analytic report found synchronization showing uptrend of the natural fibre composite market size forecasts for 2018–2024, shown in Figure 4.10 [52]. The report claims promising reduction of component mass, energy consumption and overall costing upon the replacement of the glass fibre by natural fibres.

The woven natural fibre thermoset composite materials preserved good performances as conventional materials with additional biodegradability. This shall ease the saturated status of solid waste accumulation in all countries. The hydrophilic natural fibres absorb large amount of water, which normally containing microorganisms and promoting breakdown on cellulose components [53]. Hence, lesser amount of rubbish will be located when natural fibre composites being used. On the other hand, natural fibres processing required lesser energy. Flax woven in PLA composite recorded more than three times lesser energy needed compared to glass fibre fabric production, under

void in matrix pocket   tow-matrix separation

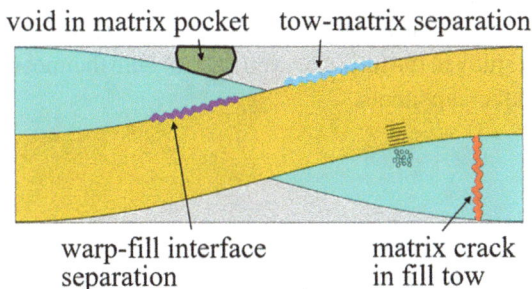

warp-fill interface separation   matrix crack in fill tow

Figure 4.9: Weave textile defects [46].

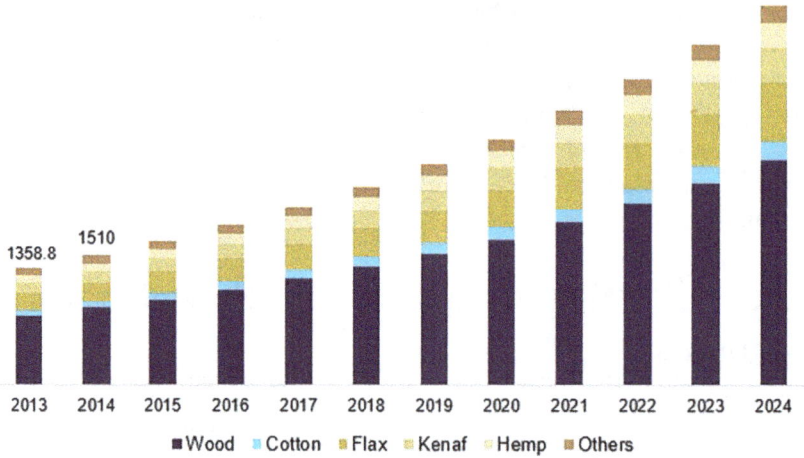

Figure 4.10: US natural fibre composites market revenue, by raw material, 2013–2024 (USD million) [52].

same manufacturing conditions [54]. It resulted in showing improvement in the development of composite material and environmental impact.

Unfortunately, the application of woven natural fibre thermoset composites was found very challenging in the IR4.0 (Industrial Revolution 4.0). Additives manufacturing (AM) or commonly referred to 3D printing technology, has variety of printing mechanism which covered the use of thermoset resin. The 3D printing provides minimum waste, labor monitoring and extremely high flexibility in product design [55]. Yet woven natural fibre seems did not considered in existing AM technology nor major research currently. From a future perspective, woven natural fibres should be involved in AM to produce strong green composites with high flexibility and lower cost.

Despite of intense research on the woven natural fibre thermoset composites, the gate of aircraft components is hardly open. Strict regulations for the aircraft components required high consistency, durability, performance, especially flammability. Although research and development improved flame retardancy of natural fibre composites [56]. The inconsistency properties of the natural fibres from batch-to-batch or even every single fibre, had made it off from the aircraft material considerations. Fortunately, the awareness on environmental issue had made us reconsidered the beauty of natural fibres. With further study in the future, woven natural fibre thermoset composites should applied for aircraft components.

4.6 Conclusions

According to the COP27: UN climate change conference 2022, majority of countries agreed to work together to accelerate global efforts to confront the climate crisis. Woven natural fibre composites offer better strength than short reinforcement natural filler in composites systems. Thermoset plastics are a material that strengthen when cured but cannot be reshape after formed. This is due to its high degree of cross-linking bonding inside the polymer, making it difficult to melt for remould. However, it gives high thermally stable, strength and geometry integrity to the polymer products. Unfortunately, hydrophilic nature of natural fibre is always incompatible with hydrophobic polymer matrix. Although treatments could enhance the interfacial strength, it may increase the overall production cost.

Despite of the disadvantages mentioned above. The oven natural fibre thermoset composites are still widely employed in a variety of advanced materials, including composite component, biomedical, army, and protective helmets due to their excellent mechanical strengths. Higher energy absorption capable of woven natural fibre composite had made used in variety of applications. This could be led by the warp-weft yarn overlap area or yarn interlacement and the gap between warp and weft acts as a mechanical interlock among the polymer matrix provides additional strong fibre/matrix bonding conditions.

There are numerous of the fabrication methods available for the woven thermoset composites (classified into open- and closed-molding technology). Those methods provide certain advantages with some limitation. Hence, selection of processing method mainly depends on the skilled worker availability, product's profile and production conditions. However, processing errors or manufacturing defects often occur by many factors, especially thermoset composites with natural fibre reinforcement. It is challenging to produce zero defect when mass producing. Yet minimizing defects and maintaining product's consistency are achievable. Processing parameters, compatibility of matrix/fibre combination, yarn production and woven waiving skills may be the reasons of composite's defects.

Lastly, the use of the woven natural fibre reinforced thermoset composites was well-fitted in the scheme in UN climate change conference 2022, to combat with global warming. The analytic report also found up-trend market size for the natural fibre composites due to the promising reduction of component mass, energy consumption and overall costing. Unfortunately, the application of woven natural fibre thermoset composites was found very challenging in the IR4.0 (Industrial Revolution 4.0) and aircraft interior components. With further study in the future, the woven natural fibre thermoset composites should extend its influences in these area. The authors would like to express their gratitude towards the Universiti Putra Malaysia for grant 9300426 (Development of New Hybrid Material for Blast-Proof Aircraft Cargo Container.

References

1. Ngo TT, Kohl JG, Paradise T, Khalily A, Simonson DL. Improving mechanical properties of thermoset biocomposites by fiber coating or organic oil addition. Int J Polym Sci 2015;2015:840823.
2. Al-Maharma AY, Al-Huniti N. Critical review of the parameters affecting the effectiveness of moisture absorption treatments used for natural composites. J Compos Sci 2019;3:27.
3. Symington M, Banks W, West O, Pethrick R. Tensile testing of cellulose based natural fibers for structural composite applications. J Compos Mater 2009;43:1083–108.
4. Chaudhary V, Bajpai PK, Maheshwari S. Effect of moisture absorption on the mechanical performance of natural fiber reinforced woven hybrid bio-composites. J Nat Fibers 2020;17: 84–100.
5. Guzman VA, Brøndsted P. Effects of moisture on glass fiber-reinforced polymer composites. J Compos Mater 2014;49:911–20.
6. Lee CH, Khalina A, Lee SH. Importance of interfacial adhesion condition on characterization of plant-fiber-reinforced polymer composites: a review. Polymers 2021;13:438.
7. Arumugam S, Kandasamy J, Venkatesan S, Murugan R, Lakshmi Narayanan V, Sultan MTH, et al. A review on the effect of fabric reinforcement on strength enhancement of natural fiber composites. Materials 2022;15:3025.
8. Li M, Pu Y, Thomas VM, Yoo CG, Ozcan S, Deng Y, et al. Recent advancements of plant-based natural fiber–reinforced composites and their applications. Compos B Eng 2020;200:108254.
9. Cruz J, Fangueiro R. Surface modification of natural fibers: a review. Procedia Eng 2016;155:285–8.
10. May-Pat A, Valadez-González A, Herrera-Franco PJ. Effect of fiber surface treatments on the essential work of fracture of HDPE-continuous henequen fiber-reinforced composites. Polym Test 2013;32:1114–22.
11. Farokhi Nejad A, Bin Salim MY, Rahimian Koloor SS, Petrik S, Yahya MY, Abu Hassan S, et al. Hybrid and synthetic FRP composites under different strain rates: a review. Polymers 2021;13:3400.
12. Osborne. Uses & applications of thermosetting plastics. Available from: https://www. osborneindustries.com/news/thermosetting-plastic-uses-applications/ [Accessed 18 Feb 2022].
13. Aisyah HA, Paridah MT, Sapuan SM, Khalina A, Berkalp OB, Lee SH, et al. Thermal properties of woven kenaf/carbon fibre-reinforced epoxy hybrid composite panels. Int J Polym Sci 2019;2019: 5258621.
14. Cihan M, Sobey A, Blake J. Mechanical and dynamic performance of woven flax/E-glass hybrid composites. Compos Sci Technol 2018:172. https://doi.org/10.1016/j.compscitech.2018.12.030.
15. Djafar Z, Renreng I, Jannah M. Tensile and bending strength analysis of ramie fiber and woven ramie reinforced epoxy composite. J Nat Fibers 2020;18:1–12.
16. Jawaid M, Abdul Khalil HPS, Bakar A. Woven hybrid composites: tensile and flexural properties of oil palm-woven jute fibres based epoxy composites. Mater Sci Eng 2011;528:5190–5.
17. Kunze W, Möhler H. Characterization of processing parameters for injection molding of thermoplastics and thermosetting plastics using thermal analysis. Thermochim Acta 1985;83: 47–58.
18. Aisyah H, Tahir MP, Sapuan S, Khalina A, Ilyas RA, Norizan MN. A review of biocomposites in biomedical application. In: Sapuan SM, Nukman Y, Abu Osman NA, Ilyas RA, editors. Composites in Biomedical Applications, 1st ed. CRC Press; 2020.
19. Khandai S, Nayak R, Kumar A, Das D, Kumar R. Assessment of mechanical and tribological properties of flax/kenaf/glass/carbon fiber reinforced polymer composites. Mater Today Proc 2019;18:3835–41.
20. Morris RH, Geraldi NR, Stafford JL, Spicer A, Hall J, Bradley C, et al. Woven natural fibre reinforced composite materials for medical imaging. Materials 2020;13:1684.

21. Salman S, Leman ZB. Physical mechanical and ballistic properties of kenaf fiber reinforced poly vinyl butyral and its hybrid composites. In: Sapuan SM, Ismail H, Zainudin ES, editors. Woodhead publishing series in composites science and engineering, natural fibre reinforced vinyl ester and vinyl polymer composites, Woodhead Publishing; 2018.
22. Sekaran ASJ, Kumar KP. Study on drilling of woven sisal and aloevera natural fibre polymer composite. Mater Today Proc 2019;16:640–6.
23. Yorseng K, Mavinkere Rangappa S, Parameswaranpillai J, Siengchin S. Influence of accelerated weathering on the mechanical, fracture morphology, thermal stability, contact angle, and water absorption properties of natural fiber fabric-based epoxy hybrid composites. Polymers 2020;12: 2254.
24. Ghoushji MJ, Eshkoor RA, Zulkifli R, Sulong AB, Abdullah S, Azhari CH. Energy absorption capability of axially compressed woven natural ramie/green epoxy square composite tubes. J Reinf Plast Compos 2017;36:1028–37.
25. Salman S, Leman Z, Hameed Sultan MT, Ishak M, Cardona F. The effects of orientation on the mechanical and morphological properties of woven kenaf-reinforced poly vinyl butyral film. Bioresources 2016;11:1176–88.
26. McDaniels K, Downs RJ, Meldner H, Beach C, Adams C. High strength-to-weight ratio non-woven technical fabrics for aerospace applications. AIAA balloon system conference, Seattle, Washington, 4–7 May 2009.
27. Sepe R, Pozzi A. Mechanical properties of woven natural fiber reinforced composites. ECCM15 – 15th European conference on composite materials, Venice, Italy, 24–28 June 2012.
28. Liu Q, Hughes M. The fracture behaviour and toughness of woven flax fibre reinforced epoxy composites. Compos Appl Sci Manuf 2008;39:1644–52.
29. Lee CH, Khalina A, Nurazzi NM, Norli A, Harussani MM, Rafiqah SA, et al. The challenges and future perspective of woven kenaf reinforcement in thermoset polymer composites in Malaysia: a review. Polymers 2021;13:1390.
30. Marya R, Abdellaoui H, Essabir H, Kakou C-A, Bouhfid R, Qaiss A. Prediction of the cyclic durability of woven-hybrid composites. In: Jawaid M, Thariq M, Saba N, editors. Woodhead publishing series in composites science and engineering, durability and life prediction in biocomposites, fibre-reinforced composites and hybrid composites. Woodhead Publishing; 2018, vol 35.
31. Laurenzi S, Marchetti M. Advanced composite materials by resin transfer molding for aerospace applications. Hu N, editor. Composites and their properties. IntechOpen; 2012.
32. Bhat P, Merotte J, Simacek P, Advani S. Process analysis of compression resin transfer molding. Compos Appl Sci Manuf 2009;40:431–41.
33. Simacek P, Advani S. Simulating three-dimensional flow in compression resin transfer molding process. Rev Eur Des Eléments Finis 2012;2005:777–802.
34. Advani S, Sozer E, Mishnaevsky L Jr. Process modeling in composite manufacturing. Appl Mech Rev 2003;56. https://doi.org/10.1115/1.1584418.
35. Hang X, Li Y, Hao X, Li N, Wen Y. Effects of temperature profiles of microwave curing processes on mechanical properties of carbon fibre–reinforced composites. Proc Inst Mech Eng B J Eng Manuf 2017;231:1332–40.
36. Hao L-C, Li Z-X, Sun F, Ding K, Zhou X-N, Song Z-X, et al. High-performance epoxy composites reinforced with three-dimensional Al_2O_3 ceramic framework. Compos Appl Sci Manuf 2019;127: 105648.
37. Hu Y, Du G, Chen N. A novel approach for Al2O3/epoxy composites with high strength and thermal conductivity. Compos Sci Technol 2016;124:36–43.
38. Umamheshwar rao RS, Mahender T. Mechanical properties and optimization of processing parameters for epoxy/glass fiber reinforced composites. Mater Today Proc 2019;19:489–92.

39. Ramírez-Herrera CA, Cruz-Cruz I, Jiménez-Cedeño IH, Martínez-Romero O, Elías-Zúñiga A. Influence of the epoxy resin process parameters on the mechanical properties of produced bidirectional [±45°] carbon/epoxy woven composites. Polymers 2021;13:1273.

40. Raghavendra G, Ojha S, Acharya S, Pal S. Jute fiber reinforced epoxy composites and comparison with the glass and neat epoxy composites. J Compos Mater 2014;48:2537–47.

41. Bhudolia SK, Perrotey P, Gohel G, Joshi SC, Gerard P, Leong KF. Optimizing bladder resin transfer molding process to manufacture complex, thin-ply thermoplastic tubular composite structures: an experimental case study. 2021;13:4093.

42. Azman Mohammad Taib MN, Julkapli NM. 4 – dimensional stability of natural fiber-based and hybrid composites. In: Jawaid M, Thariq M, Saba N, editors. Mechanical and physical testing of biocomposites, fibre-reinforced composites and hybrid composites. Woodhead Publishing; 2019: 61–79 pp.

43. Tatsuno D, Yoneyama T, Kawamoto K, Okamoto M. Hot press forming of thermoplastic CFRP sheets. Procedia Manuf 2018;15:1730–7.

44. Peerzada M, Abbasi S, Lau KT, Hameed N. Additive manufacturing of epoxy resins: materials, methods, and latest trends. Ind Eng Chem Res 2020;59:6375–90.

45. Latif R, Wakeel S, Khan N, Siddiquee A, Verma S, Khan Z. Surface treatments of plant fibers and their effects on mechanical properties of fiber-reinforced composites: a review. J Reinforc Plast Compos 2018;38. https://doi.org/10.1177/0731684418802022.

46. Woo K, Lim JH, Han C. Effect of defects on progressive failure behavior of plain weave textile composites. 2021;14:4363.

47. Beaugrand J, Guessasma S, Maigret J-E. Damage mechanisms in defected natural fibers. Sci Rep 2017;7:14041.

48. Yan L, Yuan X. Improving the mechanical properties of natural fibre fabric reinforced epoxy composites by alkali treatment. J Reinforc Plast Compos 2012;31:425–37.

49. Plenco. Basic thermoset part design suggestions. Wisconsin, USA: Plenco: Plastics Engineering Company; 2015.

50. WWF (n.d.). Time to shift from pledges to action. Available from: https://wwf.panda.org/discover/our_focus/climate_and_energy_practice/cop27/ [Accessed 17 May 2022].

51. Nations U. (n.d.). Envision2030: 17 goals to transform the world for persons with disabilities. Available from: https://www.un.org/development/desa/disabilities/envision2030.html [Accessed 17 May 2022].

52. Research GV. GVR Report cover Natural Fiber Composites (NFC) Market Size, Share & Trends Report. Natural Fiber Composites (NFC) market size, share & trends analysis report by raw material, by matrix, by technology, by application, and segment forecasts 2018–2024; 2017:139 p.

53. Luthra P, Vimal KK, Goel V, Singh R, Kapur GS. Biodegradation studies of polypropylene/natural fiber composites. SN Appl Sci 2020;2:512.

54. Tchana Toffe G, Oluwarotimi Ismail S, Montalvão D, Knight J, Ren G.. A scale-up of energy-cycle analysis on processing non-woven flax/PLA tape and triaxial glass fibre fabric for composites. J Manuf Mater Process 2019;3:92.

55. Nugroho WT, Dong Y, Pramanik A. Chapter 4 – 3D printing composite materials: a comprehensive review. In: Low I-M, Dong Y, editors. Compos Mater. Elsevier; 2021:65–115 pp.

56. Lee CH, Sapuan SM, Hassan MR. Thermal analysis of kenaf fiber reinforced floreon biocomposites with magnesium hydroxide flame retardant filler. Polym Compos 2018;39:869–75.

Fathin Sakinah Mohd Radzi*, Anuar Abu Bakar,
Mohd Azman Asyraf, Nik Adib Nik Abdullah and Mat Jusoh Suriani*

5 Manufacturing defects and interfacial adhesion of *Arenga Pinnata* and kenaf fibre reinforced fibreglass/kevlar hybrid composite in boat construction application

Abstract: In recent years, *Arenga Pinnata* and kenaf fibres have been discovered to have a high potential for usage as fibre reinforcement in material matrix composites for a several of application. The scope for this study is to encourage widespread use of eco hybrid composite in various applications specifically in the maritime field. The purpose of this study is to look into the influence of fibre loading on manufacturing defects and interfacial adhesion of *Arenga Pinnata* and kenaf fibre reinforced fiberglass/kevlar hybrid composite materials used in boat construction. The hybridization of natural fibre with fiberglass/kevlar is recommended as a solution to overcome the disadvantages of natural fibre which can give balanced strength and stiffness, enhances fatigue resistance, fracture toughness and impact resistance. General conditions in green composites are proposed, along with some preliminary data on the mechanical hybrid composites. In conclusion, the percentage of *Arenga Pinnata* and kenaf fibre contents that show reduces manufacturing defects and excellent interfacial adhesion will be proposed for boat construction.

Keywords: *Arenga Pinnata*; interfacial adhesion; Kenaf; manufacturing defects; natural fibre composites.

*Corresponding authors: Fathin Sakinah Mohd Radzi, Faculty of Ocean Engineering Technology and Informatics, Universiti Malaysia Terengganu, 21030 Kuala Nerus, Terengganu, Malaysia, E-mail: fathinsakinah96@gmail.com. https://orcid.org/0000-0002-5348-3101; and Mat Jusoh Suriani, Faculty of Ocean Engineering Technology and Informatics, Universiti Malaysia Terengganu, 21030 Kuala Nerus, Terengganu, Malaysia; and Marine Materials Research Group, Faculty of Ocean Engineering Technology and Informatics, Universiti Malaysia Terengganu, 21030 Kuala Nerus, Terengganu, Malaysia, E-mail: surianimatjusoh@umt.edu.my
Anuar Abu Bakar, Faculty of Ocean Engineering Technology and Informatics, Universiti Malaysia Terengganu, 21030 Kuala Nerus, Terengganu, Malaysia; and Marine Materials Research Group, Faculty of Ocean Engineering Technology and Informatics, Universiti Malaysia Terengganu, 21030 Kuala Nerus, Terengganu, Malaysia, E-mail: anuarbakar@umt.edu.my
Mohd Azman Asyraf, Faculty of Ocean Engineering Technology and Informatics, Universiti Malaysia Terengganu, 21030 Kuala Nerus, Terengganu, Malaysia, E-mail: asyrafazman23@yahoo.com
Nik Adib Nik Abdullah, MSET Inflamable Composit Sdn. Bhd. PT 7976K Gong Badak Industrial Zone, 21030 Kuala Nerus, Terengganu, Malaysia

As per De Gruyter's policy this article has previously been published in the journal Physical Sciences Reviews. Please cite as:
F. S. M. Radzi, A. Abu Bakar, M. A. Asyraf, N. A. Nik Abdullah and M. J. Suriani "Manufacturing defects and interfacial adhesion of *Arenga Pinnata* and kenaf fibre reinforced fibreglass/kevlar hybrid composite in boat construction application" *Physical Sciences Reviews* [Online] 2022. DOI: 10.1515/psr-2022-0078 | https://doi.org/10.1515/9783110769227-005

5.1 Introduction

In the last few years, composite materials are one of the rising materials. The application of composite materials has developed relentlessly prevailing conquering new markets and supplanting traditional metal alloys. Metal composites, ceramic composites, polymer composites, and composite building materials are the most common types of composite materials [1]. Composite materials, in specific, are being constructed and reinvented with the goal of improving and adapting traditional products while also developing innovative product in a better and responsible manner [2]. Nowadays, because of the growing global social awareness about environmental effect, renewable energy sources, sustainability and polymer natural fibre composites have recently piqued the interest of researchers due to the fact that they are biodegradable [3]. Basically, natural fibres can be divided into two categories which are animal fibres and plant fibres [4]. Recently, natural fibres have been used in conjunction with plastics. Natural fibres are completely biodegradable [5, 6], environmentally friendly, abundantly, reduced health hazard, renewable [6], inexpensive [7] and have a low density [8]. Regrettably, natural fibres' proneness to moisture and humidity, as well as the resulting deterioration in mechanical properties, can make it difficult for them to be accepted for exterior applications in the industry [9]. Thus, many researchers try to overcome the problem by implementing the hybridization of composites within the synthetic and natural fibre as well as it can reduce the cost of material and increased the physical, mechanical and thermal properties of the produced composites [10, 11]. Hybrid composite materials are one of the burgeoning disciplines in composite science research, and they're gaining traction in a variety of industries [12]. The hybridization enhances the environmental friendliness of synthetic fibre composites as well [13]. Generally, mechanical characteristics of hybrid composites were studied using natural/natural fibre, natural/synthetic fibre, and natural/natural or natural/synthetic/ additive enhanced reinforced polymeric materials [14]. Natural fibres are frequently used in conjunction with synthetic fibre-based polymer composites in current times. Kenaf and *Arenga Pinnata* fibres are among the most promising in this study, and they have lately gained increased interest for combining with synthetic fibres to generate hybridised composites [12]. There are many advantages of hybridization natural fibre composite with synthetic fibre that can be proposed in the maritime field in boat construction.

5.2 Materials and methods

The fibres were provided in random and long fibre form of *Arenga Pinnata* and kenaf fibres that purchased from Hafiz Adha Enterprise (Industri Enau Malaysia) and Innovative Pultrusion Sdn. Bhd., Negeri Sembilan, Malaysia, respectively. The fibreglass (CSM 225), kevlar and polyester resin were supplied by MSET Inflatable Composite Corporation Sdn. Bhd., Terengganu.

5.2.1 Preparation of materials

Arenga Pinnata and kenaf that have been treated with (NaOH) were gathered and cut into 23 cm lengths. The *Arenga Pinnata* and kenaf were combed to unravel the fibres' tight connection. Besides, the mechanical properties of *Arenga Pinnata* and kenaf fibres show increment from the combed fibre [15].

The fabrication of this hybrid composites were prepared by using *Arenga Pinnata* and kenaf fibres and the polyester resin act as the matrix. The sample were prepared by different weight percentage composition of *Arenga Pinnata* and kenaf fibres reinforced hybrid fiberglass/kevlar polymeric composites such as 0, 30, 45, 60 and 75%. This research was used the traditional method which is hand lay-up technique in a mould steel. During the fabrication process, the mould steel was waxed and cleaned before being covered with numerous layers of kevlar, fibreglass, and natural fibres, *Arenga Pinnata* and kenaf fibres, to prevent materials adhering to the mould after drying. To overcome the bubbles or voids in the samples, they were all finished with a roller, which assists the matrix that extends beyond the region being rolled, so pushing air and bubbles out of the samples [15]. Then, the moulds was compressed and leaved for 24–48 h in room temperature to dry.

5.2.2 Impact test

For the determination of strength of *Arenga Pinnata* and kenaf fibres reinforced hybrid fiberglass/kevlar polymeric composite samples were performed Charpy low-velocity by using LS-22006-50 Charpy Impact tester followed ASTM D6110. The test specimens were cut by following the dimension of 15×10 (cm²) $\times 0.5$ cm (thickness). A pendulum with a specified mass linked to a spinning arm attached to the machine body is used in the Charpy low-velocity impact test. The pendulum swing descends from a height at an angle of 131° and strikes the test samples, absorbing some of the pendulum's kinetic energy [15].

5.2.3 Scanning electron microscopy

The performance during the impact test of the specimen depends on the manufacturing defects and interfacial bonding of *Arenga Pinnata* and kenaf fibres reinforced hybrid fibreglass/kevlar polymeric composite that can be observed through the scanning electron microscopy (SEM) model Hitachi TM-1000 software as shown in Figure 5.2.1 below. The specimens must be dried off by using an oven and then coated with a thin layer of pure gold using sputter-coated equipment.

5.3 Result and discussion

The impact properties of *Arenga Pinnata* and Kenaf fibres reinforced hybrid fiberglass/kevlar polymeric composite materials were investigated by using the Charpy low-velocity impact test. The absorption of energy and impact strength was studied. To compare the energy absorption capabilities of different *Arenga Pinnata* and kenaf fibres content under the same impact energies the impact load was setting to 30.25 kg and the angle is 131° at room temperature. The total energy required to break the specimen is represented by the absorbed impact energy (in Joules). The impact strength

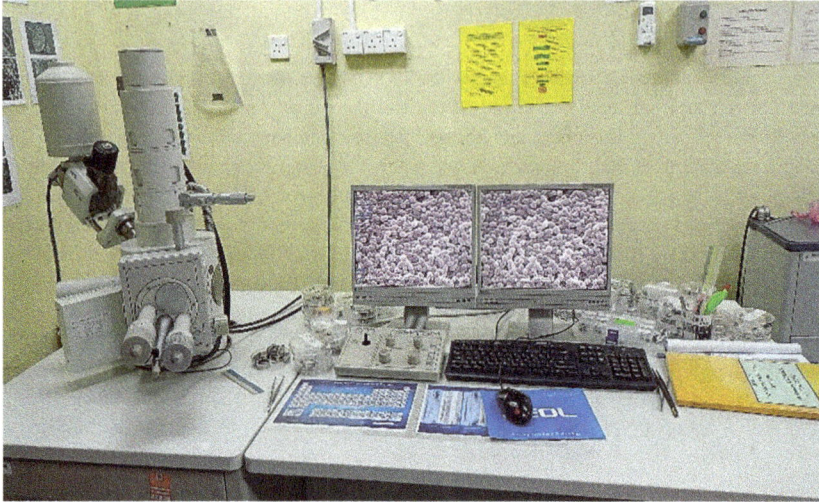

Figure 5.2.1: Scanning electron microscopy (SEM).

(J/mm^2) was estimated by dividing the absorbed impact energy by the sample cross-section area [16].

Figure 5.3.1 shows the absorbed energy values of specimens with the *Arenga Pinnata* and kenaf fibre content. The result shows the increment of fibre content in the specimens exhibit the higher impact strength and shows more energy absorbs. It observed that the increment of energy absorbs in the specimen until 60% of fibre contents and 75% of fibre contents shows the decrement absorbed of energy. In Figure 5.3.1 shows the absorption energy values with respect to *Arenga Pinnata* fibre content that indicates energy absorbed values of specimens in 0 (control sample), 30, 45, 60 and 75% are 3.05, 6.99, 7.13, 7.56 and 6.85 J. For kenaf fibre content the values of absorption energy of specimens 0 (control specimen), 30, 45, 60 and 75% are 3.05, 3.97, 5.57, 8.71 and 5.00 J. The standard deviation also have been calculated for *Arenga Pinnata* and kenaf fibres in specimen 0 (control specimen), 30, 45, 60 and 75%. For the *Arenga Pinnata* fibres it shows 0.005774, 0.005774, 0.01, 0.005774, 0.01 and for Kenaf fibres shows 0.005774, 0.011547, 0.005859, 0.005132 and 0.015275 respectively. From both results energy absorbs of *Arenga Pinnata* and kenaf with different of fibre content it show that the absorption of energy kenaf fibre in 60% of fibre content is higher than the *Arenga Pinnata* fibre. This enhancement was made possible by a greater number of high-strength fibres capable of successfully transferring impact stress [16]. Many factors determine the impact characteristics of fibre composites, including interfacial bond strength, matrix, and fibre properties. A composite's impact failure is caused by variables such as fibre/matrix debonding, fibre or matrix fracture and fibre pull out [17]. From Figure 5.3.1, the absorption of energy decreases in 75% of fibre content due to uneven polyester resin impregnated with the fibre. This may cause weak interfacial

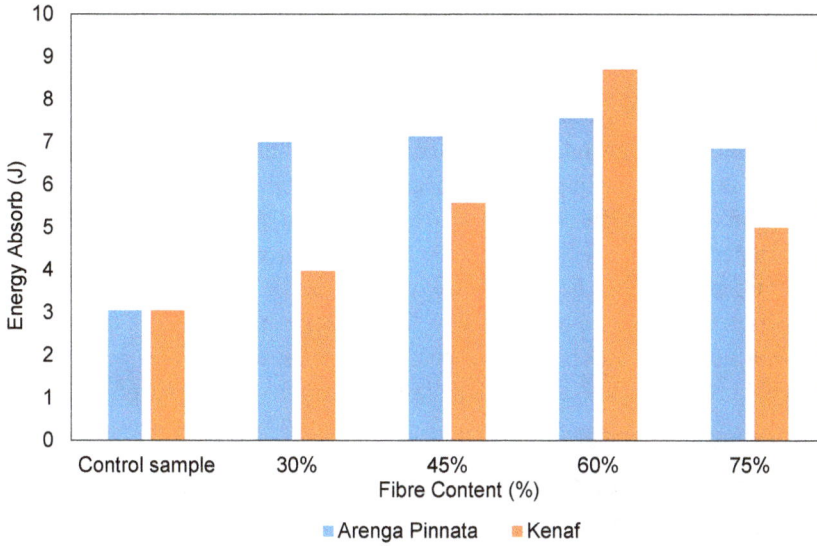

Figure 5.3.1: Energy absorbs of *Arenga Pinnata* and kenaf fibres reinforced hybrid fibreglass/kevlar polymeric composite.

adhesion between the fibres with resin and delamination which affect the energy values for this *Arenga Pinnata* and kenaf fibres content.

In Figure 5.3.2 depicts the impact strength of *Arenga Pinnata* and kenaf fibre reinforced hybrid fibreglass/kevlar polymeric composites. The result shows that 60% of

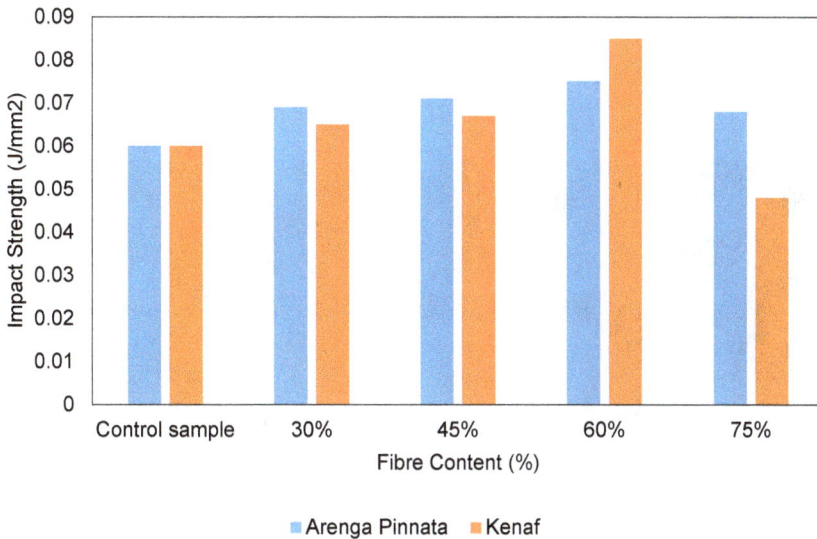

Figure 5.3.2: Impact strength of *Arenga Pinnata* and kenaf fibres reinforced hybrid fibreglass/kevlar polymeric composite.

kenaf fibre reinforced hybrid fibreglass/kevlar polymeric composite has higher impact strength than *Arenga Pinnata* fibre content. The association between *Arenga Pinnata* and kenaf fibre loading with impact strength of hybrid polymeric composites has a significant dispersion, with a tendency of decreasing impact strength as *Arenga Pinnata* and kenaf fibre loading increases [16]. It observed that the 60% of *Arenga Pinnata* and kenaf fibre is the highest impact strength among their type of fibre which are 0.075 and 0.085 J/mm^2, respectively. This is due to an increment in fibre content exceeding 60% causes a decrease in energy absorption. This is because the *Arenga Pinnata* and kenaf fibre content of 75% exceeds the threshold amount [15]. The higher fibre content in the composite which is 75% of fibre, the volume of resin in the composite might be too small that affect the mechanical properties of composites [17]. In general, the proportion of fibre content has a significant influence on the impact energy [15].

5.3.1 Determination of manufacturing defect by using scanning electron microscopy

The Charpy low-velocity impact test fracture surface was determined by scanning electron microscopy (SEM) to understand the manufacturing defects and interfacial adhesion correlation impact behaviour of *Arenga Pinnata* and kenaf fibres reinforced hybrid fibreglass/kevlar polymeric composite. Figure 5.3.3a and b show 75% of *Arenga Pinnata* and kenaf fibre reinforced hybrid fibreglass/kevlar polymeric composite manufacturing defects.

Figure 5.3.3a depicts voids that occur in the specimen 75% *Arenga Pinnata* fibre reinforced hybrid fibreglass/kevlar polymeric composite when there are air bubbles within the framework that happen during the fabricating process. There are other elements that might generate voids, such as the surroundings condition and the resin system. For Figure 5.3.3b observed that the fibre was pulled out and void was clearly seen

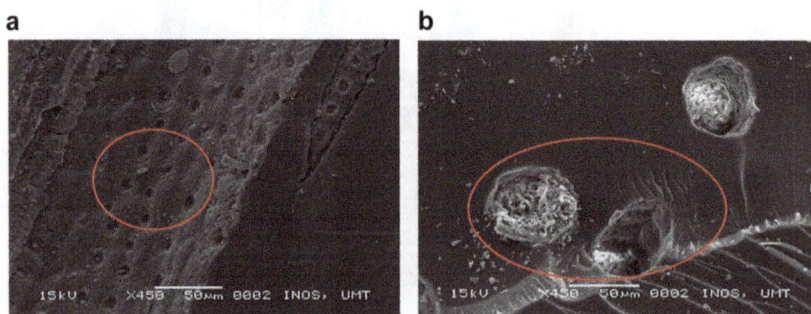

Figure 5.3.3: a. Voids of 75% *Arenga Pinnata* fibre reinforced hybrid fiberglass/kevlar polymeric composite. b. Fibre was pulled out and void clearly seen of 75% *Arenga Pinnata* fibre reinforced hybrid fibreglass/kevlar polymeric composite.

a

b

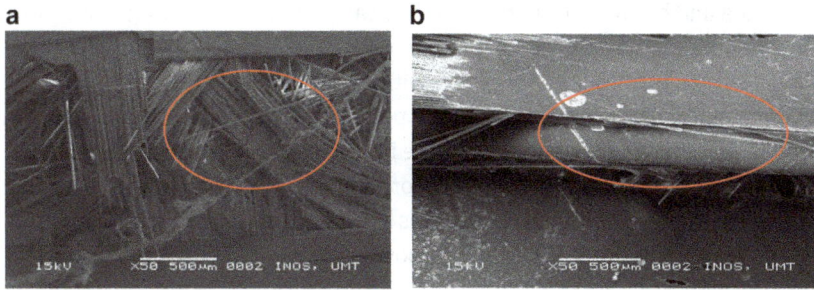

Figure 5.3.4: a. Fibre bending and misalignment of 75% kenaf fibre reinforced hybrid fibreglass/ kevlar polymeric composite. b. Resin-rich zone and matrix crack of 30% kenaf fibre reinforced hybrid fiberglass/kevlar polymeric composite.

in the specimen. Poor fibre dispersion and fibre matrix interfacial adhesion are the factors of fibre pull out. Furthermore, one of disadvantage natural fibre is low compatibility with the matrix, which has resulted in fibre matrix debonding and can cause fibre pull out.

Figure 5.3.4a show fibre bending and misalignment of 75% kenaf fibre hybrid specimen. The manufacturing defects that occur might affect low mechanical properties of hybrid specimen. Poor dispersion of fibre matrix during fabrication process could be the factor of fibre bending and misalignment composite. From Figure 5.3.4b it can be observed that resin-rich zone and matrix crack that occur in 30% of kenaf fibre reinforced hybrid fibreglass/kevlar polymeric composite. It was also claimed that because of the presence of moisture, the swelling of the resin induced deformation in the resin-rich areas, implying that the poorer interfaces near those areas were susceptible to greater loads and hence more prone to break or matrix split [18].

5.3.2 Correlation of manufacturing defects and interfacial adhesion to impact properties

Manufacturing defects that exhibit in the specimens can give effect to the impact properties. Figure 5.3.3a and b depicts manufacturing defects, voids and fibre pull out in 75% of *Arenga Pinnata* fibre reinforced hybrid fibreglass/kevlar polymeric composite observed from scanning electron microscopy. From the result it is clear that the absorption of energy and impact strength of specimen of 75% of *Arenga Pinnata* tend to decrease due to manufacturing defects and poor compatibility of fibres with the matrix. The fibres that are not fully impregnated with resin might cause weak interfacial adhesion between fibres and matrix, which usually affects the total absorption of energy values. The uneven load transfer due to poor dispersion of fibres also led to the decrement in the impact properties.

Figure 5.3.4a and b show fibre bending and misalignment of 75% of kenaf fibre and resin-rich zone and matrix crack in 30% of kenaf fibre reinforced hybrid fiberglass/kevlar polymeric composites. The specimen with high fibre content have weak interfacial adhesion fibre matrix due to the lower percentage of polyester resin that can cause the fibre bending and misalignment fibre in the composites. The resin-rich zone and matrix crack also can affect the values absorption energy of impact strength in the specimen. Hence, from the obtained result it is important to determine the optimum ratio of fibre to matrix in fabricating the composite samples to get high impact properties value.

5.4 Conclusions

In this study, the absorption of energy and impact strength of the *Arenga Pinnata* and kenaf fibres reinforced hybrid fibreglass/kevlar polymeric composite increase with the increment of the fibre content percentage from 0 (control sample), 30, 45 and 60% of each type of natural fibres. The 75% of *Arenga Pinnata* and kenaf fibres reinforced hybrid fibreglass/Kevlar polymeric composite decreased in the impact properties caused by the manufacturing defects in the specimen and poor interfacial adhesion of fibre matrix composites. The highest values absorption of energy and impact strength in *Arenga Pinnata* and kenaf fibres was in 60% of fibre content which are 7.56 J and 0.075 J/mm^2 for *Arenga Pinnata* and 8.71 J and 0.085 J/mm^2 for kenaf fibre, respectively. The values of impact properties kenaf fibre reinforced hybrid fibreglass/kevlar polymeric composite is higher compared to *Arenga Pinnata* fibres. The manufacturing defects such as voids, fibre pull out, misalignment and resin-rich zone that occur due to weak interfacial adhesion fibre matrix directly influenced low performance of impact properties. Thus, the least manufacturing defects in the composites show the good impact behaviour and composites.

References

1. Ramasamy M, Daniel AA, Nithya M, Kumar SS, Pugazhenthi R. Characterization of natural – synthetic fiber reinforced epoxy based composite - hybridization of kenaf fiber and kevlar fiber. Mater Today Proc 2020;37:1699–705.
2. Peças P, Carvalho H, Salman H, Leite M. Natural fibre composites and their applications: a review. J. Compos. Sci 2018;2:66.
3. Zamri MH, Akil HM, MohdIshak ZA. Pultruded kenaf fibre reinforced composites: effect of different kenaf fibre yarn tex. Procedia Chem 2016;19:577–85.
4. Nurazzi NM, Asyraf MRM, Rayung M, Norrrahim MNF, Shazleen SS, Rani MSA, et al. Thermogravimetric analysis properties of cellulosic natural fiber polymer composites: a review on influence of chemical treatments. Polymers 2021;13. https://doi.org/10.3390/polym13162710.

5. Atiqah A, Jawaid M, Sapuan SM, Ishak MR, Ansari MNM, Ilyas RA. Physical and thermal properties of treated sugar palm/glass fibre reinforced thermoplastic polyurethane hybrid composites. J Mater Res Technol 2019;8:3726–32.

6. Mohd Nurazzi N, Khalina A, Sapuan SM, Ilyas RA. Mechanical properties of sugar palm yarn/woven glass fiber reinforced unsaturated polyester composites: effect of fiber loadings and alkaline treatment. Polimery/Polymers 2019;64:665–75.

7. Asyraf MRM, Ishak MR, Norrrahim MNF, Nurazzi NM, Shazleen SS, Ilyas RA, et al. Recent advances of thermal properties of sugar palm lignocellulosic fibre reinforced polymer composites. Int J Biol Macromol 2021;193:1587–99.

8. Bhambure S, Rao AS. Experimental investigation of impact strength of kenaf fiber reinforced polyester composite. Mater Today Proc 2021;46:1134–8.

9. Tamrakar S, Kiziltas A, Mielewski D, Zander R. Characterization of kenaf and glass fiber reinforced hybrid composites for underbody shield applications. Compos Part B 2021;216:108805.

10. Suriani MJ, Sapuan SM, Ruzaidi CM, Nair DS, Ilyas RA. Flammability, morphological and mechanical properties of sugar palm fiber/polyester yarn-reinforced epoxy hybrid biocomposites with magnesium hydroxide flame retardant filler. Textil Res J 2021;91:2600–11.

11. Nurazzi NM, Khalina A, Sapuan SM, Ilyas RA, Rafiqah SA, Hanafee ZM. Thermal properties of treated sugar palm yarn/glass fiber reinforced unsaturated polyester hybrid composites. J Mater Res Technol 2020;9:1606–18.

12. Fernando G. Hybrid composites a review. Acta Mater 2009;23:314–45.

13. Rakesh P, Diwakar V, Venkatesh K, Savannananavar RN. A concise review on processing of hybrid composites produced by the combination of glass and natural fibers. Mater Today Proc 2019;22: 2016–24.

14. Senthilkumar K, Saba N, Rajini N, Chandrasekar M, Jawaid M, Siengchin S, et al. Mechanical properties evaluation of sisal fibre reinforced polymer composites: a review. Construct Build Mater 2018;174:713–29.

15. Suriani MJ, Sapuan SM, Ruzaidi CM, Naveen J, Syukriyah H, Ziefarina M. Correlation of manufacturing defects and impact behaviors of kenaf fiber reinforced hybrid fiberglass/Kevlar polyester composite. Polimery/Polymers 2021;66:30–5.

16. Yahaya R, Sapuan SM, Jawaid M, Leman Z, Zainudin ES. Mechanical performance of woven kenaf-Kevlar hybrid composites. J Reinf Plast Compos 2014;33:2242–54.

17. Öztürk S. Effect of fiber loading on the mechanical properties of kenaf and fiberfrax fiber-reinforced phenol-formaldehyde composites. J Compos Mater 2010;44:2265–88.

18. Suriani MJ, Ruzaidi CM, Wan Nik WB, Zulkifli F, Sapuan SM, Ilyas RA, et al. Effect of fibre contents toward manufacturing defects and interfacial adhesion of arenga pinnata fibre reinforced fibreglass/kevlar hybrid composite in boat construction. J Phys Conf Ser 1960;1:2021.

Mohd Khairun Anwar Uyup*, Lee Seng Hua, Alia Syahirah Yusoh,
Asniza Mustapha and Siti Norasmah Surip

6 Wettability of keruing (*Dipterocarpus* spp.) wood after weathering under tropical climate

Abstract: Exposure of unprotected wood to weathering can increase the wettability of the wood and the exposure period should be carefully monitored to preserve surface quality from severe deterioration. This study investigated the wettability of keruing (*Dipterocarpus* spp.) wood after weathering exposure for 1–4 weeks. The keruing samples were first planed and coated at the edges prior to expose under tropical climate. Contact angle, crack formation and lignin content of the samples were recorded on weekly basis. The results showed that the wettability of weathered keruing wood increased tremendously (i.e., 32%) after 4 weeks of exposure compared to control samples. The increase in wettability can be associated with the increase in crackformation and reduction of lignin content after exposure. It is hereby recommended that keruing wood is not suitable to be exposed to weather for more than 3 weeks before finishing or coating is applied on its surface.

Keywords: keruing; lignin; pre-weathering; surface quality; tropical climate; wettability.

6.1 Introduction

Keruing (*Dipterocarpus* spp.) is a medium hardwood with an air-dried density of 690–945 kg/m³. Generally, keruing timber is moderately durable to non-durable especially under exposed condition in tropics condition such as Malaysia [1]. Keruing falls into strength group B [2] or SG 5 [3] and is suitable for heavy construction. In Malaysia, keruing is one of the commonly used timbers for outdoor applications and also known as a material of choice for structural and building material. However, its uses are highly dependent on its properties i.e., strength and stiffness, finishing characteristics as well

*Corresponding author: Mohd Khairun Anwar Uyup,** Forest Products Division, Forest Research Institute Malaysia, 52109 Kepong, Selangor, Malaysia, E-mail: mkanwar@frim.gov.my. https://orcid.org/0000-0002-3787-9907
Lee Seng Hua, Institute of Tropical Forestry and Forest Products (INTROP), Universiti Putra Malaysia, 43400 UPM, Serdang, Selangor, Malaysia
Alia Syahirah Yusoh and Asniza Mustapha, Forest Products Division, Forest Research Institute Malaysia, 52109 Kepong, Selangor, Malaysia
Siti Norasmah Surip, Faculty of Applied Sciences, Universiti Teknologi MARA (UiTM), 40450 Shah Alam, Selangor, Malaysia

As per De Gruyter's policy this article has previously been published in the journal Physical Sciences Reviews. Please cite as: M. K. A. Uyup, L. Seng Hua, A. S. Yusoh, A. Mustapha and S. N. Surip "Wettability of keruing (*Dipterocarpus* spp.) wood after weathering under tropical climate" *Physical Sciences Reviews* [Online] 2022. DOI: 10.1515/psr-2022-0079 | https://doi.org/10.1515/9783110769227-006

as maintenance requirements. It is a well-known fact that weathering negatively influences overall quality of the wood and wood products used for exterior purposes. Long term outdoor or natural exposure of unprotected wood can cause severe surface degradation especially before being coated with paint or other finishes [4]. Besides, it is important to note that weathering before finishing (pre-weathering) will result in mechanical stress, chemical and physical changes on the wood surfaces that weakens the future applied paints and coating on the wood surface [5].

On the other hand, the wettability of wood is an important criterion in resulting optimal adhesive bond strength. It provides a series of information on the interaction between the wood surface and liquids i.e., water, coating, and adhesives [6, 7]. A good adhesive bonding is favourable in the production of engineered wood products with superior performance [8]. Liquid adhesive must be able to wet the wood surface, flowing over the surface of wood as well as penetrating into the wood to attain a good adhesive bonding. In most cases, poor surface quality of wood could negatively affect the adhesive bonding. For example, rougher wood surfaces with highly distributed peaks, valleys and crevices over the wood could lead to air pockets and blockages, which prevented complete adhesive wetting and introduced stress concentrations when the adhesive had cured [9]. A study by Hiziroglu et al. [10] reported that the surface roughness of five Malaysian wood species increased as a result of outdoor exposure.

Wettability of wood is positively correlated to the surface roughness of wood as the higher the surface roughness, the higher the wettability of wood [11]. Therefore, the wettability of post-weathering wood is worth investigating to understand it changes over exposure periods. Surface wettability of wood is reported to increase when wood is exposed to weathering conditions. Studies by Kalnins et al. [12] stated that the wettability of western red cedar and southern pine wood increased with increasing weathering time, as indicates by the continuously decreasing contact angle over exposure time. Authors also mentioned that wood tends to lose its natural water repellence ability due to increase in wettability [13]. Nevertheless, the losing rate and its relation to the wood chemical constituents are not well known as limited studies have been conducted on this topic. Therefore, this study aims to investigate the effect of pre-weathering of keruing wood after outdoor exposure for a period of 1, 2, 3 and 4 weeks. The adhesion characteristics of keruing wood were determined by evaluating its wettability through contact angle measurement. On top of that, crack formation and the changes in the lignin content as a result of outdoor exposure were also determined.

6.2 Materials and methods

6.2.1 Preparation of samples

Keruing (*Dipterocarpus* spp.) timbers were obtained from a local sawmill. The timbers were planed into dimensions of 300 × 100 × 25 mm (length × width × thickness) using single-planner machine and were end coated at edges with epoxy paint before being exposed to the weather. A total of 40 samples were pre-weathered at the Forest Research Institute Malaysia (FRIM), Kepong, Selangor, Malaysia. Wood samples were secured on a steel rack oriented and exposed at 45° facing south as described by Hiziroglu et al. [10] and Anwar et al. [14] for 1–4 weeks. Meanwhile, 8 controls samples were stored in a conditioning room at 27 °C and 65% of relative humidity for comparison purpose. Both control and weathered samples were then tested for wettability and chemical constituents after weathering.

6.2.2 Observation of crack formation

The assessment of the samples was done visually at weathering site and in the lab from week one to week 4. The cracking was visibly observed by the naked eyes when there was a break extending through to the surface of the samples. The result of the observation was made based on photographic reference standards (ASTM D661-93) [15].

6.2.3 Evaluation of wettability

Five samples were collected on weekly basis for determination of the contact angle. The samples were trimmed into dimensions of 100 × 20 × 5 mm (length × width × thickness) and were conditioned at 27 ± 2 °C and 65 ± 3% of relative humidity for a week. Upon reached constant moisture content, 0.04 mg of distilled water was deposited on the surface of each sample using an injection tube. The apparent contact angle was measured immediately after the deposition on the sample surface. The measurement was made for every 1 min for a period of 10 min by using First Ten Angstroms (FTA 1000) wettability tester. The final time required for the water droplet to reach 0° was also recorded.

6.2.4 Evaluation of chemical analysis

Lignin content of the control and pre-weathered samples were determined according to the procedure specified in TAPPI T 222 om-88: Acid-insoluble lignin in wood and pulp [16]. Total of 15 samples were collected on weekly basis and planed from 1 mm of the top surface of the samples. The samples were then grounded into powder form prior to lignin content determination. The lignin content was calculated as followed:

$$\text{Lignin (\%)} = \frac{Q_3}{WX} \times 100$$

where,
- $Q_3 = Q_2 - Q_1$
- Q_2 = filter paper dry weight + lignin residue
- Q_1 = filter paper dry weight
- WX = sample oven dried from extraction test

6.2.5 Statistical analysis

The data for each test were statistically analyzed by analysis of variance (ANOVA) using the Statistical Analysis Software (SAS 9.4). The least significant difference test was used to determine which of the means were significantly different from one another. This method ranks the means by denoting different letters (a, b and c) to the factors to mark significance and calculates the least significant difference that occurs between the group means. Means followed by the same letters are not significantly different at $\propto\ <\ 0.05$. A regression analysis was carried out between percentage of crack and contact angle.

6.3 Results and discussion

6.3.1 Visual observation

Figure 6.1 shows the surface of keruing wood before (control) and after exposure to weather for 4 weeks. The surface of pre-weathered wood sample appeared to be darker than the control sample. Based on the observations, the wood surface will start to crack after being exposed for a week. Wood is prone to crack when unfinished wood is exposed to outdoor for longer duration. The crack size will also increase gradually over time and become more severe after 4 weeks exposure. According to Williams and Feist [17], the phenomenon of photo-chemical degradation is initially manifested by colour changes, followed by loosening of timber fibers, and will gradually erode the wood surface.

6.3.2 Wettability

Changes in wettability of both control and pre-weathered woods is confirmed by the contact angle measurement. Figure 6.2 illustrates the wettability of keruing wood after pre-weathered for 1–4 weeks. For comparison, samples without pre-weathering were used as controls. From the figure, it shows that the control samples exhibited lowest wettability as indicated by high contact angle. The initial contact angle of the control samples was 65°

Figure 6.1: Surface of keruing wood before and after outdoor exposure.

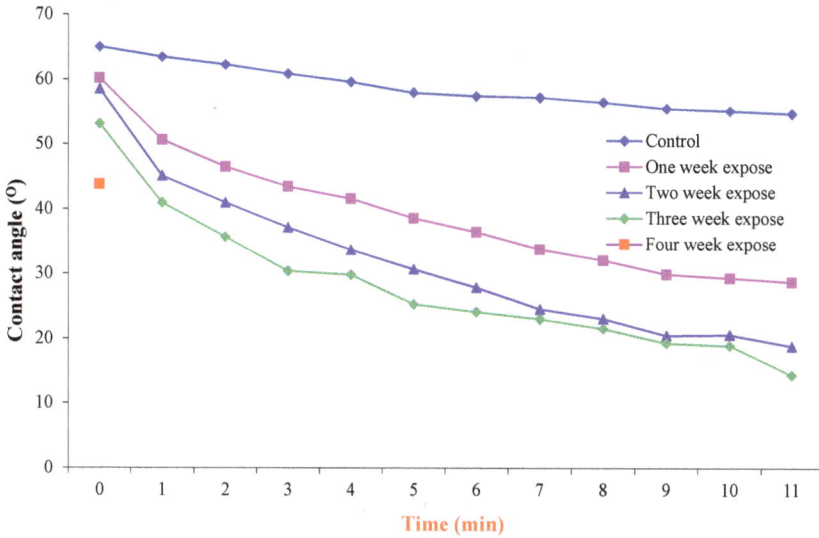

Figure 6.2: Wettability of keruing wood after exposure.

and reduced slightly to 55° after 11 min. Conversely, the wettability of all pre-weathered samples reduced markedly from approximately 60° to less than 30° after 11 min.

Figure 6.3 shows the mean of time required to complete surface wetting. In this study, the contact angle of control samples was relatively high and required the longest time to reach 0°. As illustrated in the figure, the water droplet reached 0° after 41 min as compared to an average of 12 min for the pre-weathered samples. The effect was much greater when the samples were pre-weathered for 4 weeks where the keruing wood

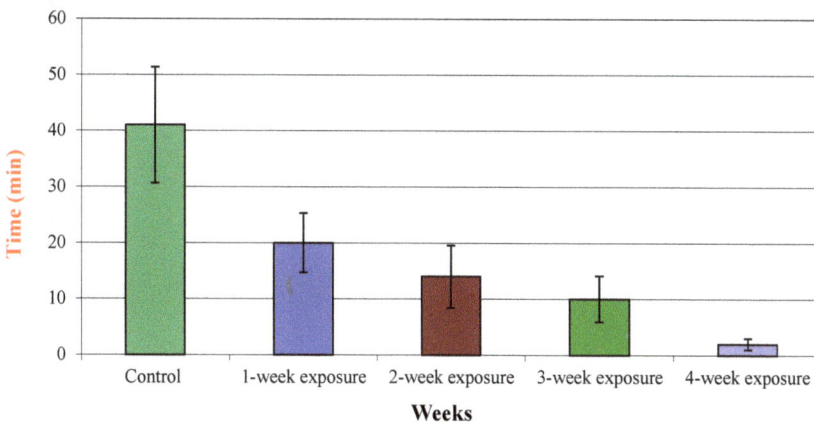

Figure 6.3: Time required to complete surface wetting.

Table 6.1: LSD analysis of contact angle (<10 s) and time to complete surface wetting.

Samples	Contact angle (<10 s)	Time to complete surface wetting (min)
Control	65[a]	41[a]
1 week pre-weathered	60[b]	20[b]
2 week pre-weathered	58[b]	14[bc]
3 week pre-weathered	53[c]	10[cd]
4 week pre-weathered	44[d]	2[d]

Means with the same letter are not significantly different ($\alpha < 0.05$). Number of specimens = 45

displayed severe degradation in surface quality as it needed only 2 min for complete wetting. The remarkably high wettability for samples exposed to 4 weeks exposure could be attributed to the physical changes such as cracks and chemical changes, i.e., reduction of lignin percentage on the wood surface. Feist and Hon [18] noticed that wood exposed to the weather undergoes physical and chemical degradation due to the combined effects of light, water, heat, environmental pollutants and certain microorganisms.

The least significant difference (LSD) analysis also shows that the contact angle of control samples was significantly different with pre-weathered samples. Table 6.1 shows the analysis of LSD for contact angle and time required for complete surface wetting. The initial contact angle showed that pre-weathering of wood gives a significant effect on the wettability of wood. Control samples recorded an initial contact angle of 65° and the value decreased as the wood were pre-weathered. The extent of decrement increased with increasing exposure period. The lowest initial contact angle of 44° was recorded in the samples pre-weathered for 4 weeks. The time required for complete surface wetting also showed similar trend. However, the difference was not significant between the samples exposed for 1 and 2 weeks. The effects became more obvious when the samples were exposed for up to 3 weeks.

The fitted regression (Figure 6.4) line indicates high inverses linear relationship between contact angle (within <10 s) and crack percentage. The predictive equation for crack percentage was significant with a coefficient of determination (r^2) of 0.86. Therefore, the wettability of the surfaces can be predicted by determining the contact angle of the water droplet (within 10 s) using the following formula:

$$y = 76.22 - 0.4543x$$

where,
- y = contact angle.
- x = crack percentage.

6.3.3 Lignin content

The lignin percentage of control and pre-weathered samples is presented in Figure 6.5. It appeared that the pre-weathered samples lost higher amount of lignin compared to

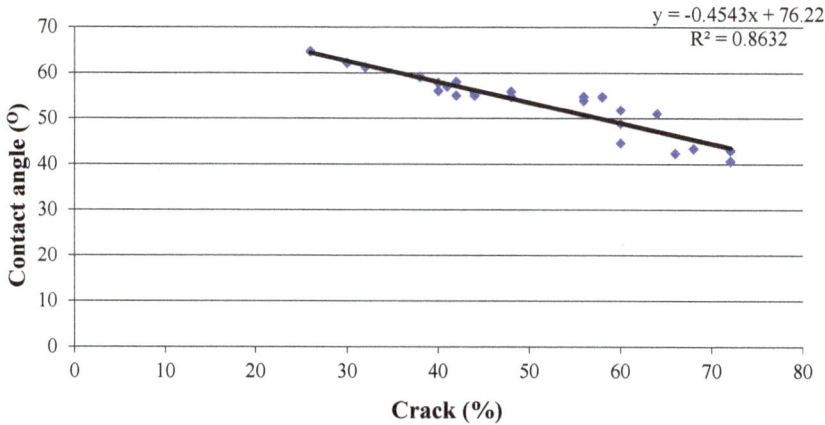

$$y = -0.4543x + 76.22$$
$$R^2 = 0.8632$$

Figure 6.4: Relationship between percentage of crack and contact angle (after 10 s water dropped).

control samples. The results indicated that the lignin was degraded after pre-weathering. After 1 week exposure, the lignin in the samples degrade at approximately 3.2%. The lowest lignin content was found in the samples pre-weathered for 4 weeks. Other researchers also observed the degradation of wood surfaces of pre-weathered wood. Lignin, in according to previous observations of the chemical changes in wood during natural weathering [18], appeared to be rapidly degraded during exterior exposure [19]. Also, lignin is an important component responsible for the water repellency of wood [20]. As the amount of lignin decreased over exposure time, the wettability and surface

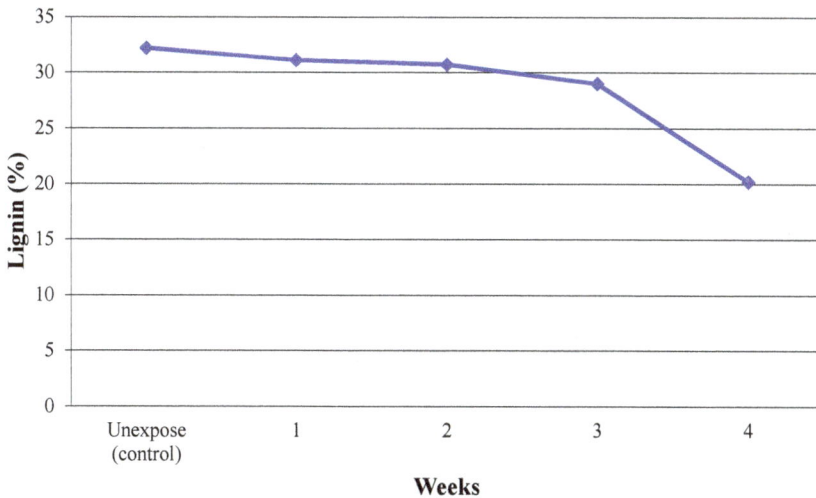

Figure 6.5: Lignin content of keruing after 1–4 weeks of exposure.

crack of the pre-weathered samples increased correspondingly. Haygreen and Bowyer [21] reported that lignin present between individual cellulose and within the cell walls. Between cells, it serves as a binding agent to hold the cells together. Within cell walls, lignin is very intimately associated with cellulose serves in imparting rigidity to the cells.

6.4 Conclusions

The findings of this study revealed that exposure for 4 weeks significantly increased the wettability of keruing wood. It was confirmed that the wettability of wood surfaces increased by exposing the wood to natural weathering prior to finishing. Significant relationships were found between contact angle (<10 s) and crack percentage. The duration of pre-weathered was found to have a direct effect on the wettability, crack percentage and lignin content of keruing wood. Based on the overall findings, the duration of the pre-weathering must be limited to a maximum of 3 weeks to avoid severe checks and cracks that will cause over penetration by a liquid adhesive.

References

1. MTC Wood Wizard. Keruing. 2021. Available from: http://mtc.com.my/wizards/mtc_tud/items/report(227).php [Accessed 29 Dec 2021].
2. Engku ABC. Basic and grade stresses for strength groups of Malaysian timbers. In: Malayan forest service trade leaflet no. 38. Kuala Lumpur: The Malaysian Timber Industry Board and Forest Research Institute Malaysia; 1998.
3. MS 544: Part 2:2001. Code of practice for the structural use of timber: permissible stress design of solid timber. Cyberjaya: The Department of Standards Malaysia; 2001.
4. Williams RS. Weathering of wood. In: Handbook of wood chemistry and wood composites. Madison: CRC Press; 2005, vol 7:139–85 pp.
5. William RS, Feist WC. Durability of paint or solid colour stain applied to preweathered wood. For Prod J 1993;43:8–14.
6. Gindl M, Reiterer A, Sinn G, Stanzl-Tschegg SE. Effects of surface ageing on wettability, surface chemistry, and adhesion of wood. Holz als Roh- Werkst 2004;62:273–80.
7. Rathke J, Sinn G. Evaluating the wettability of MUF resins and pMDI on two different OSB raw materials. Eur J Wood Wood Prod 2013;71:335–42.
8. Alia-Syahirah Y, Paridah MT, Hamdan H, Anwar UMK, Nordahlia AS, Lee SH. Effects of anatomical characteristics and wood density on surface roughness and their relation to surface wettability of hardwood. J Trop For Sci 2019;31:269–77.
9. Frihart CR, Hunt CG. Adhesives with wood materials: bond formation and performance. In: Wood handbook: wood as an engineering material, centennial ed. General technical report FPL, GTR-190. Madison, WI: US Dept of Agriculture, Forest Service, Forest Products Laboratory; 2010.
10. Hiziroglu S, Anwar UMK, Hamdan H, Tahir PM. Evaluation of surface quality of some Malaysian species as function of outdoor exposure. J Mater Process Technol 2008;199:156–62.

11. Anwar UMK, Lee SH, Husain H, Siam NA, Jasmani L. Effects of weathering on mechanical properties and surface roughness of endospermum malaccense wood modified with propionic anhydride. Drewno 2021;63:207.
12. Kalnins MA, Knaebe MT. Wettability of weathered wood. J Adhes Sci Technol 1992;6:1325–30.
13. Kalnins MA, Feist WC. Increase in wettability of wood with weathering. For Prod J 1993;43:55.
14. Anwar UMK, Hiziroglu S, Hamdan H, Latif MA. Effect of outdoor exposure on some properties of resin-treated plybamboo. Ind Crop Prod 2011;33:140–5.
15. ASTM D661-93. Standard test method for evaluating degree of cracking of exterior paints. West Conshohocken, Pennsylvania, United States: ASTM International; 2011.
16. Tappi method T222 om-88. In: Acid-Insoluble Lignin in wood and pulp. Atlanta, GA: Tappi Press; 1988.
17. Williams RS, Feist WC. Water repellents and water-repellent preservatives for wood. Gen Tech Rep. FPL-GTR-109, Forest Products Laboratory. WI: U.S. Department of Agriculture; 1999:12 p.
18. Feist WC, Hon DNS. Chemistry of weathering and protection. In: Rowel RM, editor. Chemistry of solid wood. Washington D.C, USA: A C S; 1984:401–51 pp.
19. Evans PD, Wallis AFA, Owen NL. Weathering of chemically modified wood surfaces. Wood Sci Technol 2000;34:151–65.
20. Rowell RM, Banks WB. Water repellency and dimensional stability of wood. General Technical Report FPL-GTR-50. Madison, Wis: USDA Forest Service, Forest Prod Lab; 1984:24 p.
21. Haygreen JG, Bowyer JL. Forest products and wood science. Iowa: The Iowa State University Press; 1988:196–202 pp.

Alia Syahirah Yusoh*, Mohd Khairun Anwar Uyup,
Paridah Md Tahir, Lee Seng Hua and Ong Chee Beng

7 Mechanical performance and failure characteristics of cross laminated timber (CLT) manufactured from tropical hardwoods species

Abstract: The study on the mechanical properties of cross laminated timber (CLT) panels made from tropical hardwood species is essential in order to promote the use of CLT as buildings material in Malaysia. The objective of this study were to evaluate the mechanical performance and failure characteristics of CLT fabricated from tropical timbers species, namely, batai (*Paraserianthes falcataria*), sesendok (*Endospermum malacensis*), rubberwood (*Hevea brasiliensis*) and kedondong (family Burceraceae). The modulus of rupture (MOR), modulus of elasticity (MOE), and compressive strength were determined. The failure characteristics of each samples were visually examined and recorded. The results indicated that CLT made from kedondong (KKK) had the highest value of MOR (82.63 N/mm^2) and MOE (11,371.33 N/mm^2) compared to other species. For compressive strength, CLT made from kedondong (KKK) and rubberwood (RRR) were not significantly different. The failure characteristics observed from bending test were tension, rolling shear and glue line failure while the crushing, shearing and splitting failure were found during compression test. Based on the results obtained, it showed that, the tropical hardwood is suitable to be used as raw material to produce CLT. However, more study should be conducted to observe the performance of CLT on durability and outdoor weathering.

Keywords: bending failure; compression failure; mechanical properties.

7.1 Introduction

Cross laminated timber (CLT) is the latest addition to the "mass timber" family that was first proposed in the mid-1990s in Austria and developed at the Graz University

*Corresponding author: Alia Syahirah Yusoh,** Forest Products Division, Forest Research Institute Malaysia, Kepong, Selangor, Malaysia, E-mail: aliasyahirah@frim.gov.my
Mohd Khairun Anwar Uyup and Ong Chee Beng, Forest Products Division, Forest Research Institute Malaysia, Kepong, Selangor, Malaysia
Paridah Md Tahir and Lee Seng Hua, Institute of Tropical Forestry and Forest Product (INTROP), Universiti Putra Malaysia, Serdang, Selangor, Malaysia

As per De Gruyter's policy this article has previously been published in the journal Physical Sciences Reviews. Please cite as: A. S. Yusoh, M. K. A. Uyup, P. Md Tahir, L. S. Hua, O. C. Beng "Mechanical performance and failure characteristics of cross laminated timber (CLT) manufactured from tropical hardwoods species" *Physical Sciences Reviews* [Online] 2022. DOI: 10.1515/psr-2022-0080 | https://doi.org/10.1515/9783110769227-007

of Technology, Austria [1, 2]. The grains of two consecutives lamella in CLT are oriented at right angles to each other. The cross-wise directions of each layer of timber in CLT give the product strength in both directions [3]. This also gives CLT a high degree of dimensional stability, as the cross-wise construction prevents the following layer from changing dimensions at right angles to it, as happens with single boards owing moisture to shifts. Consequently, CLT retains its dimensions much better than glulam or solid wood in response to change in moisture. According to the properties of this mass timber product, CLT is often used as wall elements [4].

The study on mechanical properties of timber species as CLT components is important to overcome the inherently weak mechanical properties and make them enable to be structural members for construction. CLT can be made from any species or the mixed-species mixture by providing its physical and mechanical properties to ensure the suitability of the timbers to be glued together. The mechanical properties of CLT have been examined in a number of prior researches. However, studies on the structural use of hardwood especially fast-grown tropical timber in CLT panel products are limited. There was a study found that fast-grown hardwood species such as yellow-poplar (*Liriodendron tulipifera*) have an acceptable mechanical performance that needed in structural engineering applications [5].

Failure of the mechanical strength evaluation must be recognized as one of the most important qualities while designing CLT. Failure is not only a requirement for safety, but also for the resistance and serviceability of the timber structure. The majority of materials engineering is predicted on producing a cost-effective design that is not prone to failure [6]. Normalized expressions, which describe the material's strength surface, are widely used to define CLT failure mode from failure criteria. Structures must adapt to external loads, transfer them to the ground and deal with the result in internal loads, which include normal force, shear force and moment [7]. As a result, the structure is subjected to loads and deformations that must not exceed the design strength and deformation limits.

In elastic phenomena such as damage or plastic strains result from a stress state that reaches surpass the failure surface. Once the related failure load has been determined, the dominant failure mode can be determined [4]. The strength of wood, on the other hand, is determined by the species and the influence of particular growth features [8]. Nevertheless, these characteristics vary within a species due to the large inherent material variations which consequently yield various types of failure characteristics. Therefore, the purpose of this study was to evaluate the mechanical properties as well as the failure characteristics of CLT manufactured from four species of Malaysian tropical hardwoods.

7.2 Material and methods

7.2.1 Samples preparation

Sawn timber from four different tropical hardwood wood species namely batai (*Paraserianthes falcataria*), sesendok (*Endospermum malaccensis*), rubberwood (*Hevea brasiliensis*) and kedondong (family Burceraceae) were used in this study. The lumbers were obtained from a local supplier and were kiln dried until they reached about 12% moisture content prior to the production of CLT.

7.2.2 Production of CLT

The dried lumbers were first cut, trimmed and planed into desired dimensions. Phenol resorcinol formaldehyde (PRF) adhesive was used together with a hardener at a ratio of 100 to 12 parts by weight of PRF to hardener. The steps of the production process of CLT were summarized in Figure 7.1. The CLT was fabricated into a three-layered board (54 mm thick × 1072 mm length × 1072 mm width) by arranging the lamina perpendicular to each other into size. The adhesive was applied onto the surface of the lamina using a roller-coater at glue spread rate of 250 g/m^2 (rubberwood and kedondong) and 300 g/m^2 (batai and sesendok) based on Alia et al. [9]. The three-layered CLT board was clamped under hydraulic pressure of 0.7 N/mm^2 for 6 h. The CLTs consisted of three-layered batai, sesendok, rubberwood and kedondong were denoted as BBB, SSS, RRR and KKK, respectively.

7.2.3 Evaluation of mechanical properties and observation on failure characteristics of CLT

The CLT were trimmed and cut into test specimens of appropriate dimensions based on the standards requirement. The density and moisture content were evaluated based on BS EN 384 (2016) [10]. The

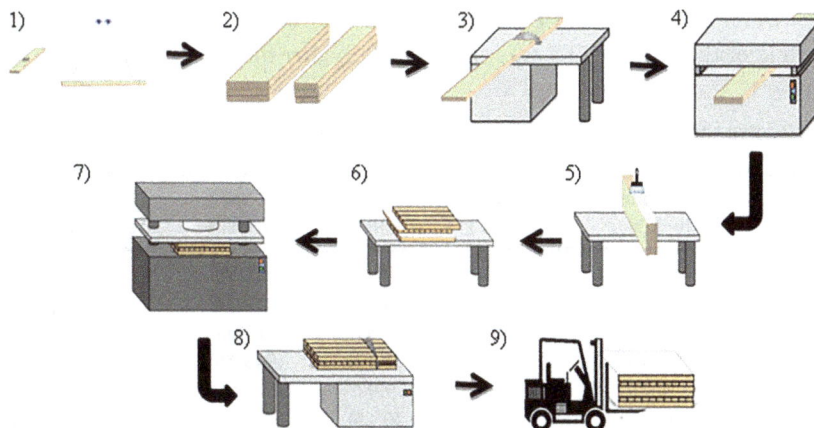

Figure 7.1: Production process of CLT; (1) Lumber selection, (2) lumber grouping, (3) lumber planing, (4) lumbers cutting to length, (5) adhesive application, (6) panel lay-up, (7) assembly pressing, (8) machining and cutting and (9) marking and conditioning.

bending and compression tests were determined based on BS EN 408 (2010) [11]. A total of 30 specimens were used for each test, i.e., 3-point bending and compression test. The test was performed using a Shimadzu Universal Testing machine with a max load of 100 kN using a speed crosshead of 0.6 mm/min. The test was ended when no further increment in strain and complete failure occurred.

7.2.4 Statistical analysis

Statistical analysis was carried out using a statistical analysis system (SAS 9.4). The means were separated by the least significant difference (LSD). Means which are followed by the same letter(s) indicate that the studied factor has no significant effect on the observed values at $p \leq 0.05$.

7.3 Results and discussion

7.3.1 Mechanical properties

The analysis of variance for the moisture content (MC), density and strength properties (MOR, MOE and compression strength) of the CLT from four wood species are shown in Table 7.1. The moisture content of the CLT ranged from 14.6% (SSS) to 14.9% (KKK). The board densities of the CLT were significantly different among the wood species. The highest density was KKK (769 kg/m³) followed by RRR (638 kg/m³), SSS (474 kg/m³) and BBB (354 kg/m³). Batai has a very low wood density [12–15], hence naturally produced equally low density CLT panel. The same observation was found in kedondong that have the highest wood density, as well as the resulting CLT panel.

For the strength properties, the CLT made from 100% kedondong (KKK) was the strongest which has significantly higher MOR than that of CLTs from other species, i.e., 82.63 N/mm². MOR of both BBB and SSS CLTs were at the lower end of the four species while of the MOR of RRR was significantly higher than these two CLTs. The

Table 7.1: Bending and compressive strengths of CLT manufactured from tropical hardwoods species.

Species of CLT	Moisture content (%)	CLT Density (kg/m³)	MOR (N/mm²)	MOE (N/mm²)	Compression strength (N/mm²)
BBB	14.83ᵃ	354ᵈ	40.05ᶜ	6165.43ᶜ	18.47ᶜ
	(0.12)	(5.23)	(6.25)	(809.59)	(2.15)
SSS	14.63ᵃ	474ᶜ	48.51ᶜ	11275.64ᵃ	27.79ᵇ
	(0.29)	(7.32)	(8.17)	(945.76)	(2.64)
RRR	14.70ᵃ	638ᵇ	59.91ᵇ	8707.95ᵇ	35.65ᵃ
	(0.16)	(6.33)	(6.19)	(795.73)	(1.36)
KKK	14.90ᵃ	769ᵃ	82.63ᵃ	11371.33ᵃ	38.22ᵃ
	(0.11)	(8.35)	(11.35)	(2789.01)	(0.79)

Values are an average of 30 specimens. Values in parentheses are standard deviations. Means followed with the same letters a,b,c,d in the same column are not significantly different at $p \leq 0.05$ according to LSD.

trend in MOR of CLTs followed that in solid wood. The MOR of solid wood from kedondong, rubberwood, sesendok and batai were 81, 66, 60 and 51 N/mm^2, respectively (Table 7.2). The trend remained the same when they were converted into CLT.

The MOE values, however, showed a slightly different trend. While CLTs from kedondong and sesendok have comparable MOE (as shown by the same LSD ranking 'a'), those of rubberwood and batai are significantly lower (LSD ranking b and c respectively). As shown in Table 7.2, the stiffness trend in CLT is the same for solid wood. This behavior is to be expected as the CLT contain about the same properties in each lamella thus would behave exactly the same when subjected to loading.

In the case of compressive strength, it was found that CLT from BBB and SSS did not follow the trend of compressive strength of the solid wood from the previous studies as shown in Table 7.2. The CLT from SSS obtained higher compressive strength compared to BBB. Meanwhile, the compressive strength of KKK shows slightly higher than RRR but was not significantly different. The compressive strength is linearly correlated with density of the wood species [16]. Thus, it might be the cause of the greatest compressive strength obtained by KKK since CLT from kedondong had the highest density in this study.

Previously, a three-layer CLT from sesendok had been produced and found that the compressive strength of the CLT was 28.10 N/mm^2 which was lower compared to the results obtained from this study [17]. In other study, the compressive strength value obtained from three-layer CLT from batai was 13.75 N/mm^2 [18]. Generally, a low density wood species would produce low compressive strength. Nevertheless, the compressive strength of CLT might also be influenced by other factors such as size of the area to which the load was applied, the grain direction of the outermost lamina of the CLT and the distance between the compressed area and the end-plane of the specimen to which the load was applied [19].

7.3.2 Failure characteristics

In this study, different types of failure characteristics were observed in the three-point bending tests of the CLT samples. Tension failure typically occurred at the bottom layer

Table 7.2: Strength properties of solid wood from batai, sesendok, rubberwood and kedondong.

Species	MOR (N/mm^2)	MOE (N/mm^2)	Compression strength (N/mm^2)	References
Batai	51	6800	28	Emmanuel & Arnaldo (1997)
Sesendok	60	11,280	27	Muhammad Bazli & Zakiah (2017)
Rubberwood	66	9240	32	MTC (2020) [24]
Kedondong	81	11200–12900	31–43	MTC (2020) [23]

All samples were tested in dry condition.

of CLT timber. Whereas, rolling shear failure occurred when there is shear stress leading to shear strains in a plane perpendicular to grain direction while glue line failure occurred when there is a failure in the bonding. The failure characteristics for the CLT were tabulated in Table 7.3. It has been noted that, in addition to the type of wood species, grain angle is one of the factor affecting load bearing capacities. Shear and rolling shear failures in the transverse layer, tensile failure in the bottommost layer, delamination or glue line failure of longitudinal bonding lines between the transverse and the bottom face layer while compression and tension failures of longitudinal fibres in face layers were the most common failure characteristics of CLT panels. Hence, the failure characteristic of a CLT panel is influenced by the grain angle of the lamina [20].

Table 7.3: Bending failure of three-layered CLT.

Types of CLT	Bending failure
BBB	
SSS	
RRR	
KKK	

BBB and SSS were mostly failed when force was applied to the specimen, initiated by rolling shear at the crosswise layer of the CLT panel. Observation conducted during testing showed tension failure happened after rolling shear failure initiated. The failure in SSS was similar to BBB, except that the SSS took a longer time to fail when maximum force was applied. Rolling shear failure in SSS occurred due to the anatomical features in the transverse layer of the CLT. The rolling shear strength and modulus are relatively low in the radial-tangential plane compared to the longitudinal-radial and longitudinal-tangential [21]. This phenomenon led to the rotation of small bundles of fibres, so-called rolling shear from shear forces which finally associated the tension failure in the bottom layer of CLT.

The fracture sound of the wood fibre in RRR could be clearly heard when the samples during the loading process start to fail. During the observation, tension failure was heard during the loading process after the glue-line failed. However, the tension failure observed was not severe as occurred in BBB and SSS since RRR can resist the higher ultimate load. Tension failure was observed in the bottom longitudinal layer. Meanwhile, there was only glue-line failure observed in KKK. The glue-line abruptly burst when the test reached its maximum load. There was no tension failure occurred in KKK which might be due to its high density, so that the sample can resist high load without failed in tension.

The fracture sound of the wood fibre in RRR could be clearly heard when the samples during the loading process start to fail. Tension failure was heard during the loading process. The glue line abruptly burst in KKK when the test reached its maximum load. This can be related to the high ductility of KKK. Generally, the most prominent failures after bending test were categorized as (i) tension failure (70%), (ii) glue line failure (23%) and rolling shear failure (6%) based on the total number of failed specimens.

For compressive failure parallel to the grain test, crushing, shearing and splitting occurred on the CLT. The crushing failure is the failure that occurs at the supports, mainly by clamp crushing the samples while shearing is the failure caused by shearing at the weak bonded area. Meanwhile, splitting is the failure that occurred along the length of the sample. The compressive failure of the CLT was tabulated in the table below (Table 7.4).

The crushing and splitting between layers near to the glue line of BBB and SSS occurred due to the low density of the wood species that were categorized as weak materials. Crushing failure is often defined by the folding of cellulose microfibrils, which can begin at modest applied loads. At high-stress levels, folding could also occur at the cell wall, resulting in a complete failure of the examined samples [22]. When BBB was compressed along the grain, the damage associated wedge splitting failure was close to the glue line of the sample. According to Buck et al. [3], splitting failure was found in the study of CLT panels with different directions of a transverse layer using Norway spruce. It was mentioned that the splitting failure that occurred in the tested samples was due to the stiffness of adhesive which caused higher stress concentration in the wood close to the glue line.

Table 7.4: Compressive failure of three-layered CLT.

Types of CLT	Compressive failure
BBB	
SSS	
RRR	
KKK	

In addition, it was also stated that stress concentrations can be reduced the increased in load-bearing capacity with a more flexible glue line that allows active distribution of the load through the entire bonded area. Shearing failure observed in SSS, RRR and KKK can be described as a pattern in which associated with a plane of weakness such as edge joint parts particularly along the glue line as for laminates at an angle to the loaded surface. Most of the compression test has experienced shearing failure (73%), crushing failure (23%) and splitting failure (5%).

7.4 Conclusions

In conclusions, the three-layered CLT manufactured using tropical hardwood species with 250–300 g/m² glue spread rate and 0.7 N/mm² clamping pressure were sufficient to produce good mechanical properties. The mechanical performances of CLT manufactured from tropical hardwood species made from kedondong were the highest, followed by rubberwood, sesendok and batai. The high density of the hardwood material possessed a great mechanical performance of CLT panel. The rolling shear failure were found mostly on CLT from low density timber (i.e., batai and sesendok) while tension failure does not occurred on CLT from high density timber such as kedondong after bending. The crushing and splitting failure after compression only happened on CLT from weak material; batai and sesendok.

References

1. Crovella P, Smith W, Bartczak J. Experimental verification of shear analogy approach to predict bending stiffness for softwood and hardwood cross-laminated timber panels. Construct Build Mater 2019;229:116895.
2. Wei P, Wang BJ, Li Z, Ju R. Development of cross-laminated timber (CLT) products from stress graded Canadian hem-fir. Wood Res 2020;65:335–46.
3. Buck D, Wang XA, Hagman O, Gustafsson A. Bending properties of cross laminated timber (CLT) with a 45 alternating layer configuration. Bioresources 2016;11:4633–44.
4. Franke S, Franke B, Harte AM. Failure modes and reinforcement techniques for timber beams–State of the art. Construct Build Mater 2015;97:2–13.
5. Mohamadzadeh M, Hindman D. Mechanical performance of yellow-poplar cross laminated timber. Blacksburg: Virginia Polytechnic Institute and State University, Department of Civil and Environmental Engineering; 2015, Report No. CE/VPI-ST-15-13.
6. Bodig J, Jayne BA. Mechanics of wood and wood composites. Melbourne, FL: Krieger Pub Co.; 1982: 712 p.
7. Franzoni L, Lebée A, Lyon F, Foret G. Influence of orientation and number of layers on the elastic response and failure modes on CLT floors: modeling and parameter studies. Eur J Wood Wood Prod 2016;74:671–84.
8. Cruz H, Yeomans D, Macchioni N, Jorissen A, Touza M, Lourenco PB. Guidelines for on-site assessment of historic timber structures. Int J Archit Heritage 2015;9:277–89.

9. Alia Syahirah Y, Paridah MT, Anwar UMK, Lee SH, Hamdan H, Omar MK. Effect of wood species, clamping pressure and glue spread rate on the bonding properties of cross laminated timber (CLT) manufactured from tropical hardwoods. Constr Build Mater 2021;273:121721.
10. BS EN 384. Structural timber. determination of characteristic values of mechanicalproperties and density. London, UK: BSI; 2016.
11. BS EN 408. Timber structures - structural timber and glue-laminated timber- determination of some physical and mechanical properties. London, UK: BSI; 2010.
12. Nordahlia AS, Lim SC, Hamdan H, Anwar UMK. Wood properties of selected plantation species: *Tectona grandis* (teak), *Neolamarckia cadamba* (kelempayan/laran), *Octomeles sumatrana* (binuang) and *Paraserianthes falcataria* (batai). Timber Technology Bulletin 2014;54:1–4.
13. Muhammad-Fitri S, Suffian M, Wan-Mohd-Nazri WAR, Nor-Yuzia Y. Mechanical properties of plywood from batai (paraserianthes falcataria), Eucalyptus (Eucalyptus pellita) and kelempayan (neolamarckia cadamba) with different layer and species arrangement. J Trop For Sci 2018;30: 58–66.
14. Feng TY, Chiang LK. Effects of densification on low-density plantation species for cross-laminated timber. In: AIP conference proceedings [AIP Publishing 4th International Conference on the Science and Engineering of Materials: ICoSEM- Kuala Lumpur, Malaysia 2020, 2284, No. 1. AIP Publishing LLC; 2020:020001 p.
15. Rahman WMNWA, Sa'ad MF, Suffian M, Sofian NYMY. Wood and veneer properties of fast growing species from batai, eucalyptus and kelampayan. Int J Law Manag Humanities 2020;4:65–71.
16. Tian Z, Gong Y, Xu J, Li M, Wang Z, Ren H. Predicting the average compression strength of CLT by using the average density or compressive strength of lamina. Forests 2022;13:591.
17. Hamdan H, Anwar UMK, Izani IM, Iskandar M, Tamizi MM, Gan KS, et al. Cross laminated timber (CLT) structure: the first in Malaysia. Timber Technology Bulletin 2017;70:1–7.
18. Liew KC, Maining ES. Mechanical and physical properties of cross-laminated timber made from batai using different glue spread amounts. In: Journal of Physics: Conference Series, 1st ed. Bristol, UK: IOP Publishing; 2021, vol. 2129:012087 p.
19. Serrano E, Enquist B. Compression strength perpendicular to grain in cross-laminated timber (CLT) In: 11th World conference on timber engineering. Trentino, Italy; 2010:441–8 pp.
20. Rostampou Haftkhani A, Hematabadi H. Effect of layer arrangement on bending strength of cross-laminated timber (CLT) manufactured from poplar (populus deltoides L.). Buildings 2022;12: 608.
21. Hosseinzadeh S, Mohebby B, Elyasi M. Bending performances and rolling shear strength of nail-cross-laminated timber. Wood Mater Sci Eng 2020;17:1–8.
22. Crespo J, Majano-Majano A, Lara-Bocanegra AJ, Guaita M. Mechanical properties of small clear specimens of eucalyptus globulus Labill. Materials 2020;13:906.
23. MTC. MTC Wood Wizard - Malaysian Timber Council (Kedondong) 2020. Available from: http://mtc.com.my/wizards/mtc_tud/items/report(45).php.
24. MTC. MTC Wood Wizard - Malaysian Timber Council (Rubberwood) 2020. Available from: http://mtc.com.my/wizards/mtc_tud/items/report(105).php.

Nurul Ain Maidin, Salit Mohd Sapuan*, Mastura Mohammad Taha
and Zuhri Mohamed Yusoff Mohd

8 Constructing a framework for selecting natural fibres as reinforcements composites based on grey relational analysis

Abstract: Material selection is crucial in product development, especially when material from a composites process application is involved. Numerous multi-criteria decision-making (MCDM) tools each have their own set of advantages and disadvantages. Using grey relational analysis (GRA), this research proposes a systematic framework evaluation approach for generating a sensible rank for material selection of natural fibre as reinforcement composites. The framework was created using the GRA technique, a robust evaluation tool that employs the grade of relation to determine the degree of similarity or difference between two sequences. The MCDM approach can be straightforward for the material selection problem. A GRA technique is used to investigate the performance of the potential material, which includes grey relational sequence creation, reference sequence definition, grey relational coefficient calculation and grey relational grade determination. This framework is applied with a case study to identify the optimum natural fibres composites material for a bike helmet. End results revealed that pineapple is the best candidate for construction of safety gear (cyclist helmet). The best possible evaluation model for material selection of the composite can be referred by design engineer in composite industry for multiple applications. Moreover, the proposed framework is an aid to help engineers and designers to choose most suitable material.

Keywords: natural fibre-reinforced composites (NFRCs), biocomposite products, grey relational analysis (GRA), multiple criteria decision making (MCDM)

*Corresponding author: Salit Mohd Sapuan, Laboratory of Biocomposite Technology, Institute of Tropical Forestry and Forest Products, Universiti Putra Malaysia, 43400 UPM Serdang, Selangor, Malaysia; and Department of Mechanical and Manufacturing, Faculty of Engineering, Universiti Putra Malaysia, 43400 UPM Serdang, Selangor, Malaysia, E-mail: sapuan@upm.edu.my
Nurul Ain Maidin, Department of Mechanical and Manufacturing, Faculty of Engineering, Universiti Putra Malaysia, 43400 UPM Serdang, Selangor, Malaysia; and Department of Manufacturing Engineering Technology, Faculty of Mechanical and Manufacturing, Engineering Technology, Universiti Teknikal Malaysia Melaka, 76100 Durian Tunggal, Melaka, Malaysia
Mastura Mohammad Taha, Department of Manufacturing Engineering Technology, Faculty of Mechanical and Manufacturing, Engineering Technology, Universiti Teknikal Malaysia Melaka, 76100 Durian Tunggal, Melaka, Malaysia
Zuhri Mohamed Yusoff Mohd, Department of Mechanical and Manufacturing, Faculty of Engineering, Universiti Putra Malaysia, 43400 UPM Serdang, Selangor, Malaysia

As per De Gruyter's policy this article has previously been published in the journal Physical Sciences Reviews. Please cite as: N. A. Maidin, S. M. Sapuan, M. M. Taha, Z. M. Y. Mohd "Constructing a framework for selecting natural fibres as reinforcements composites based on grey relational analysis" *Physical Sciences Reviews* [Online] 2022. DOI: 10.1515/psr-2022-0081 | https://doi.org/10.1515/9783110769227-008

8.1 Introduction

Natural fibres are in high demand as a base material replacement due to the rising use of biocomposites in consumer goods. Natural raw materials such as kenaf, oil palm, abaca, sisal, coir, sugar palm, bamboo, pineapple, jute and banana are being generated, increasing as the global population grows. The natural fibre sector reported a 60% rise in demand for natural fibres. Furthermore, the predicted annual growth rate is estimated to be between 10 and 20% [1, 2]. Natural fibre–reinforced composites (NFRCs) are gaining appeal as an alternative for metals and synthetic fibre–reinforced composites due to increased demand for lighter materials with a lower environmental imprint [3, 4]. Natural fibres come in two varieties: animal-based fibres and plant-based fibres. Animal-based fibres, including cocoon silk, chicken feathers, wool, and spider silks, are widely used in biomedical applications such as implants. These bio-products must be bio-resorbable (also known as "bio-degradable" in certain publications), which means they must be able to break down and assimilate back into the body, or biocompatible, which means they must not be hazardous to humans [5]. Plant-based fibres like jute, hemp, sisal kenaf, coir, flax, bamboo and banana are frequently combined with polymers to create NFRP composites for household products. Cost and strength are essential considerations. These fibres are made from sustainable resources that do not affect the environment. Much research has been published in the last year that investigates the ability of NFRCs to substitute metal-based materials by displaying a high mechanical property score [6–12].

There are various ways to evaluate the performance of a material, and numerous factors can affect its performance. The design engineer must select the optimal composites mixture for the design function of the intended product. It is important to select the proper materials when constructing something [13]. In the industrial sector, there is no standard method for selecting materials. Diverse picking instruments employ distinct methods and dimensions [14]. Numerous researchers have examined MCDM problems before [15]. The scoring models proposed by [16], the axiomatic design proposed by [17], and the analytic hierarchy process (AHP) proposed by [18] are examples of these techniques. Almost all of the methods mentioned above determine the value of each criterion using weights. According to [19], the outcome depends on the weight. For instance, different weights produce distinct results. When the weight of a decision-maker (DM) changes, the entire calculation must be redone. This can be difficult and inefficient for DMs who struggle with mathematics. Grey system theory is a method for dealing with incorrect, incomplete or unclear data. The first user was [20], who studied uncertainty in system models to make more accurate predictions and decisions. In 1982, Deng introduced the Grey System Theory. It was about making decisions without sufficient information or knowing what would occur. The category of information that is neither known nor unknown is grey information. The Grey System Theory is utilised to solve difficult problems involving complex data. The GRA method can solve problems with multiple criteria and intricate relationships [21]. Additionally, it can determine the optimal process parameter for two or more response variables [22]. Grey Relationship analysis is effective in situations with multiple criteria [23].

Therefore, designers and material engineers would benefit from a straightforward and systematic technique selection procedure [24]. This work proposes a model based on grey relational sequence generation, a well-known GRA method, to select the best natural fibres as reinforcements in biocomposites for bicycle helmets. This will aid the industry's design engineer and manufacturing team in selecting the best materials.

8.2 Methodology

This section presents the methodology with the proposed framework in Figure 8.1, and the detailed grey relational analysis (GRA) approach is given in a subsection.

8.2.1 Proposed Framework

A subfield of grey system theory known as grey relational analysis (GRA) can be used to investigate the complex interrelationships between various variables and components. GRA employs the grade of relation as a model for impact assessment to assess the similarity of comparability and reference sequences. Designers and material scientists would benefit from a straightforward and systematic method of selecting techniques. GRA is a four-step process that must be followed to arrive at sound decisions. Developing a grey relational sequence begins with constructing a comparability sequence using all of the alternatives' properties as a basis. All performance attributes can be compared to the newly created grey relational series by transforming them into a single comparability sequence. Afterwards, the grey correlation coefficient is computed for each of the attributes. The estimated coefficients and weights assigned to the performance attributes are then used to calculate the GRG (grey relational grade) [21, 23, 25]. The resulting GRG is then used to order the available choices.

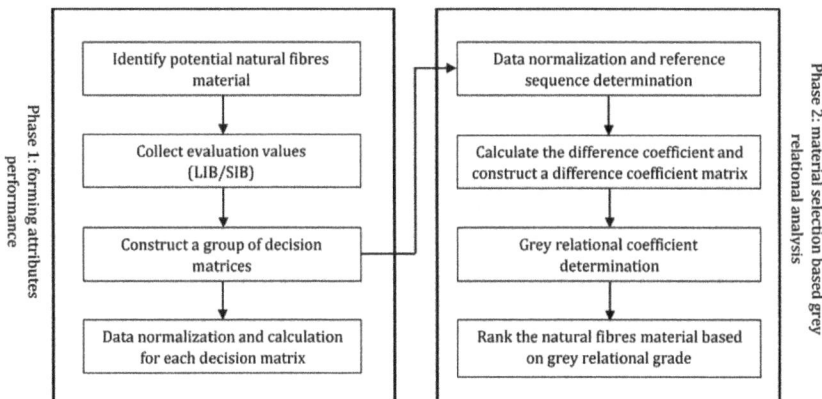

Figure 8.1: Framework for the selection of the best natural fibre composites.

For larger the better quality characteristic, normalized value (X_i) is given by

$$X_i = \left(\frac{x_i - x_{min}}{x_{max} - x_{min}}\right) \tag{8.1}$$

For smaller the better quality characteristic, normalized value (X_i) is given by

$$X_i = \left(\frac{x_{max} - x_i}{x_{max} - x_{min}}\right) \tag{8.2}$$

where $(x_i)_{max}$ and $(x_i)_{min}$ are the maximum and minimum values of the original sequence x_i.

Grey coefficient, y_i is given by

$$y_i = \left\{\frac{(\delta_i)_{max} + (\delta_i)_{min}}{(\delta_i)_{max} + (\delta_i)}\right\} \tag{8.3}$$

where, Delta value, δ_i is given by

$$\delta_i = (X_i)_{max} - X_i \tag{8.4}$$

8.2.2 Data collection

Many studies on natural fibre–reinforced composites (NFRCs) [26–30] are merely data collection and do not provide new insights. There are seven (7) natural fibre properties listed in Table 8.1. These specifications were arranged following the general requirements for cycling helmet materials and production aspects. One of the most important requirements for materials was "strength," which was associated with both material characteristics and natural fibre limitations. The following seven characteristics of natural fibres were selected to determine their relative importance. Natural

Table 8.1: Selection criteria and alternatives developed to find the most suitable natural fibre in biocomposites for cyclist helmet application [31–35].

Criteria	Alternatives	Description
Strength	Tensile strength	To determine the durability and strength of materials
	Young's modulus	To measure elastic constant to resist deformation when stress is applied
	Elongation at break	To determine the strength and toughness level Young's
	Density	To identify lighter natural fibres
	Cellulose	To determine the degree of polymerisation and reflect the mechanical properties
	Lignin	To predict the rigidity of the fibres
	Fibre's length	To predict impact strength of the fibres and bonding condition at the interface

Figure 8.2: Natural fibre of banana (a), pineapple (b), kenaf (c), jute (d) and bamboo (e).

fibre composites were rated based on a recent and well-known literature review. Equation (8.1) is used if the desired value of the performance attribute is greater (LIB), and Equation (8.2) is used if the desired value of the attribute is lower (LIB) (SIB). This evaluation is required to build the attribute comparability sequence (Equations (8.3) and (8.4)) and normalise the choice matrix. For the calculations, Microsoft Excel 2016 was used as well. This approach makes it possible to select the best natural fibre–reinforced composite model for use in cycling helmets.

Information gleaned from secondary sources about the mechanical and physical properties of (NFRC) helmets can be used to make better decisions about fit and comfort. Mechanical properties and material constraints are the most critical considerations when choosing a cycling helmet material [36, 37]. Data from the literature led to selection of the following materials for this study: banana, pineapple, kenaf, jute, and bamboo (Figure 8.2). Table 8.2 shows data from recent literature on (5) five natural fibre substitutes from 2017 to 2021 that are comparable and complete. Data dispersion can be reduced by using the average number in this case.

8.3 Results and discussion

Detecting a trend or relationship between input factors and desired outcomes is difficult because of the lack of complete information in grey systems. This real-time modelling approach relies on minimal data rather than a statistical model, which would otherwise have to deal with enormous data. By defining the relationships between variables and identifying the critical variables that have the greatest impact on desired outcomes, grey relational analysis (GRA) is used to identify GRG. There are many competing criteria in the material selection problem, similar to the grey system. The GRA procedure is as follows, based on Table 8.2.

Table 8.2: Qualitative measures of natural fibres alternatives in biocomposites proposed for cyclist helmet application [4, 11, 31–35].

No.	Fibre	Density (g/cm³)	Tensile strength (MPa)	Young's modulus (GPa)	Performance				
					Elongation to break (%)	Fibre length (mm)	Lignin (%)	Cellulose (%)	
1	Bamboo	0.6–1.11	140–800	30–50	1.4	90	21–31	26–43	
2	Pineapple	0.8–1.6	170–1627	1.44–82	2.4	30	8.3–12.7	80.5	
3	Kenaf	1.4	930	11–60	1.5	500	17–21.5	53.5	
4	Jute	1.3–1.5	200–773	9–31	1.5–1.8	60.75	9–13	59–71.5	
5	Banana	1.35	529–914	8–32	3	10	5–10	62–64	
	GRA GENERATING	SIB	LIB	LIB	LIB	LIB	LIB	LIB	

Table 8.3: Results of grey relational generating.

Alternative no.	Step 1 and 2-Sequence generation						
	Performance						
	Density (g/cm³)	Tensile strength (MPa)	Young's modulus (GPa)	Elongation to break (%)	Fibre's Length (mm)	Lignin (%)	Cellulose (%)
1	1.0000	0.0000	0.9208	0.0000	0.1633	1.0000	0.0000
2	0.3704	0.9315	1.0000	0.6250	0.0408	0.1622	1.0000
3	0.0000	1.0000	0.7136	0.0625	1.0000	0.6351	0.4152
4	0.0000	0.0359	0.0000	0.1563	0.1036	0.1892	0.6685
5	0.0926	0.5467	0.0000	1.0000	0.0000	0.0000	0.5978

8.3.1 Grey relational sequence generation

The primary goal of grey relational generation is to generate identical sequences from raw data. The desired performance attributes were greater for tensile strength, Young's modulus, elongation to break, fibre length, lignin and cellulose (LIB). Concerning tensile strength, alternative No. 1 yielded a minimum of 470, and alternative No. 3 yielded a maximum of 930 (470 – 470)/(930 – 470) = 0 was the grey relational produced by Alternative No. 1 using Equation (8.1). On the other hand, the tensile attribute's alternative No. 2 is calculated as (898.5 – 470)/(930 – 470) = 0.9315. The grey relational generation process, on the other hand, utilised Equation (8.2) for attributes with a lower desired value (SIB). Density, for example, had a maximum value of 1.4 from alternatives 3 and 4 and a minimum value of 0.86 from options 1. Equivalent to (1.4 – 0.86)/(1.4 – 0.86) = 1, the grey relational produced by alternative no. 1 was calculated using Equation (8.2). For all grey relational generating findings, see Table 8.3.

8.3.2 Derivation of reference sequence

Grey relational sequence X_i values were scaled into the range [0, 1] following the generation of the comparability sequence X_i. Other alternatives were beaten by the one with the highest X_i value (one). As a result, this kind of performance criterion was rare. A sequence with similar values to one (X_0) was defined and compared to the generated sequence. Because of the series' similarity, the better option was selected. As a starting point, the following is the reference sequence:

$$X_0 = (x_{01}, x_{02}, \ldots, x_{0j}, \ldots, x_{0n}) = (1, 1, \ldots, 1\ldots, 1)$$

After calculating Δ_{ij}, and Δ_{max}, all grey relational coefficients can be determined from Equation (8.3). In this study, the distinguishing coefficient was assumed to be 0.5. For

Table 8.4: Results of grey relational coefficient.

Step 3-Grey relation coeeficient						
Performance						
Density (g/cm³)	Tensile strength (MPa)	Young's modulus (GPa)	Elongation to break (%)	Fibre's length (mm)	Lignin (%)	Cellulose (%)
1.0000	0.3333	0.8633	0.3333	0.3740	1.0000	0.3333
0.4426	0.8795	1.0000	0.5714	0.3427	0.3737	1.0000
0.3333	1.0000	0.6358	0.3478	1.0000	0.5781	0.4609
0.3333	0.3415	0.3333	0.3721	0.3581	0.3814	0.6013
0.3553	0.5245	0.3333	1.0000	0.3333	0.3333	0.5542

example, $\Delta_{i1} = |1 - 0.2326| = 0.7674$, $\Delta_{max} = 1$, and $\Delta_{min} = 0$, if $\delta = 0.5$, then $\gamma(x_{0j}, x_{ij}) = (0 + 0.5 \times 1)/(0.7674 + 0.5 \times 1) = 0.3945$. $\Delta_{i1} = |1 - 1| = 0$, $\Delta_{max} = 1$, and $\Delta_{min} = 0$, if $\delta = 0.5$, then $\gamma(x_{0j}, x_{ij}) = (0 + 0.5 \times 1)/(0 + 0.5 \times 1) = 1$. Whereas, $\Delta_{i2} = |1 - 0.3704| = 0.6296$, $\Delta_{max} = 1$, and $\Delta_{min} = 0$, if $\delta = 0.5$, then $\gamma(x_{0j}, x_{ij}) = (0 + 0.5 \times 1)/(0.6296 + 0.5 \times 1) = 0.4426$. The entire results for the grey relational coefficient are shown in Table 8.4.

8.3.3 Calculation of GRG

In this scenario, it was presumed that every aspect of the performance was equally important. Because of this, the seven aspects of performance received the same weighting (1/10). Table 8.5's column 2 shows the GRG, calculated using Equation (8.4). Column 3 of Table 8.5 displays the GRA's results.

The GRA grade scores calculated by Microsoft Excel 2016 software with the corresponding formula were used to rank five natural fibre alternatives according to the synthesised results for the primary criteria. The outcomes are shown in Figure 8.3. It

Table 8.5: Results of grey relational grade and rank.

Step 4-Grey relation grade		
Alternative material	GRG	Rank
Pineapple	0.6586	1
Kenaf	0.6223	2
Bamboo	0.6053	3
Banana	0.4906	4
Jute	0.3887	5

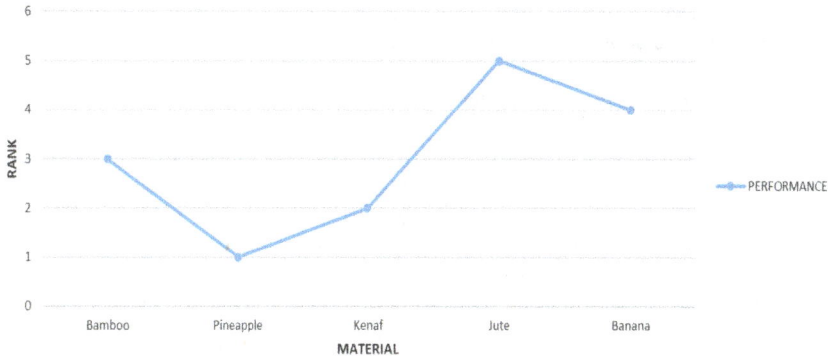

Figure 8.3: The GRA final synthesis results.

received the highest grade of 0.6586, which placed Pineapple at the top of the list. Following bamboo, banana and jute with grades of 0.6053, 0.4906 and 0.4885 were kenaf, which had the second-highest grade of 0.6232. Identifying natural fibres' chemical composition and morphology is essential to understanding their unique properties [28–40]. Despite the highly validated results, the authors believe that the natural fibre selection method would have been more complete if additional details from other criteria had been included. When making a decision, several factors must be taken into consideration. Due to the importance of selecting natural fibres that meet a specific demand, those in charge of the selection process must devise criteria as specific as possible about that need.

8.4 Conclusions

Natural fibre–reinforced composites (NFRCs) with relevant mechanical and physical properties were selected using the GRA technique. For bicycle helmet design, the grade of relation is used to measure how similar or different two sequences are in terms of material performance and is proposed to use this tool. A potential material's performance can be evaluated using this GRA technique, which includes the creation of a reference sequence, the calculation of a grey relational coefficient and the determination of a grey relational grade. The new method for material selection was shown to be logical in this study.

References

1. Akil H, Santulli C, Sarasini F, Tirillo J, Valente T. Environmental effects on the mechanical behaviour of pultruded jute/glass fibre-reinforced polyester hybrid composites. Compos Sci Technol 2014; 94:62.

2. Noryani M, Sapuan SM, Mastura MT, Zuhri MYM, Zainudin ES. A statistical framework for selecting natural fibre reinforced polymer composites based on regression model. Fibers Polym 2018;19: 1039–49.
3. Wilson A. Vehicle weight is the key driver for automotive composites. Reinforc Plast 2017;61: 100–2.
4. Gholampour A, Ozbakkaloglu T. A review of natural fiber composites: properties, modification and processing techniques, characterization, applications. J Mater Sci 2020;55:1–64.
5. Vaisanen T, Haapala A, Lappalainen R, Tomppo L. Utilization of agricultural and forest industry waste and residues in natural fibre-polymer composites: a review. Waste Manag 2016;54:62–73.
6. Sahari J, Sapuan SM, Zainudin ES, Maleque MA. Mechanical and thermal properties of environmentally friendly composites derived from sugar palm tree. Mater Des 2013;49:285.
7. Mansor MR, Sapuan SM, Zainudin ES, Nuraini AA, Hambali A. Hybrid natural and glass fibers reinforced polymer composites material selection using Analytical Hierarchy Process for automotive brake lever design. Mater Des 2013;51:484.
8. Petrone G, Meruane V. Mechanical properties updating of a non-uniform natural fibre composite panel by means of a parallel genetic algorithm. Compos Appl Sci Manuf 2017;94:226.
9. Mastura MT, Sapuan SM, Mansor MR, Nuraini AA. Environmentally conscious hybrid bio-composite material selection for automotive anti-roll bar. Int J Adv Manuf Technol 2017;91:2031.
10. Anandkumar R, Ramesh Babu S, Sathyamurthy R. Investigations on the mechanical properties of natural fiber granulated composite using hybrid additive manufacturing: a novel approach. Adv Mater Sci Eng 2021. https://doi.org/10.1155/2021/5536171.
11. Li M, Pu Y, Thomas VM, Yoo CG, Ozcan S, Deng Y, et al. Recent advancements of plant-based natural fiber–reinforced composites and their applications. https://doi.org/10.1016/j.compositesb. 2020.108254.
12. Lotfi A, Li H, Dao DV, Prusty G. Natural fiber–reinforced composites: a review on material, manufacturing, and machinability. J Thermoplast Compos Mater 2021;34:238–84.
13. Shah DU. Natural fibre composites: comprehensive Ashby-type materials selection charts. Mater Des 2014;62:21.
14. Mardani A, Jusoh A, Nor K, Khalifah Z, Zakwan N, Valipour A. Multiple criteria decision-making techniques and their applications – a review of the literature from 2000 to 2014. Econ Res Ekon Istraz 2015;28:516.
15. Wang P, Meng P, Zhai J, Zhu Z. A hybrid method using experiment design and grey relational analysis for multiple criteria decision making problems. Knowl Base Syst 2013;53:100–7.
16. Nelson CA. A scoring model for flexible manufacturing systems project selection. Eur J Oper Res 1986;24:346–59.
17. Kulak O, Kahraman C. Fuzzy multi attribute selection among transportation companies using axiomatic design and analytic hierarchy process. Inf Sci 2005;170:191–210.
18. Pohekar SD, Ramachandran M. Application of multi-criteria decision making to sustainable energy planning - a review. Renew Sustain Energy Rev 2004;8:365–81.
19. Yurdakul M, Tansel Y. Application of correlation test to criteria selection for multi criteria decision making (MCDM) models. Int J Adv Manuf Technol 2009;40:403–12.
20. Deng JL. Grey information space. J Grey Syst 1989;1:103–17.
21. Geum Y, Cho Y, Park Y. A systematic approach for diagnosing service failure: service-specific FMEA and grey relational analysis approach. Math Comput Model 2011;54:3126–42.
22. Patel GM, Krishna P, Parappagoudar MB. Optimization of squeeze cast process parameters using Taguchi and grey relational analysis. Procedia Technol 2014;14:157–64.
23. Patil AN, Walke G, Gawkhare M. Grey relation analysis methodology and its application. 2019;4: 409–11.

24. Mora´n J, Granada E, Mı´guez JL, Porteiro J. Use of grey relational analysis to assess and optimize small biomass boilers. Fuel Process Technol 2006;87:123–7.
25. Jayakrishna K, Vinodh S. Application of grey relational analysis for material and end of life strategy selection with multiple criteria. Int J Mater Eng Innovat 2017;8:250–72.
26. Ku H, Wang H, Pattarachaiyakoop N, Trada M. A review on the tensile properties of natural fiber reinforced polymer composites. Compos B Eng 2011;42:856.
27. Yan CH, Po HM, Lau K, Cardona F, Hui D. Natural fibre-reinforced composites for bioengineering and environmental engineering applications. Compos B Eng 2009;40:655–63.
28. JohanssonBras CJ, Mondragon I, Nechita P, Plackett D, Šimon P, Svetec DG, et al. Renewable fibers and bio-based materials for packaging applications - a review of recent developments. Bioresources 2012;7:2506–52.
29. Puglia D, Biagiotti J, Kenny JM. A review on natural fibre-based composites—part II. J Nat Fibers 2005;1:23–65.
30. Syduzzaman M, Faruque MAA, Bilisik K, Naebe M. Plant-based natural fibre reinforced composites: a review on fabrication, properties and applications. Coatings 2020;10:1–34.
31. Al-Oqla FM, Salit MS, AL-Oqla FM, Salit MS. Material selection for composites. In: Materials selection for natural fiber composites. Cambridge, UK: Woodhead Publishing; 2017:73–105 pp.
32. Mastura MT, Sapuan SM, Mansor MR, Nuraini AA. Environmentally conscious hybrid bio-composite material selection for automotive anti-roll bar. Int J Adv Manuf Technol 2017;89:2203–19.
33. Salwa HN, Sapuan SM, Mastura MT, Zuhri MYM. Analytic hierarchy process (AHP)-based materials selection system for natural fiber as reinforcement in biopolymer composites for food packaging. Bioresources 2019;14:10014–46.
34. Sahu P, Gupta MK. A review on the properties of natural fibres and its bio-composites: effect of alkali treatment. Proc Inst Mech Eng, Part L: J Mater: Des Appl 2020;234:198–217.
35. Zwawi M. A review on natural fiber bio-composites, surface modifications and applications. Molecules 2021;26:404.
36. Alam F, Chowdhury H, Elmir Z, Sayogo A, Love J, Subic A. An experimental study of thermal comfort and aerodynamic efficiency of recreational and racing bicycle helmets. Procedia Eng 2010;2: 2413–8.
37. Bahrami M, Abenojar J, Martínez MÁ. Recent progress in hybrid biocomposites: mechanical properties, water absorption, and flame retardancy. Materials 2020;13:5145.
38. Huzaifah MRM, Sapuan SM, Leman Z, Ishak MR. Comparative study on chemical composition, physical, tensile, and thermal properties of sugar palm fiber (arenga pinnata) obtained from different geographical locations. Bioresources.Com 2017;12:9366–82.
39. Hazrati KZ, Sapuan SM, Zuhri MYM, Jumaidin R. Extraction and characterization of potential biodegradable materials based on dioscorea hispida tubers. Polymers 2021;13:1–19.
40. Harussani MM, Sapuan SM, Rashid U, Khalina A. Development and characterization of polypropylene waste from personal protective equipment (Ppe)-derived char-filled sugar palm starch biocomposite briquettes. Polymers 2021;13:1707.

Norihan Abdullah*, Khalina Abdan, Ching Hao Lee,
Muhammad Huzaifah Mohd Roslim, Mohd Nazren Radzuan and
Ayu Rafiqah shafi

9 Thermal properties of wood flour reinforced polyamide 6 biocomposites by twin screw extrusion

Abstract: The use of waste wood flour as polymer reinforcements has recently gained popularity because of its environmental benefits. The goal of this research is to determine the thermal properties of a waste wood flour/polyamide 6 composite made via extrusion. The fillers were melt compounded with polyamide 6 at filler concentrations of 5%, 10%, and 15% using a twin screw extruder, followed by compression molding. The processability of waste wood flour/polyamide 6 composite was evaluated using thermogravimetric analysis (TGA), differential scanning calorimeter (DSC), and dynamic thermomechanical analysis (DMA). According to the TGA analysis, the thermal stability of the composites decreases as the natural fiber content increases. The onset temperature of rapid thermal deterioration was reduced somewhat from 425 °C (neat PA6) to 405 °C (15 wt% wood flour). According to the DSC results, the addition of natural fibers resulted in quantify changes in the glass transition (T_g), melting (T_m), and crystallization temperature (T_c) of the PA6 composites. The storage modulus from the DMA study increased from 1177 MPa (neat PA6) to 1531 MPa due to the reinforcing effects of wood flour (15 wt%). Waste wood flour/polyamide 6 composites offer advantageous thermal properties, enabling us to profit from the strengthening potential of such cellulosic reinforcements while remaining recyclable and generally renewable.

Keywords: composites; extrusion; polyamide 6; thermal properties; wood flour.

---

*Corresponding author: Norihan Abdullah, Laboratory of Biocomposite, Institute of Tropical Forestry and Forest Products (INTROP), Universiti Putra Malaysia, 43400 Serdang, Selangor, Malaysia, E-mail: GS60118@student.upm.edu.my

Khalina Abdan, Ching Hao Lee and Ayu Rafiqah shafi, Laboratory of Biocomposite, Institute of Tropical Forestry and Forest Products (INTROP), Universiti Putra Malaysia, 43400 Serdang, Selangor, Malaysia

Muhammad Huzaifah Mohd Roslim, Department of Crop Science, Faculty of Agricultural Science and Forestry, Universiti Putra Malaysia Bintulu Campus, 97008 Bintulu, Sarawak, Malaysia

Mohd Nazren Radzuan, Department of Biological and Agricultural Engineering, Faculty of Engineering, Universiti Putra Malaysia, 43400 Serdang, Selangor, Malaysia

As per De Gruyter's policy this article has previously been published in the journal Physical Sciences Reviews. Please cite as: N. Abdullah, K. Abdan, C. H. Lee, M. H. Mohd Roslim, M. N. Radzuan and A. R. shafi "Thermal properties of wood flour reinforced polyamide 6 biocomposites by twin screw extrusion" *Physical Sciences Reviews* [Online] 2022. DOI: 10.1515/psr-2022-0082 | https://doi.org/10.1515/9783110769227-009

9.1 Introduction

Several sectors, including transportation, automotive, outdoor building and decoration, and others, have shown interest in the use of natural fibers as reinforcement in polymer-based composites [1]. Natural fibers are growing in popularity as people become more concerned about environment, notably fuel depletion and climate change [2]. Because of their superior mechanical properties, glass and carbon fibers are the most often utilized man-made fibers for reinforcing polymer composites. However, they are nonrenewable resources that need enormous amounts of energy to produce [3]. Due to their many advantages over inorganic fillers and petroleum-based synthetic fibers, natural renewable plant fibers such as wood, jute, coir, flax, hemp, bamboo, and sisal are commonly employed in the manufacturing of fiber-reinforced composites [4]. These natural fibers are used because of their bio-based origin, high specific mechanical qualities, low density, sustainability, global availability, good thermal insulation, biodegradability, cheap price, durability, and simplicity of processing [5, 6]. Natural plant fibers are becoming more popular in engineering applications as an ecofriendly alternative to typical synthetic fibers. Multiple life cycle assessment (LCA) studies conclude that natural fibers are more environmentally friendly than synthetic fibers throughout the entire composite material life cycle (can be easily converted into thermal energy without residues, allowing for less pollution and providing additional carbon credits) [7, 8].

Excessive exploitation of natural resources has focused the world's attention on the need of properly protecting and using such resources. Wood waste is by far the most substantial component of the waste stream generated by construction and demolition procedures. The Malaysian timber company generates 3.4 million m^3 of wood waste per year, as stated by the Forest Research Institute Malaysia (FRIM) in 2016. Urban wood waste includes sawdust, wood chips, bark, slab, trees, branches, and other wood debris. Thus, increasing the use of wood waste will result in favorable incentives while also aiding in the resolution of the landfill issue [8, 9]. Economic competitiveness may be attained by employing less costly reinforcement than the matrix. Compared to unfilled thermoplastics, lignocellulosic material filled thermoplastics exhibit higher bending strength, stiffness, high elasticity (both flexural and tensile modulus), lower thermal expansion, and are less costly [7, 10]. Natural fiber/polymer composites consist of natural fibers incorporated in thermoplastic resins such as polypropylene (PP), polyethylene (PE), poly (vinyl chloride), and polystyrene. They have significant promise as an innovative alternative engineering material for residential and industrial applications. Nonetheless, their applicability in structural and construction applications is still severely restricted because of their poor mechanical strength and stability [11]. As a result, many methods for improving the mechanical properties and stability of natural fiber/polymer composites have been attempted, such as the use of compatibilizers and alkali treatment to modify the surface of the natural fibers, thereby promoting interfacial bonding between the natural fiber and the matrix [12]. However, owing to the greater processing costs for these technologies, future industrial expansion is actually difficult.

Polyamides (6, 6/6, 11, and 12) are the most often used engineering thermo-plastics in a wide variety of important engineering applications. PA6 has sparked substantial attention as a viable matrix for fiber-reinforced composites due to its outstanding mechanical, thermal, and chemical properties [13, 14]. When compared to thermoset polymers (epoxy, polyester, polyurethane), they have higher me-chanical properties than polyolefins (PE, PP, etc.) [15] and easier recyclability, which is important for end-of-life materials. Furthermore, because of their similar hydro-philic nature, PA6 and lignocellulosic fillers have great compatibility, enabling the need of coupling agents and surface modification of natural fibers to be avoided [11]. Recent research on lignocellulosic filler reinforced PA6 has been extensively pub-lished [16–19]. However, their high melting temperatures (220 °C for polyamide 6) are usually hard in the manufacturing of natural fiber composites (NFCs) [20], and thermal deterioration is a key source of difficulty when working with natural fibers, leading to lower mechanical characteristics [21]. Despite the increasing interest in polyamide-based composites, few studies on polyamide-based NFCs have been done, and the present literature focuses on mechanical and morphological charac-teristics. There has been little investigation into the rheological and thermal prop-erties of PA-based NFCs.

The influence of a natural fiber (wood flour) on the thermal characteristics of PA6-based NFCs is discussed. The thermal properties of materials are examined using differential scanning calorimetry (DSC), dynamic mechanical analysis (DMA), and thermogravimetric analysis (TGA).

9.2 Experimental

9.2.1 Materials

Polyamide 6 was supplied by the Shanghai King Chemicals Co., Ltd. with a density of 1.13 g cm^{-3} as the reinforcement, and waste wood flour were supplied by SWCorp, Pahang. The wood flour waste was dried in the industrial oven for 24 h at 60 °C. The waste wood flour that passed through a 40 mesh (400 μm) screen was used in this study.

9.2.2 Fabrication of waste wood flour/polyamide 6 composites

Waste wood flour and PA6 particles were oven-dried for 6 h at 103 °C. The waste wood flour/PA6 composites were then prepared in two steps: extrusion and pelleting, followed by compression molding based on composite compositions of 5%, 10%, and 15% (Table 9.1). Before feeding the components into the extruder hopper, all of the components were manually mixed. A counter-rotating intermeshing twin screw extruder with temperature profiles of 220 °C, 225 °C, 230 °C, and 235 °C from hopper to die and 50 rpm, respectively. The extruded compounds were cooled in a water pool (23 ± 2 °C) before being granulated into pellets at a rate of 10 mm/s via a 2 mm gauge strand die. Further, before compression molding, the pellets were dried in an oven at 103 °C for 6 h to minimize the moisture content. All

Table 9.1: Formulations of wood flour/polyamide 6 composites.

Samples	PA6 (wt%)	Wood flour (wt%)
PA6	100	0
5% WF	95	5
10% WF	90	10
15% WF	85	15

composite pellet formulations resulting from the compounding procedure were compression molded for 5 min using a hot press machine at 220 °C under a pressure of 400 kN with dimensions of 300 mm × 300 mm × 3 mm (width × length × thickness).

9.2.3 Thermal analysis

9.2.3.1 Thermal gravimetric analysis (TGA): Thermal gravimetric analysis (TGA) is to measure the thermal stability of composites and provides quantitative measurement of mass change in materials associated with transition and thermal degradation. Thermal gravimetric analysis was performed using an ASTM E1131 TA instrument Q500 (New Castle, DE, USA). The temperature varied from 30 °C to 600 °C, with a heating rate of 10 °C/min. To avoid sample oxidation, the samples weighing 5–6 mg were placed in an open platinum pan inside a nitrogen flow environment (50 mL/min). The TA universal analysis software was used to record all of the results.

9.2.3.2 Different scanning calorimeter (DSC): In addition to TGA, a differential scanning calorimeter (DSC) test was carried out with the TA instrument Q20 in accordance with ASTM D3418. DSC is widely used to characterize the thermo-physical properties of polymer composites. In this study, the DSC testing was to investigate the melting temperatures and enthalpies of the waste wood flour/polyamide 6 composites. For each formulation, the DSC was performed three times, and dry samples weighing 5–6 mg were placed in aluminum pans. The samples were subjected to a heating cycle ranging from 30 °C to 250 °C at a rate of 10 °C/min. Then, the TA universal analysis software is reporting all of the data. The DSC thermograph was used to investigate the melting and crystallization behavior of materials. The degree of crystallinity (X_c) was calculated corresponding to the melting enthalpy of PA6 in the sample, ΔH_m and $\Delta H_{100\%}$ is the theoretical enthalpy of 100% crystalline PA6, which is equivalent to 230 J/g [22].

$$X_c = (\Delta H_m / \Delta H_{100\%})100 \tag{9.1}$$

9.2.3.3 Dynamic thermomechanical analysis (DMA): Dynamic mechanical analysis (DMA) is a technique used to study the viscoelastic behaviour of polymer composites. A three point bending mode was used under nitrogen protection to conduct DMA on the wood flour/polyamide 6 (WF/PA6) composites using the TA instrument Q800 in accordance with ASTM D4065 to determine the glass transition temperature (Tg), storage modulus (E'), loss modulus (E''), and loss factor (Tan). The temperature varied from 30 °C to 200 °C, with a heating rate of 3 °C/min, a frequency of 1 Hz, an amplitude of 15 m, and specimens of 60 mm × 12 mm × 3 mm.

For DSC, DMA, and TGA tests, at least three samples were tested for each formulation, and the findings are reported as an average of tested samples.

9.3 Result and discussion

9.3.1 Thermal Gravimetric Analysis Results

Thermogravimetric analysis is one of the important analysis techniques to determine the thermal behavior of the composites. The TGA and DTG (derivative thermogravimetric analysis) curves and TGA data of the samples are shown in Figures 9.1 and 9.2 and Table 9.1. The TGA curves of wood flour reinforced PA6 composites are shown in Figure 9.1 were classified into three zones. The initial weight loss of <4% in the first area was below 250 °C, which might be attributable to the evaporation of residual moisture as well as small molecular components during sample storage and processing [16, 23]. The second zone exhibited 90 wt% loss up to 500 °C, whereas the third region extended to 600 °C and showed >2 wt% loss. As a result, the degradation temperatures of neat PA6 and WF/PA6 composites varied. The thermal stability of the 5%, 10% and 15% of WF/PA6 composites reduced as the wood flour content increased. The results in Figure 9.2 indicate that as the fiber content increased, the main DTG peak for WF/PA6 composites shifted to a lower temperature. Because of the higher onset temperatures of PA6, it was established that wood flour had lower fast thermal degradation onset temperatures than PA6 composites [24].

The TGA data of the studied samples are presented in Table 9.1. In contrast, the residue weight of the composite samples at the end of TGA measurement increased proportionally from around 0.8% (neat PA6) to 6% (15 wt% wood flour content), showing that as the wood flour content of the composite materials increased, the

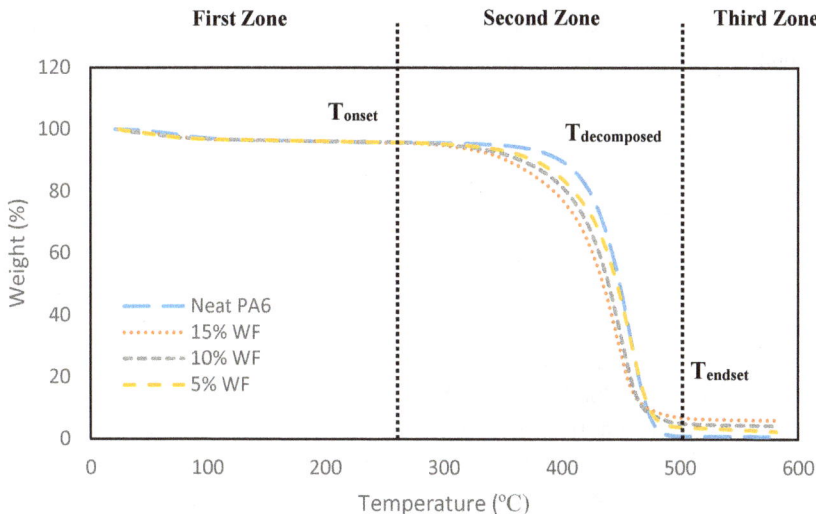

Figure 9.1: TGA curves of neat PA6 and WF/PA6 composites.

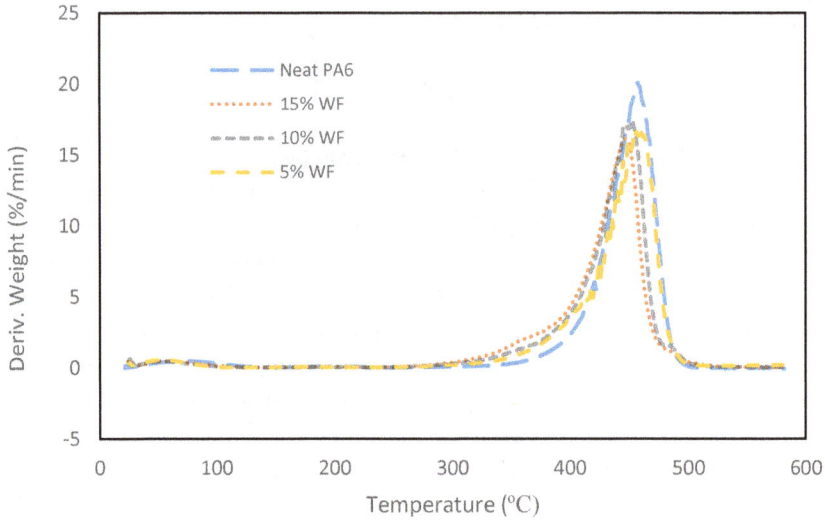

Figure 9.2: DTG peak for neat PA6 and WF/PA6 composites.

composite materials were more resistant in extreme environments, even though rapid thermal degradation onset occurred earlier [24]. It has higher lignin contents, which help in char formation to aid for structural integrity [25]. The T_{onset} of 5% and 15% were lower compared to neat PA6. 5% wood flour started to decomposed much earlier than that of 15%. However, at 10% the T_{onset} is higher before drop again at 15% indicate that the 10% was found dispersed in the PA6 matrix to delay the degradation of the wood flour in blend. The decomposition temperature of the WF/PA6 composites fell from 425 °C for neat PA6 to 405 °C for the addition of 15% wood flour. The lower decomposition temperature of WF/PA6 composites indicates that PA6 has greater thermal stability than the composites.

9.3.2 Differential Scanning Calorimetry Results

The DSC curves of the neat PA6 and different % of WF/PA6 composites are shown in Figure 9.3. Table 9.2 shows the crystallization temperature (T_c) of neat PA6 and WF/PA6 composites. T_c values of neat PA6 was 208 °C and WF/PA6 composites were between 203 °C and 208 °C. As a result, the addition of wood flour had little to no influence on the T_c of WF/PA6 composites. This result is supported with prior study on PA6-based NFCs [24]. The T_g reflected the amorphous region's movement of the PA6 molecular chain. Since neat PA6 had no compatibility issues and the molecular chain motion was not impeded, T_g was detected at a low temperature. In the amorphous region, which needed a high energy and free volume to accomplish the glass transition, the wood flour particles inhibited the mobility of the neat PA6 molecular chain. This limitation

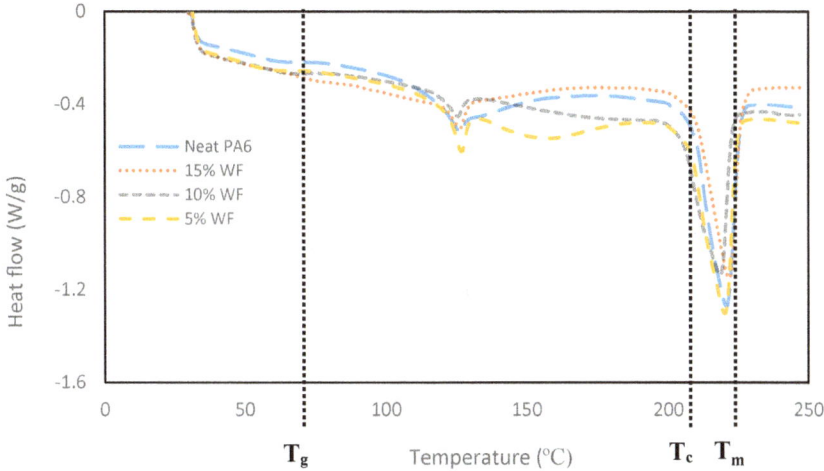

Figure 9.3: DSC melting curves of the neat PA6 and WF/PA6 composites.

increased as the wood flour percentage increased [26]. As a result, T_g rises in proportion to the amount of wood flour present.

The neat PA6 melting point from the exothermic melting peak was found to be 221 °C. Melting temperatures for 5%, 10% and 15% WF/PA6 composites ranged between 218 °C and 221 °C, implying that the wood flour had no influence on the T_m of the composites, wood flour has a negligible effect on T_m because of a small change in crystallite size. Result showed that the addition of 5% and 10% wood flour in PA6 matrix reduced the T_m, however at 15% the T_m increase again indicate that there are greater intermolecular forces between the wood flour and PA6 matrix. The wood flour prevented the mobility of the PA6 molecular chains, resulting in partial crystals, which explains the 5% and 15% decreases in crystallinity. Similar findings were reported by [24, 26]. Wood flour can act as a nucleating site for PA6 crystallization, but it can also act as a barrier to crystal development. However, at 10% the melting enthalpy and X_c increased due to nucleating ability of the wood flour. At 5% most of it lower than neat PA6. The results showed that thermal stability of 5% was lower than that of neat PA6. Table 9.3 shows the DSC data of neat PA6 and WF/PA6 composites.

Table 9.2: TGA data of neat PA6 and WF/PA6 composites.

Sample	T_{onset}	T_{endset}	T_{decomp}	Weight loss (%)	Residue at 600 °C (%)
PA6	249.81	457.29	425.72	95.03	0.808
5% WF	235.63	458.39	419.28	93.53	2.475
10% WF	241.30	452.24	413.72	91.62	4.348
15% WF	235.24	446.68	405.95	90.40	6.106

Table 9.3: DSC data of neat PA6 and WF/PA6 composites.

Sample	Melting temperature, T_m (ºC)	Crystallization temperature, T_c (ºC)	Glass Transition, T_g (ºC)	Melting enthalpy, ΔH_m (J/g)	Crystallinity Index, X_c (%)
PA6	220.52	208.40	52.71	60.66	26.37
5% WF	219.98	207.02	46.12	53.43	23.23
10% WF	218.26	203.22	59.43	75.39	32.78
15% WF	221.58	208.40	62.95	54.00	23.48

9.3.3 Dynamic Mechanical Analysis Results

Figure 9.4 shows the change in storage modulus (E'), loss modulus and loss factor (tan delta) for neat PA6 and wood flour/PA6 composites. The decrease in the E' curves proceeded until the rubbery plateau was reached, at which point the E' curves leveled out. The transition from the glassy plateau to the rubbery plateau through the visco-elastic transition of neat PA6 is reflected by a significant step on the E' about 51 °C. The addition of wood flour to the glassy and rubbery phases induces a small change in the E'. Because of the occurrence of static hydrogen bonds between hydrophilic filler and matrix amide groups, this increase is correlated to the transmission of stresses between fibers and matrix in the glassy state [27]. In the rubbery state, neat PA6 and WF/PA6 composites behave differently; the PA6 exhibits the flow of a linear polymer, while the composites behave as a network. In composites, wood flour functions as additional topological nodes. Figure 9.4a shows the storage modulus of neat PA6 was 1177 MPa while for the WF/PA6 composite increased to 1532 MPa with a 15 wt% increase in wood flour content. This result was attributed to better stress transfer at the fiber interface, which results in a higher modulus with the addition of wood flour over the neat polymer [26]. In the rubbery zone, the 15 wt% wood flour reinforced PA6 showed improved temperature stability than the neat polymer.

Meanwhile, Figure 9.4b indicates that the composite's loss modulus increased from 111 MPa for neat PA6 to 130 MPa for 15 wt% WF/PA6 composites. The mobility and relaxation of the neat PA6 molecular chain were limited due to the good mechanical adhesion between the PA6 molecular chain and the wood flour particles. When the wood flour concentration was higher, the restriction was more severe (15 wt%). Under external stress, the wood flour particles and the PA6 matrix rubbed against one another across the interface, increasing the energy consumption during the neat PA6 molecular chain's movement as well as the loss modulus of the composite [26].

Figure 9.4c shows the addition of wood flour slightly affected the glass transition temperature of the WF/PA6 composites in the tan delta curves as a function of temperature. Furthermore, there was no apparent movement in the peak of the tan delta

(a)

(b)

(c)

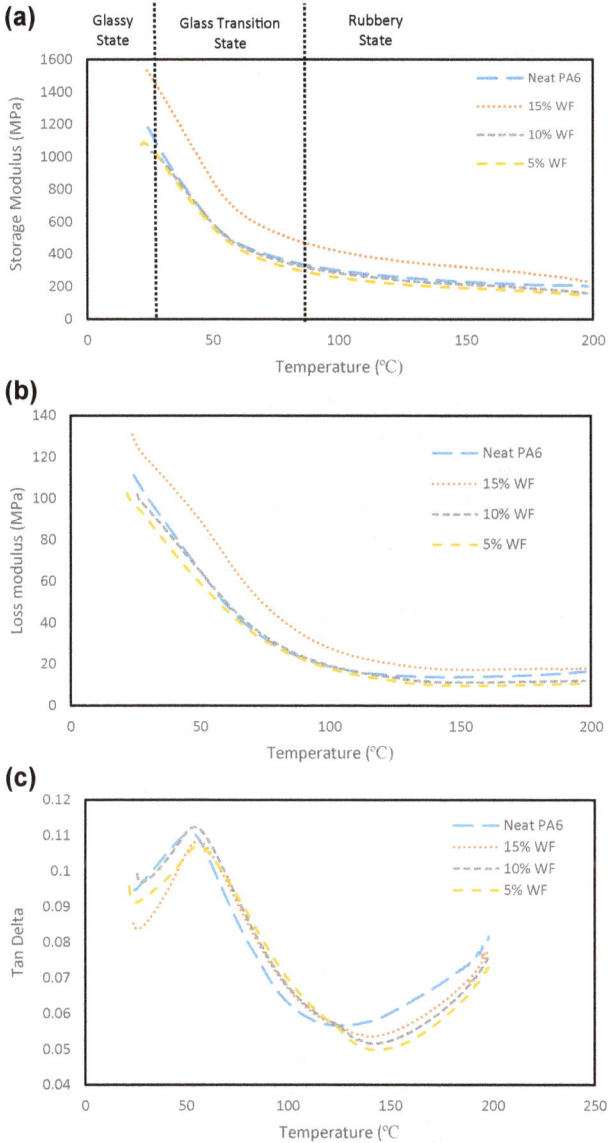

Figure 9.4: DMA curves of neat PA6 and WF/PA6 composites, (a) storage modulus, E' (b) loss modulus, E'' (c) tan delta, δ.

maximum. Wood four networks and/or strong interactions amongst wood flour in the PA6 matrix are most likely responsible for the reinforcing effect [28]. The neat PA6 showed a tan delta max peak temperature of around 52 °C, whereas the WF/PA6 composites had values ranging from 53 °C to 55 °C. With the presence of the 15 wt%

wood flour composition, the most notable effect of the wood flour was a shift in T_g to a higher temperature. The decreased mobility of the PA6 chains caused by wood flour mixing may explain T_g's shift to higher temperatures. Natural fibers may have limited polymer melt mobility, resulting in a reduction in the damping of the composite material at about the glass transition temperature (tan delta max peak height values decreased from 0.11 for PA6 to 0.108 for the 15% WF/PA6 composites).

9.4 Conclusions

In this study, WF/PA6 composites with different contents of WF were prepared by extrusion and the thermal properties of WF/PA6 composites were investigated. The thermal stability of the WF/PA6 composites decreased as the wood flour content increased. However, at 600 °C, the residual weight increased from 0.81% for neat PA6 to 6.11% with the addition of 15 wt% wood flour. There were no significant variations in the T_g, T_m, and T_c temperatures of the waste wood flour reinforced PA6 composites as the natural fiber content increased. Moreover, the storage modulus (E') of the WF/PA6 composites increased as the wood flour content increased, with the highest E' observed in the case of 15 wt% wood flour. The maximum tan delta and peak temperature from DMA did not change significantly when the wood flour content increased.

Acknowledgement: The authors would like to thank the Putra Grant Berimpak (GBP/ 2020/9691300) for providing research funding that made this research possible.

References

1. Sanjay MR, Madhu P, Jawaid M, Senthamaraikannan P, Senthil S, Pradeep S. Characterization and properties of natural fiber polymer composites: a comprehensive review. J Clean Prod 2018;172: 566–81.
2. Atiqah A, Muthukumar C, Thiagamani SMK, Krishnasamy S, Ansari MNM. Characterization and interface of natural and synthetic hybrid composites. Encyclopedia of Renewable and Sustainable Materials. Amsterdam: Elsevier, 2020;4:389–400 pp.
3. Kona S, Lakshumu Naidu A, Bahubalendruni MVAR. A review on chemical and mechanical properties of natural fiber reinforced polymer composites. Int J Perform Eng 2017;13:189–200.
4. Taj S, Munawar M, Khan S. Natural fiber-reinforced polymer composites. Proc Pakistan Acad Sci 2007;44:129–44.
5. Keya KN, Kona NA, Koly FA, Maraz KM, Islam MN, Khan RA. Natural fiber reinforced polymer composites: history, types, advantages and applications. Mater Eng Res 2019;1:69–85.
6. Karthi N, Kumaresan K, Sathish S, Gokulkumar S, Prabhu L, Vigneshkumar N. An overview: natural fiber reinforced hybrid composites, chemical treatments and application areas. Mater Today Proc 2019;27:2828–34.

7. Islam Md. A. Natural fiber as a substitute to synthetic fiber in polymer composites: a review. Res J Eng Sci 2013;2:46–53.
8. Omran A, Mahmood A, Aziz H, Robinson G. Investigating households attitude toward recycling of solid waste in Malaysia: a case study. Int J Environ Res 2009. (ISSN: 1735–6865);3:3.
9. Gulitah V, Liew KC. Morpho-mechanical properties of wood fiber plastic composite (WFPC) based on three different recycled plastic codes. Int J Biobased Plast. 2019;1:22–30.
10. Mohammed L, Ansari MNM, Pua G, Jawaid M, Islam MS. A review on natural fiber reinforced polymer composite and its applications. Int J Polym Sci 2015;2015:2–16.
11. Zierdt P, Theumer T, Kulkarni G, Däumlich V, Klehm J, Hirsch U, et al. Sustainable materials and technologies sustainable wood-plastic composites from bio-based polyamide 11 and chemically modified beech fibers. SUSMAT 2015;6:6–14.
12. Ku H, Wang H, Pattarachaiyakoop N, Trada M. A review on the tensile properties of natural fiber reinforced polymer composites. Compos B Eng 2011;42:856–73.
13. Singh R, Kumar R, Ranjan N, Penna R, Fraternali F. On the recyclability of polyamide for sustainable composite structures in civil engineering. Compos Struct 2017;184:704–13.
14. Kunchimon SZ, Tausif M, Goswami P, Cheung V. Polyamide 6 and thermoplastic polyurethane recycled hybrid fibres via twin-screw melt extrusion. J Polym Res 2019;26:2–14.
15. Ashori A. Wood – plastic composites as promising green-composites for automotive industries!. Bioresour Technol 2008;99:4661–7.
16. Xu K, Zheng Z, Du G, Zhang Y, Wang Z, Zhong T, et al. Effects of polyamide 6 reinforcement on the compatibility of high-density polyethylene/environmental-friendly modified wood fiber composites. J Appl Polym Sci 2019;136:16–8.
17. Alonso-Montemayor FJ, Tarres Q, Oliver-Ortega H, Espinach FX, Narro-Cespedes RI, Castaneda-Facio AO, et al. Enhancing the mechanical performance of bleached hemp fibers reinforced polyamide 6 composites: a competitive alternative to commodity composites. Polymers 2020;12: 2–17.
18. Santos PA, Spinacé MAS, Fermoselli KKG, De Paoli MA. Polyamide-6/vegetal fiber composite prepared by extrusion and injection molding. Compos Part A Appl Sci Manuf 2007;38:2404–11.
19. Singh R, Kumar R, Ranjan N. Sustainability of recycled ABS and PA6 by banana fiber reinforcement: thermal, mechanical and morphological properties. J Inst Eng Ser C 2019;100:351–60.
20. Andrzejewski J, Ani J, Szulc J. A comparative study of biocarbon reinforced polyoxymethylene and polyamide: materials performance and durability. Compos Part A Appl Sci 2022;152. https://doi.org/10.1016/j.compositesa.2021.106715.
21. Pereira de Melo R. Development of polyamide composites with natural fibers for automotive applications. Ecole Nationale Supérieure des Mines de Paris: Universidade federal do Rio de Janeiro; 2015.
22. Millot C, Fillot LA, Lame O, Sotta P, Seguela R. Assessment of polyamide-6 crystallinity by DSC: temperature dependence of the melting enthalpy. J Therm Anal Calorim 2015;122:307–14.
23. Şeker Hirçin B, Yörür H, Mengeloğlu F. Effects of filler type and content on the mechanical, morphological, and thermal properties of waste casting polyamide 6 (W-PA6G)-based wood plastic composites. Bioresources 2022;16:655–68.
24. Kiziltas EE, Yang HS, Kiziltas A, Boran S, Ozen E, Gardner DJ. Thermal analysis of polyamide 6 composites filled by natural fiber blend. Bioresources 2016;11:4758–69.
25. Monteiro SN, Calado V, Rodriguez RJS, Margem FM. Thermogravimetric behavior of natural fibers reinforced polymer composites-an overview. Mater Sci Eng A 2012;557:17–28.
26. Zhu S, Guo Y, Chen Y, Liu S. Low water absorption, high-strength polyamide 6 composites blended with sustainable bamboo-based biochar. Nanomaterials 2020;10:1–15.

27. Lods L, Richmond T, Jany D, Dantras E, Lacabanne C, Durand JM, et al. Thermal stability and mechanical behavior of technical bamboo fibers/bio-based polyamide composites. J Therm Anal Calorim 2022;147:1097–106.
28. Buchenauer A. Wood fiber polyamide composites for automotive applications [Master's thesis]: University of Waterloo; 2016.

Ahmad Fazil Nur Diyana*, Abdan Khalina, Mohd Sapuan Sali,
Ching Hao Lee*, Humaira Alias Aisyah, Mohd Nurazzi Norizan and
Rafiqah Shafi Ayu

10 Characterization of lignocellulosic *S. persica* fibre and its composites: a review

Abstract: As the demand for renewable, cost-effective, and environmentally accept-able materials in a variety of applications has developed, natural fibres have become more popular as reinforcement in composite materials. *Salvadora persica* L. is the most common traditional source of chewing stick (miswak) advised by Prophet Muhammad. It is also known as Arak in Arabic and Peelu in Urdu. A lot of research has been done in the last few years to investigate if its traditional applications in dental care are still valid. For this review, a variety of databases (Science Direct, PubMed, Wiley Online Library, and Google Scholar), books and primary sources were examined, surveyed, and analysed. Miswak fibre qualities and attributes were addressed in this review study to evaluate if the fibre may be used as an alternative to natural fibre reinforcing in composites. The history and uses of the miswak tree, as well as the structure of the miswak tree, are presented first, followed by a discussion of fibre characterization, with a focus on fibre structure and composition. Finally, the effect of miswak on the physical, mechanical, and thermal properties of composites is discussed. Miswak fibre and its composites present considerable challenges and potential as a reinforcement or filler alternative in a variety of applications, including dentistry.

Keywords: miswak; natural fiber composite; *Salvadora persica*.

*Corresponding authors: Nur Diyana Ahmad Fazil and Lee Ching Hao, Laboratory of Bio Composite Technology, Institute of Tropical Forestry and Forest Products (INTROP), Universiti Putra Malaysia, 43400 Serdang, Malaysia, E-mail: nur.diyana.ahmad.fazil@gmail.com, leechinghao@upm.edu.my
Abdan Khalina, Laboratory of Bio composite Technology, Institute of Tropical Forestry and Forest Products (INTROP), Universiti Putra Malaysia, 43400 Serdang, Selangor, Malaysia; and Department of Agriculture and Biotechnological Engineering, Universiti Putra Malaysia, Sedang 43400, Selangor, Serdang, Malaysia
Mohd Sapuan Sali, Department of Mechanical Engineering, Universiti Putra Malaysia, 43400, Selangor, Serdang, Malaysia
Humaira Alias Aisyah, Mohd Nurazzi Norizan and Rafiqah Shafi Ayu, Laboratory of Bio composite Technology, Institute of Tropical Forestry and Forest Products (INTROP), Universiti Putra Malaysia, 43400 Serdang, Selangor, Malaysia

As per De Gruyter's policy this article has previously been published in the journal Physical Sciences Reviews. Please cite as: A. F. Nur Diyana, A. Khalina, M. S. Sali, C. H. Lee, H. A. Aisyah, M. N. Norizan and R. S. Ayu "Characterization of lignocellulosic *S. persica* fibre and its composites: a review" *Physical Sciences Reviews* [Online] 2022. DOI: 10.1515/psr-2022-0043 | https://doi.org/10.1515/9783110769227-010

10.1 Introduction

Environmentally friendly biodegradable polymers have gotten a lot of attention in recent decades due to the desire to reduce global pollution produced by petroleum-based synthetic plastics [1–3]. By definition, biodegradability refers to a material's ability to be entirely digested by native microorganisms in the ecosystem. Natural lignocellulosic fibres have been widely used and explored as an eco-friendly polymer that is based on renewable plant material, entirely biodegradable and low cost to tackle those issues stated by Sherif Mehanny et al. in [4]. Natural composites use 17% less energy than synthetic composites on average. By 2025, the global bio composite market is expected to increase at a 9.59% annual pace, with a net worth of USD 41 billion [5]. Also, Mohanty et al. reported by 2020, roughly 10% of fundamental chemical building blocks will originate from renewable plant resources, with that percentage predicted to climb to 50% by 2050 as barriers to their usage are removed [6]. However, the development of natural fibre reinforced polymer has several obstacles such as inadequate adhesion, poor wettability, significant moisture absorption, and debonding are the key problems [7]. These limitations, however, can be mitigated by using a coupling agent, compatibilizer and fibre pre-treatment [7].

The term biomass has been used to encompass lignocellulosic in general, however this phrase has larger meanings than lignocellulosic. Lignocellulosic materials can indeed be found in abundance. Claassen et al. calculated their annual production to be between 10 and 50 billion tonnes (about half of the biomass) [8]. Cellulose, hemicelluloses, lignin, and pectin make up lignocellulose fibres. Because they are the consequence of photosynthesis, lignocellulosic materials are also known as photomass. Wood, agricultural residues, water plants, grasses, and other plant compounds are all sources of lignocellulosic. Since wood and other lignocellulosic are inexpensive, low in processing energy, renewable, and sturdy, they have been employed as "engineering materials." However, certain schools of thought believe that lignocellulosic composites such as plywood, particleboard, fibreboard and laminated lumber are not materials since they lack consistent, predictable, reproducible, continuous, and uniform qualities. This is true for solid lignocellulosic like wood, but it is not always true for lignocellulosic composites. A lignocellulosic composite is a reconstituted product created from a mixture of one or more substances that is held together by a bonding agent. Plywood, particleboard, fibreboard and laminated lumber are the most well-known lignocellulosic composites. The proportion of these products varies depending on the type of fibre. As a result, lignocellulose fibres have a wide range of mechanical properties [9–11]. The fundamental objective is to valorize the available lignocellulose fibres, which are typically chosen in accordance with local production and governmental or industrial partnerships [12].

Natural fibres are divided into three categories based on their origins: plants, animals, and minerals. Plant fibres, in general, are utilised to reinforce polymers in the composite sector. Hairs (cotton, kapok), fibre-sheafs of dicotyl plants or vessel-sheafs of monocyclic plants (e.g., flax, hemp, jute and ramie) and hard fibres (sisal, henequen, and coir) are only a few examples of plant fibres. Wood fibres from trees are the most common, although other fibre types are beginning to develop. Decks, docks, window frames, and moulded panel components are all examples of when plastic/wood fibre composites are used [13]. According to [14]; 460 million pounds of plastic or wood fibre composites were manufactured in 1999. The manufacturing industry is likewise concerned about the availability of raw materials, and the pressures on it to utilise ever-greener technology have sparked interest in this field of research around the world [14]. In lignocellulosic fibre reinforced composites, Jacob et al. [15] has studied the characterisation of interfaces and fibre surface. The adhesion between fibre and matrix is the most essential element in composites; hence characterisation of the interface is crucial [15]. To create a strong interface on both the fibre surface and the polymer matrix surface, a variety of treatments on lignocellulosic surfaces as well as the characterization technique has been explored. Furthermore, several lignocellulosic materials such as rice and wheat straw, sugarcane bagasse, and rice and coffee husks, which were either used as household fuels or sometimes burned *in situ* in the fields as a way of disposal, have been tested as fillers/reinforcements in polymer or cement to develop new materials for the construction industry [16]. These successful endeavours have opened up a vast array of raw materials for biodegradable material creation.

Plants have been utilised to improve dental health and promote oral cleanliness for millennia, and this practise is still practised in various societies around the world. *Salvadora persica* L. (*S. persica*) is a *S. persica* plant species belonging to the Salvadoraceae family. From Rajasthan (India), Nepal, and Malaysia in the east, through Pakistan, Iran, Iraq, Saudi Arabia, and Egypt to Mauritania in the west, and from North Africa through Sudan, Ethiopia and Central Africa to southwestern Africa, *S. persica* has a vast geographic distribution [17, 18]. It is primarily grown in many parts of the world for the production of roots, stems, leaves, and fruits for medicinal and food purposes [19]. *S. persica* fibres have been identified as a material with significant potential [19], which has piqued the interest of researchers and academics interested in investigating the prospect of using it to reinforce polymers [20, 21]. Despite their numerous benefits, lignocellulosic fibres suffer from poor wettability and high moisture absorption, which prevents the hydroxyl groups from reacting with polar matrices and producing an interlock when employed as reinforcing materials [21]. Physical and perhaps chemical treatment of the fibre surfaces to improve compatibility with the polymer matrix is used to alleviate setbacks, create reactive groups, and interlock with polymer matrices. Mercerization, acetylation, benzoylation, silane coupling, graft co-polymerisation, permanganate treatment, maleate coupling and other treatments can be used to improve the interfaces between fibre and polymer matrix. The modification cleans and modifies the fibre surfaces, enhances the unevenness of the external surfaces, and either galvanises the hydroxyl groups (OH) or introduces new functionalities that interlock the fibres with the polymer matrix, improving the mechanical characteristics of composites [22].

S. persica is widely used as a toothbrush, and the roots, stems, leaves, and fruits are consumed as food in arid environments; the woods were used as firewood or charcoal, and the flowers were a good source of nectar for honey bees; and the most important of all the different parts of the plants: root, bark, stem, leaves, seeds, flowers and fruits were used in the preparation of medicine for various ailments and diseases such as digestive, circulatory [23]. As a result, it is with the mind that this plentiful supply can be transformed into a very valuable product. Wood, stems and roots, which were once thought to be low-value items, are now being employed as potential reinforcement materials in a variety of thermoplastic composites. There is currently limited evidence on the use of this fibre as a reinforcing material in polymer composites, so this review will focus on the characterization of fibre in terms of composition and structure, as well as its impact on the physical, mechanical and thermal properties of composite materials.

10.2 History of *S. persica* tree and its fibre

The chewing stick can be thought of as the forerunner of today's toothbrush. Chewing sticks made from various plants are widely used in Asia, Africa and the Middle East. "Miswak" is an Arabic word that refers to any plant component used as a chewing stick [24]. *S. persica* is a stick that is used to clean teeth and gums in the traditional sense. One end of a little pencil-sized stick is chewed until it becomes a brush, and then the frayed end is used to clean the teeth. Various plants have been employed for this purpose in various regions of the world. Arak (*S. persica*) is mostly used in the middle east; lime tree (*S. aurantafolia*) and orange tree (*Citrussinensis*) are used in West Africa, Neem (*Azadirachta indica*) in Indian subcontinent while the roots of the senna (*Cassia vinnea*) are used by American and African Negros and Laburnum (*Cassia sieberianba*) has been used in Sierra Leone [24].

Miswak is a term used in Middle Eastern countries to refer to the Arak plant, as well as the Muslim community worldwide. Miswak is typically made from the arak plant's root, but its bark and branches have also been employed [25]. The use of the arak plant has the longest documented history of any chewing stick [25]. Numerous researches have been conducted to demonstrate the efficacy of *S. persica* when compared to modern methods of oral hygiene maintenance at home. People who used *S. persica* scored lower on the community periodontal index of treatment need (CPITN) in a 1991 study by [26].

S. persica use was found to be at least as efficient as teeth brushing in the prevention of gingivitis and plaque in a study by [27]. The antibacterial action of the arak plant was also effective for the prevention/treatment of periodontal disorders, according to this study [28–30].

The value of *S. persica* is due primarily to its mechanical cleaning action [31]. When compared to the usage of conventional toothbrushes, Varma et al. [23] found that using *S. persica* as an oral hygiene aid resulted in significantly reduced plaque scores. *S. persica* is typically used for a longer amount of time than a toothbrush, and it is easier to reach the buccal/labial surfaces of the teeth than the lingual and proximal areas [23]. In Saudi Arabia, Al-Ghamdi et al. [27] reported no significant difference in buccal/labial plaque ratings between *S. persica* and toothbrush users, despite the extended duration of *S. persica* use. Chewing sticks are frequently utilised as the sole oral cleansing technique in many underdeveloped countries. The frequent eradication of dental plaque is an important part of maintaining oral hygiene and preventing dental caries and periodontal disease.

Several theories have been proposed to explain the cleansing ability of *S. persica*, including (a) the mechanical effects of its fibres, (b) the release of beneficial compounds from the chewing stick while in use, or (c) a combination of these. However, because to a dearth of studies detailing the length, frequency, and duration of *S. persica* use, it is unable to make significant judgements of its mechanical cleaning effects on dental health [23]. According to a 2018 poll, the majority of Jordanians believe that brushing their teeth with a toothbrush and a *S. persica* is the most efficient way to maintain good oral health [32]. The *S. persica* is a low-cost, widely available oral hygiene equipment that can be utilised by the vast majority of individuals [33].

10.3 Uses of *S. persica* tree

S. persica has been used for a variety of reasons in the past, including food, fuel, cosmetics, oral hygiene and, of course, medications [19, 34]. The leaves, for example, can be cooked into a sauce and served as salads or green vegetables. The fruits can be consumed fresh, cooked or dried. The wood is also used for charcoal and fuel. Furthermore, the tree's resin is said to be valuable in the production of varnish. Hair removal with a knife has also been reported to be aided by crushed leaves of *S. persica* submerged in cow urine. Furthermore, the leaves and young shoots are fed to camels, cows, goats and sheep, and the leaves have been shown to boost lactation in cows and improve animal body weight. Honey bees have found the petals to be a rich source of nectar, and the honey of *S. persica* is thought to have medicinal benefit [35, 36]. *S. persica* has been utilised for medical purposes by various ethnic groups, mainly in Asia and Africa. The root, bark, stem, leaves, seeds, flowers and fruits of the plants have all been employed in a range of formulations for internal and external use to treat a number of maladies and disorders of the digestive, musculoskeletal, circulatory, glandular and urinary systems [34, 37–39].

10.4 Characterization of *S. persica* fibre

It's a relatively recent approach to use *S. persica* fibre as reinforcement in a polymer matrix. The study of *S. persica* fibre and its composites began in the 1990s, although the properties of these materials are yet unknown. Most natural fibres can be used to create a variety of bio composite products. The composition of such fibres is the most important factor to examine because it dictates some of the bio composites' essential features. In order to design any bio composite material, it is necessary to understand the chemical composition of natural fibres as well as their adhesive surface qualities [40–42].

10.4.1 Structure and chemical composition of natural fibre

Natural fibre's chemical composition, chemical structure, and cell structure arrangement are all critical. The performance and durability of natural fibre composites in their diverse applications is determined by these characteristics [40, 43]. Figures 10.1 and 10.2 depicts the structural organisation of plant fibre. The natural fibre has a layered structure, with the major wall on the outside enclosing the secondary wall. S1, S2 and S3 are the three layers that make up the secondary wall.

Natural fibres contain cellulose, hemicellulose, lignin, pectin, ash and other naturally occurring components, according to their chemical components [46, 47]. Hemicellulose, cellulose and lignin are the three principal components of lignocellulose material, accounting for 10–25%, 31–70 wt% and 7–26 wt%, respectively [48–50].

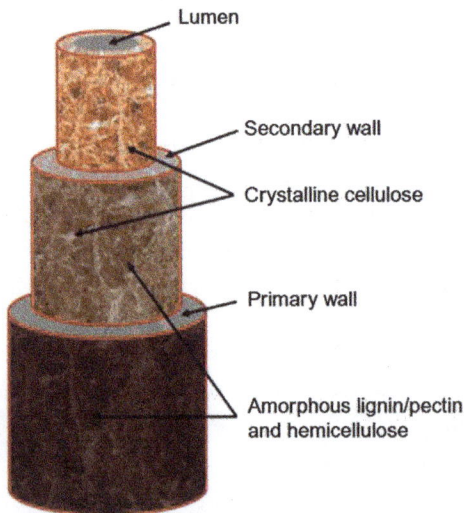

Figure 10.1: Bhattacharyya et al. [44] describes the structure of the plant fiber.

Secondary wall S3

Lumen

Helically
arranged
crystalline
microfibrils
of cellulose

Secondary wall S2

Spiral angle

Secondary wall S1

Primary wall

Amorphous
region mainly
consisting of lignin
and hemicellulose

Disorderly arranged
crystalline cellulose
microfibrils networks

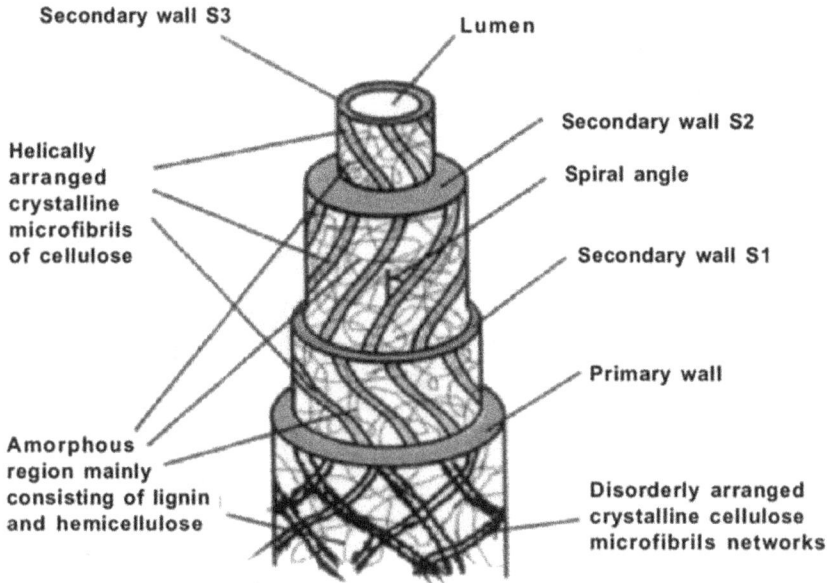

Figure 10.2: John and Thomas [45] describe the structural constitution and arrangement of a natural plant fibre cell.

Chemical energy is stored in these plant fibre constituents, which varies from plant to plant [49]. When it comes to natural fibre's composition, the most important component is cellulose, which provides mechanical strength [51, 52]. Furthermore, cellulose has a highly crystalline structure with about 80% crystalline area [50, 53].

Hemicellulose, unlike cellulose, does not assist much to the strength of natural fibres, although it does bind the fibre's microfibrils [52]. When hemicellulose burned up early in the thermal analysis, the most essential two fibre components are cellulose and lignin. Lignin is an amorphous complex of phenylpropane unit polymers that serves as an encrusting agent or a matrix [54, 55]. The features of natural fibres and composites are influenced by crystal structure, microfibril angle, lumen size, crystalline area size, porosity and void [46].

10.4.2 *S. persica* stem cell wall ultrastructure

The stem ultrastructure is made up of stem cortex, xylem, and parenchyma, according to the anatomy of the *S. persica* stem in Figure 10.3. UV radiation, diseases and insects are all protected by 3-dimensional epicuticular crystals of diverse shapes and sizes on the surface. The secondary xylem on the *S. persica* stem surface is made up of groups of xylem fibres with thick-walled small cells. Wood parenchyma and ray parenchyma are

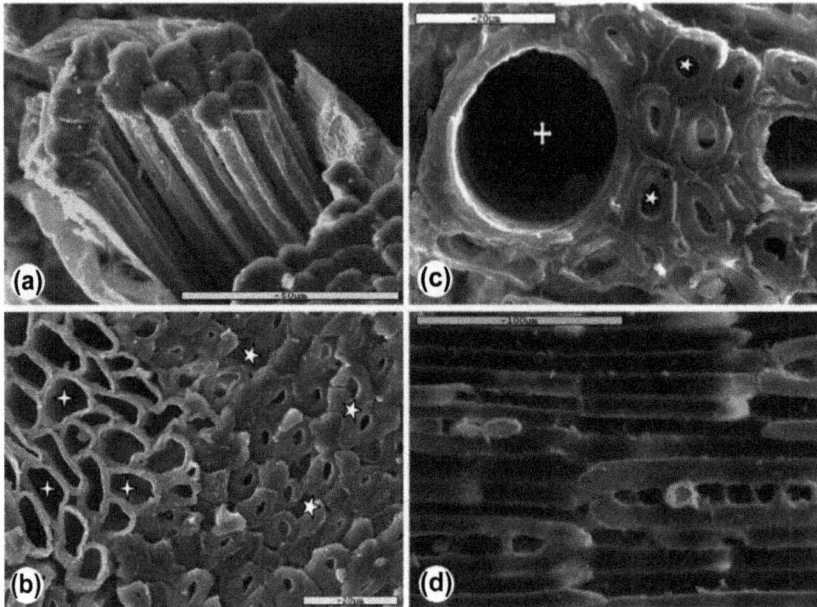

Figure 10.3: Sawidis [57] describes; (a) in the cortex, there is a band of polygonal phloem fibres with a thick lignified secondary cell wall, (b) fibres that are flattened and intimately connected with parenchyma cells (stars), (c) metaxylem vessels (quad arrow) are tightly connected with xylem fibres (stars), (d) longitudinal segment of xylem fibres with blunt ends.

two types of parenchyma cells found in the xylem. The wood parenchyma in the centre of the stem is crushed, forming a chamber that is eventually filled with amorphous inclusions or rhombohedra crystals. The stem vessels of the *S. persica* are orientated axially, with no branching for connectivity. Sulphur in wood parenchyma cells was discovered by SEM-EDX research, which likely acts as a defence against pathogenic germs, making the *S. persica* stem a good oral hygiene aid [56, 57].

10.4.3 *S. persica* stem composition and structure

S. persica fibre has the same cellulose content and cellulose molecular chain arrangement as wood cellulosic fibres. Plant cellulose is found in the woody xylem tissues, where cellulosic microfibrils are embedded in a matrix of lignin and hemi-cellulose, which is an amorphous non-cellulosic carbohydrate [57]. *S. persica* fibres are axial elongate fibres with moderate radial expansion, thick walls of 3–15 m in diameter, and mean length 100–1000 m of very long cells in libriform fibres and elementary vessels of elongated cells and roundish cells of 20–40 m in diameter, as determined by optical microscopy. The biological features and chemical contents of *S. persica* fibre have been examined by many researchers [24, 31, 58, 59]. According to [60], the properties of the toothbrushes are superior to those of traditional toothbrushes. In comparison to regular toothpaste, [28] displays good quality attributes. In comparison to conventional mouthwash, [58] found that it has good quality properties. Raw fibre can survive in arid environments [57] and has been shown to have significant anti-bacterial activity, making it perfect for tooth brushing [27, 61]. Although there has been little research on the cellulose type and microfibrillar angle, which is a role in the mechanical strength of fibres.

10.5 *S. persica* based polymer composite

It is vital to understand the physical, mechanical and thermal properties of natural fibre composites in order to recognise their strengths and shortcomings [62]. Several publications on the physical, chemical, mechanical and thermal properties of *S. persica* fibre will be rigorously examined.

10.5.1 Effect *S. persica* on the physical properties of composites

Particulate fillers are used to alter the physical and mechanical properties of polymers in a variety of ways. At both high and low temperatures, polymeric materials are known for their versatility, chemical resistance, hardness and long-term durability [63]. Commercial aviation, flooring, and the electrical industry are just a few of the industries that use polymers and polymer-matrix composites reinforced with filler particles [64]. In commercial applications, the usage of particle fillers in polymers is primarily aimed at cutting costs and enhancing stiffness [65].

Savaş [21] has investigated the physical characteristics of miswak powder with polypropylene thermoplastic composites in terms of contact angle measurements, surface free energy and water uptake.

The contact angle is an important element in composites fabrication and processing because it defines the wetting phenomenon. Savaş [21] had discovered for pure PP; the highest contact angle was found to be 111.89° in distilled water. He employed the dynamic sessile drop method to determine contact angles. The contact angles for distilled water were all larger than 90°, suggesting that it was hydrophobic, but the contact angles for the other three test liquids (diiodomethane, ethylene glycol and formamide) were all greater than 90°, indicating that they were amphiphilic. However, the contact angles tend to increase as the miswak concentration is increased to 100.48° for PM30. For a given fluid, a smaller contact angle suggests that the fluid spreads over a larger region, indicating that the fluid has a better wettability on that particular surface [66]. As a result, the miswak concentration composite has a higher wettability than the PP matrix in this circumstance. Except for PM5 and PM0, the other test liquids (diiodomethane, ethylene glycol, formamide) followed a similar pattern with some fluctuations, which is thought to be due to the attributes of the liquid used as well as geometric factors such like surface fractions contacting the liquid and pores, which end up causing the droplet to sit on the air as well as the solid, starting to affect the contact angle. Using contact angle measurements, [67] investigated the surface properties of lignocellulosic wheat straw reinforced in a polyolefin (polymer of polyethylene and polypropylene) matrix. The researchers discovered that the hydrophilic lignocellulosic filler and the hydrophobic polyolefin matrix had weak interfacial adhesion [67].

Surface energy and contact angle measurements are closely related since both are used to quantify the wetting behaviour of fibres and are directly related to adhesion potential to matrices. Savaş [21] calculated the γ_S^{tot} using the OWK method and the γ_C using the Zisman method for surface free energy. Due to the non-polar character of the PP, it is noticed that PM0 has a low polar component value, γ_L^P of surface free energy. Furthermore, the composites had lower γ_L^P values than PM0, as shown in the study. Because the surface energy of the polar portion of the miswak treated with PP-g-MA is very low, a significantly lower value of γ_L^P for the composites suggests a higher surface energy of the dispersion part at the interface. However, when the miswak concentration rises, the findings of surface free energy calculation and contact angle measurement differ slightly, with the maximum γ_S^{tot} of 35.715 mJ m^2 for PM0, and other surface free energies determined in the following order: PM10 > PM5 > PM30 > PM20. In comparison to PBAT, LCR and cellulose fibres, Le Digabel et al. [67] discovered that the hydrophobic property of PP is demonstrated by its low polar component. The dispersive components of the substrates range from 32 to 40 mJ m^{-2}, and the polar components range from 2 to 22 mJ m^{-2}. The fibres have high values, which is consistent with their well-known hydrophilic nature. These differences are due to the substrate's polar or nonpolar nature. Dhakal et al. [66] discovered that unsaturated polyester has a polar surface energy of 16.27 mN m, whereas hemp/UP has a polar energy of 32.13 mN m. According to the findings, the polar energy composite with NaOH increased in value while it was predicted to drop, resulting in little crystalline alteration of cellulose.

For the water absorption, Savaş [21] discovered that after 48 h of immersion, the samples' WU values were in the range of 0.02–0.26% for water absorption. The WU values of PP/miswak composites were found to be higher than those of pure PP, and they rose exponentially as miswak content was increased. This is most likely due to hydrogen atoms in water molecules connecting to the miswak's cellulose cell wall units' free hydroxyl groups. The fibrous structure of the miswak, which may speed up water penetration via capillary action, as well as the eventual diffusion of water molecules to the miswak and PP surfaces, are also contributing causes. Furthermore, because to its large surface area, the miswak powder has a large number of pores (average pore size: 12.56 lm), making it a natural adsorbent for dye uptake and purification applications. As a result, increasing the miswak concentration in the composites resulted in more water being adsorbed. Abdulkhani et al. [68] found a similar finding in their study of the impact of lignocellulosic material with solvents on the mechanical and physical properties of composite films. The highest WA is observed during the first immersion time of the PW and CEL film composites, according to the findings. The hydrogen bonding of water molecules to the free hydroxyl groups present in the cellulose cell wall of fibrous material and the diffusion of water molecules into the microfibrils interfaces are the reasons for the higher water uptake of PW and CMP films than CEL films. Capillary action is accelerated by the huge number of porous tube structures found in poplar fibre [68]. Wan Busu et al. [69] conducted another study on the level of water uptake of an untreated kenaf/glass hybrid composite. According to the experiments, untreated kenaf composites had a larger proportion of water uptake than untreated glass composites. The reason for this is that the materials can absorb more water due to hydrogen bonding or the creation of OH and H groups from water, resulting in poor adhesion between fibre and matrix. It may be deduced that the difference in water uptake levels as a result of the cellulosic materials' weak absorption resistance is mostly attributable to the presence of polar groups, which attract water molecules through hydrogen bonding. It has been found to be linked to the lignocellulosic fibre composition, implying that fibre composition influences water intake [69].

Surface roughness and degradation were also mentioned as factors restricting the material in damp conditions in the experiments. The final roughness value was calculated using the Ra and Rz values. The % increase in surface roughness values of the composites rose with larger miswak concentration due to the hydrophilic feature of the cellulose structure of the filler, as stated above. The average roughness (Ra) values of the composites after water immersion ranged from 0.181 to 0.649 nm for samples with 5 wt% (PM5) and 30 wt% (PM30) miswak concentrations. The percent increase in average roughness values of the composites for the samples PM5 and PM30 ranged from 5.8 to 151.6%. Surprisingly, although absorbing 52% more water than pure PP, PM5 had lower Ra and Rq (except for Rz) surface roughness values than PM0 after the test. The researcher concluded that 5 wt% miswak reinforcement might be effective in wet situations without causing substantial surface degeneration, despite the fact that the average roughness was lower than PM0. Peng et al. [42] used atomic force microscope AFM methods to investigate the changes in surface morphology of weathered samples of polypropylene composites reinforced with wood flour (WF), cellulose and lignin in various loadings, and the results revealed that the surface roughness of all samples increased after 960H weathering, owing to lignin's role in preventing photodegradation of composites [42]. The average surface roughness of sisal, hemp and nettle fibres is studied by Lila et al. [41]. Sisal fibre roughness was determined to be in the 200–220 nm range by AFM, while Hemp and Nettle fibres were not possible to perform AFM due to their short diameter (20–60 µm), since the minimum area required is in the 50 m range. Surface roughness of fibres propagates mechanical connection between fibres and matrix and aids in improving mechanical characteristics, according to studies, as evidenced by SEM pictures. As a result, fibres with a higher surface roughness have better material mechanical characteristics.

A deeper understanding of interfacial behaviours at the molecular level is required to determine the expected mechanical properties of freshly constructed composite systems at the macroscale [70]. Furthermore, Savaş [21] said that the mechanical performance of composites is influenced by interfacial contacts (or adhesion) between the polymer matrix and the filler. Increased miswak concentration reduces surface free energy, causing poor interface reinforcement/matrix adhesion and, as a result, a loss of composite tensile properties.

10.5.2 Effect *S. persica* on the mechanical properties of composites

One of the key research studies focus areas is material mechanical properties. The adjectives "strong," "tough," and "ductile" are widely used to describe polymers, and their mechanical properties include strength, toughness, and ductility. Material strength, often known as tensile strength or tensile properties, is widely evaluated in material investigations. Flexural strength is a measurement of a material's ability to bend. Flexural strength is the ability of a polymer composite sample to bend when bent. All strength indicates how much force is required to break something, but it does not indicate what happened to our sample throughout the breakage process. When something is stressed, it undergoes elongation behaviour, which is defined by a change in shape [62].

The mechanical properties of the *S. persica*-based reinforced thermoset acrylic resin composite tensile strength (TS), tensile modulus (TM), flexural strength (FS), flexural modulus (FM), and impact strengths (IS) have been studied by several researches. Khalaf [71] used the hand lay-up approach to make the *S. persica* powder/ PMMA composites. To make the composite, varied concentrations of 3, 5, and 7 wt% of fibres were employed. The addition of 5% *S. persica* powder had no significant effect on the tensile and compressive strengths, according to the findings. When compared to the control group, the addition of *S. persica* powder at a ratio of 3% had no significant influence on the experimental group's impact strength, while the addition of *S. persica* powder at a ratio of 7% revealed a significant drop in tensile strength, impact strength and compressive strength.

The composite panel was made with self-cure acrylic resin and *S. persica* fibre by Oleiwi et al. [72]. The composites were made by hand lay-up method with 3, 6 and 9 wt% *S. persica* fibre in self-cure acrylic resin with varying lengths of 2-, 6- and 12-mm. Tensile strength and young modulus increased with fibre length and concentration, with the highest values of tensile strength and young modulus for specimens reinforced with *S. persica* fibres being 71 MPa and 4.9 GPa, respectively, at an optimum weight fraction of 9% and fibre length of 12 mm, which was significantly higher than other formulations.

Using the melt blending and injection moulding method, Savaş [21] synthesised and examined the mechanical characteristics of *S. persica* powder with polypropylene thermoplastic composites. The mechanical characteristics of composites with fibre concentrations of 5, 10, 20 and 30% were investigated. The tensile strength of the materials lowers from 23.25 to 21.78 MPa for PM0 and PM30, respectively, according to the test results. Elongation at break values for PM0 and PM30 drop from 38.4% to as low as 5%, respectively. In contrast to tensile strengths and elongation at break values, the composites' tensile moduli and Shore D hardness values increase in lockstep with increasing *S. persica* content. The lowest modulus was found to be 0.57 GPa in the sample with a 5 wt% *S. persica* concentration (PM5). Pure PP has an elastic modulus of 0.91 GPa, which is between PM20 and PM30's tensile moduli of 0.84 and 1.18 GPa, respectively. Flexural properties, in contrast to tensile qualities, showed an upward trend with fibre addition. The sample with 30% *S. persica* content (PM30) had the maximum flexural strength and flexural modulus, measuring 41.37 MPa and 1.97 GPa, respectively. The flexural strength of the pure PP polymer in the control group is 34.65 MPa, which is determined to be in the middle of the flexural strengths of PM10 (33.98 MPa) and PM20 (34.65 MPa) (36.33 MPa). PM30 had the highest hardness value of 67.8 and PM5 had the lowest hardness rating of 63.2.

However, Chaaben et al. [20] found that when *S. persica* (30 wt%) was added to *S. persica* (*S. persica*) powder reinforced poly (methyl methacrylate) resin, the hardness of the specimens diminished ($p = 0.05$). The use of 5% HAP as a coupling agent in the first stage, on the other hand, significantly increased the hardness. The improvement in hardness became less noteworthy when the proportion of HAP was raised. These findings are in line with those of [73], who discovered that adding HAP particles to unfilled monomer mixes enhanced the surface hardness of the material.

10.5.3 Effect *S. persica* on the thermal properties of composite

To determine the fraction of residual monomers, Chaaben et al. [20] utilised a DSC test on PMMA/*S. persica* powder. Inflammatory symptoms and mucous membrane irritations can be caused by residual monomers in polymerized denture base polymers [70, 74]. On the DSC thermograms, the glass transition pattern (Tg) was discernible, but there was no exothermic peak. The lack of a polymerization peak could be explained by the composite's lack of leftover monomers [75]. The proportion of remaining monomers was determined using HPLC to validate the theory. The peak of MMA, according to [76], was between 7 and 8 min, hence there was no significant increase between 7 and 8 min. As a result, we can deduce that the composite under investigation has no leftover monomers.

Savaş [21] used a dynamic mechanical analysis (DMA) test to investigate the thermal characteristics of *S. persica* powder with polypropylene thermoplastic composites. The graphs of the loss modulus, elastic (or storage) modulus, and tan versus temperature of PP/*S. persica* composites with various *S. persica* concentrations were examined. Three transition peaks can be visible in the loss modulus-temperature graph for pure PP(PM0), according to him. He discovered that the large relaxation zone's maximum temperature is 11.3 °C, which corresponds to the amorphous phase's glass transition in semi-crystalline isotactic PP. He also discovered that the PP relaxes at around −50 °C due to the motions of small-chains like methyl and methylene, and that the c-transition maximum is recorded at a higher temperature (around 52.5 °C), indicating that the PP crystallites are stiffer and more amorphous.

The transition peaks of PP/*S. persica* composites with various *S. persica* concentrations are identical to those of the PM0, he claims. With the exception of PM5, which had a lower loss modulus at low temperatures, up to roughly 0 °C, PM0 had a lower loss modulus than the composites. This is thought to be due to the fact that phase inversion at low temperatures induces morphological and structural changes. He had seen that as *S. persica* content increases, the composites' αc-transition peaks broaden and shift to lower temperatures than the PM0. The loss moduli of the composites rose as the *S. persica* concentration was increased beyond roughly 70 °C, despite some temperature variations. He found a similar pattern for elastic moduli, observing that, like the loss modulus, the elastic modulus of composites increased with increasing *S. persica* content in the composites. In the elastic modulus/temperature graphs of composites, elastic moduli decline monotonically with temperature. PM0 has a lower elastic modulus (1165 MPa at 250 °C) than composites. Furthermore, the damping factor, tan d, rose with increasing temperature and was seen to rise with increasing *S. persica* concentration, especially after about 100 °C.

10.6 Limitations of *S. persica* as filler or reinforcement agent

Variable quality as a result of unpredictable weather and moisture absorption, limited maximum processing temperatures, lower strength properties, lower durability, poor fire resistance, especially at high temperatures, and price fluctuation as a result of harvest results or agricultural politics can limit their industrial application [77]. However, with natural fiber compatible resins such as unsaturated polyester, epoxy, polypropylene, and others, sufficient surface treatment with suitable binder, compatibilizing agent and coupling agent should be used to overcome these limits. Natural and synthetic fibre hybridization can also be advantageous [78].

10.7 Conclusions and perspective

The purpose of the review was to discuss the structure and composition of *S. persica* plant fibre, as well as the impact of *S. persica* fibre on the physical, mechanical and thermal properties of polymer composites. As a result of the findings, it can be concluded that *S. persica* is an important historically used medicinal plant, particularly in the field of dentistry. Natural fibre is made up of cellulose, hemicellulose and lignin components, and the cellulose molecular chain of *S. persica* is discovered to be identical to that of wood cellulosic fibre. Furthermore, the cortex, xylem, parenchyma, and sulphur contained in the *S. persica* stem wall assist the stem to act as a defence against pathogenic microorganisms, which contributes to good mouth, tooth, and gum health. *S. persica* has a high wettability because to its hydrophilic nature and the fibre composition is linked to the amount of water taken up by the composite with lignocellulosic fibre. The mechanical performance of a composite is influenced by the interfacial adhesion between the polymer matrix and the filler, and it has been found that the lower the content of *S. persica* fibre in the composite, the better the tensile properties. However, the same cannot be said for hardness, as a high hardness value was found in a composite with 30 wt% fibre content. The loss modulus and elastic modulus of *S. persica* fibre composites increased as the amount of *S. persica* fibre in the composite increased. *S. persica*, like any other lignocellulosic fibre, has some limitations, such as moisture absorption, limited durability and poor fire resistance; however these limitations can be solved with natural fibre compatible resins. Based on prior research, 5 wt% *S. persica* reinforcement is expected to have strong tensile capabilities and thermal stability, and these findings will encourage the use of sustainable and large deposits of *S. persica* fibre as filler in polymer composites. This review may also help to maximise the usage of *S. persica* in the development of cost-effective materials for engineering and dentistry applications. As a result, the use of synthetic fibres may be reduced.

References

1. Bourmaud A, Corre Y-M, Baley C. Fully biodegradable composites: use of poly-(butylene-succinate) as a matrix and to plasticize l-poly-(lactide)-flax blends. Ind Crop Prod 2015;64:251–7.
2. Fu Y, Wu G, Bian X, Zeng J, Weng Y. Biodegradation behavior of poly(butylene adipate-Co-terephthalate) (PBAT), poly(lactic acid) (PLA), and their blend in freshwater with sediment. Molecules 2020;25. https://doi.org/10.3390/molecules25173946.
3. Weng Y-X, Jin Y-J, Meng Q-Y, Wang L, Zhang M, Wang Y-Z. Material behaviour. Polym Test 2013;32: 918–26.
4. High-content lignocellulosic fibers reinforcing starch-based biodegradable composites: properties and applications; 2016. https://doi.org/10.5772/65262.
5. Zwawi M. A review on natural fiber bio-composites, surface modifications and applications. Molecules 2021;26. https://doi.org/10.3390/molecules26020404.

6. Mohanty AK, Vivekanandhan S, Pin JM, Misra M. Composites from renewable and sustainable resources: challenges and innovations. Science 2018;362:536–42.
7. Jusoh AF, Rejab R, Siregar J, Bachtiar D. Natural fiber reinforced composites: a review on potential for corrugated core of sandwich structures. MATEC Web Conf 2016;74:00033.
8. Claassen PAM, van Lier JB, Lopez Contreras AM, van Niel EWJ, Sijtsma L, Stams AJM, et al. Utilisation of biomass for the supply of energy carriers. Appl Microbiol Biotechnol 1999;52:741–55.
9. Hottle T, Bilec M, Landis A. Sustainability assessments of bio-based polymers. Polym Degrad Stabil 2013;98:1898–907.
10. Mukherjee T, Kao N. PLA based biopolymer reinforced with natural fibre: a review. J Polym Environ 2011;19:714–25.
11. Shah D, Nag R, Clifford M. Why do we observe significant differences between measured and 'back-calculated' properties of natural fibres? Cellulose 2016;23. https://doi.org/10.1007/s10570-016-0926-x.
12. Gallos A, Paës G, Allais F, Beaugrand J. Lignocellulosic fibers: a critical review of the extrusion process for enhancement of the properties of natural fiber composites. RSC Adv 2017;7:34638–54.
13. Li Q, Matuana L. Surface of cellulosic materials modified with functionalized polyethylene coupling agents. J Appl Polym Sci 2003;88:278–86.
14. Patterson J. New opportunities with wood-flour-foamed PVC. J Vinyl Addit Technol 2001;7:138–41.
15. Jacob M, Joseph S, Pothan LA, Thomas S. A study of advances in characterization of interfaces and fiber surfaces in lignocellulosic fiber-reinforced composites. Compos Interfac 2005;12:95–124.
16. Satyanarayana KG, Arizaga GGC, Wypych F. Biodegradable composites based on lignocellulosic fibers—an overview. Prog Polym Sci 2009;34:982–1021.
17. Khoory T. The use of chewing sticks in preventive oral hygiene. Clin Prev Dent 1983;5:11–4.
18. Wu CD, Darout IA, Skaug N. Chewing sticks: timeless natural toothbrushes for oral cleansing. J Periodontal Res 2001;36:275–84.
19. Aumeeruddy MZ, Zengin G, Mahomoodally MF. A review of the traditional and modern uses of Salvadora persica L. (Miswak): toothbrush tree of Prophet Muhammad. J Ethnopharmacol 2018; 213:409–44.
20. Chaaben R, Taktak R, Mnif B, Guermazi N, Elleuch K. Innovative biocomposite development based on the incorporation of Salvadora persica in acrylic resin for dental material. J Thermoplast Compos Mater 2020;1–17. https://doi.org/10.1177/0892705720939167.
21. Savaş S. Structural properties and mechanical performance of Salvadora persica L. (Miswak) reinforced polypropylene composites. Polym Compos 2018;40. https://doi.org/10.1002/pc.24939.
22. Akhtar MN, Sulong AB, Radzi MKF, Ismail NF, Raza MR, Muhamad N, et al. Influence of alkaline treatment and fiber loading on the physical and mechanical properties of kenaf/polypropylene composites for variety of applications. Prog Nat Sci: Mater Int 2016;26:657–64.
23. Varma SR, Sherif H, Serafi A, Fanas SA, Desai V, Abuhijleh E, et al. The antiplaque efficacy of two herbal-based toothpastes: a clinical intervention. J Int Soc Prev Community Dent 2018;8:21–7.
24. Aboul-Enein BH. The miswak (Salvadora persica L.) chewing stick: cultural implications in oral health promotion. Saudi J Dental Res 2014;5:9–13.
25. Sadhan Re. IA, Almas K. Miswak (chewing stick): a cultural and scientific heritage. Saudi Dental J 1999;11:80–8.
26. al-Khateeb TL, O'Mullane DM, Whelton H, Sulaiman MI. Periodontal treatment needs among Saudi Arabian adults and their relationship to the use of the Miswak. Community Dent Health 1991;8: 323–8.
27. Al-Ghamdi F, Jari N, Al-Yafi D, Redwan S, Gogandy B, Othman H. Tooth brushing behavior and its prevalence versus miswak usage among the dental students of the faculty of dentistry at king abdulaziz university. Int Dental J Student's Res 2015;2:49–56.

28. El-Desoukey RMA. Comparative microbiological study between the miswak (Salvadora persica) and the toothpaste. Int J Microbiol Res 2015;6:47–53.
29. Gupta P, Shetty H. Use of natural products for oral hygiene maintenance: revisiting traditional medicine. J Compl Integr Med 2018;15. https://doi.org/10.1515/jcim-2015-0103.
30. Amjed S, Junaid K, Jafar J, Amjad T, Maqsood W, Mukhtar N, et al. Detection of antibacterial activities of Miswak, Kalonji and Aloe vera against oral pathogens & anti-proliferative activity against cancer cell line. BMC Complement Altern Med 2017;17:265.
31. Chaurasia A, Patil R, Nagar A. Miswak in oral cavity - an update. J Oral Biol Craniofac Res 2013;3: 98–101.
32. Hamasha AA, Alshehri A, Alshubaiki A, Alssafi F, Alamam H, Alshunaiber R. Gender-specific oral health beliefs and behaviors among adult patients attending King Abdulaziz Medical City in Riyadh. Saudi Dent J 2018;30:226–31.
33. Halawany HS. A review on miswak (Salvadora persica) and its effect on various aspects of oral health. Saudi Dent J 2012;24:63–9.
34. Shah A, Rahim S. Ethnomedicinal uses of plants for the treatment of malaria in Soon Valley, Khushab, Pakistan. J Ethnopharmacol 2017;200:84–106.
35. Orwa C, Mutua A, Kindt R, Jamnadass R, Simons A. Agroforestree database: a tree reference and selection guide. Nairobi, Kenya: World Agroforestry Centre; 2009. Version 4.
36. Sher H, Alyemeni MN. Pharmaceutically important plants used in traditional system of arab medicine for the treatment of livestock ailments in the Kingdom of Saudi Arabia. Afr J Biotechnol 2011;10:9153–9.
37. Demissew T, Najma D, Kinyamario JI, Kiboi S. The utilization of medicinal plants by the Masaai community in arid lands of Kajado county. Nairobi, Kenya: University of Nairobi; 2016.
38. Katewa SS, Chaudhary BL, Jain A. Folk herbal medicines from tribal area of Rajasthan, India. J Ethnopharmacol 2004;92:41–6.
39. Mali PY, Bhadane VV. Ethno-medicinal wisdom of tribals of Aurangabad district (M.S.), India. Indian J Nat Prod Resour 2011;2:102–9.
40. Li M, Cha DJ, Lai Y, Villaruz AE, Sturdevant DE, Otto M. The antimicrobial peptide-sensing system aps of Staphylococcus aureus. Mol Microbiol 2007;66:1136–47.
41. Lila MK, Saini GK, Kannan M, Singh I. Effect of fiber type on thermal and mechanical behavior of epoxy based composites. Fibers Polym 2017;18:806–10.
42. Peng Y, Liu R, Cao J. Characterization of surface chemistry and crystallization behavior of polypropylene composites reinforced with wood flour, cellulose, and lignin during accelerated weathering. Appl Surf Sci 2015;332:253–9.
43. Chandramohan. Natural fiber reinforced polymer composites for automobile accessories. Am J Environ Sci 2013;9:494–504.
44. Bhattacharyya D, Subasinghe A, Kim NK. Chapter 4 - Natural fibers: Their composites and flammability characterizations. Multifunctionality of Polymer Science. William Andrew Publishing; 2015:102–143 pp.
45. John M, Thomas S. Biofibres and biocomposites. Carbohydr Polym 2008;71:343–64.
46. Kalia S, Dufresne A, Cherian BM, Kaith BS, Avérous L, Njuguna J, et al. Cellulose-based bio- and nanocomposites: a review. Int J Polym Sci 2011;2011:837875. https://doi.org/10.1155/2011/837875.
47. Rong M, Zhang M, Liu Y, Yang G, Zeng H. The effect of fiber treatment on the mechanical properties of unidirectional sisal-reinforced epoxy composites. Compos Sci Technol 2001;61:1437–47.
48. Li X, Tabil LG, Panigrahi S. Chemical treatments of natural fiber for use in natural fiber-reinforced composites: a review. J Polym Environ 2007;15:25–33.
49. McKendry P. Energy production from biomass (Part 2): conversion technologies. Bioresour Technol 2002;83:47–54.

50. Yang H, Yan R, Chen H, Lee DH, Zheng C. Characteristics of hemicellulose, cellulose and lignin pyrolysis. Fuel 2007;86:1781–8.
51. Dashtizadeh Z, Khalina A, Cardona F, Lee CH. Mechanical characteristics of green composites of short kenaf bast fiber reinforced in cardanol. Adv Mater Sci Eng 2019;2019:1–6.
52. Reddy N, Yiqi Y. Biofibers from agricultural byproducts for industrial applications. Trends Biotechnol 2005;23:22–7.
53. Lau K-t., Hung P-y., Zhu M-H, Hui D. Properties of natural fibre composites for structural engineering applications. Compos B Eng 2018;136:222–33.
54. Frollini E, Leao A, Mattoso L, Rowell R, Han J, Rowell J. Characterization and factors effecting fiber properties. Nat Polym Agrofibers Compos 2000;113–34.
55. Han JS. Properties of nonwood fibers. In: Proceedings of the Korean Society of wood science and technology annual meeting. Republic of Korea; 1998:3–12 pp.
56. Balto H, Al-Sanie I, Al-Beshri S, Aldrees A. Effectiveness of Salvadora persica extracts against common oral pathogens. Saudi Dent J 2017;29:1–6.
57. Sawidis T. Anatomy and ultrastructure of Salvadora persica stem: adaptive to arid conditions and beneficial for practical use. Acta Biol Cracov Ser Bot 2013;55. https://doi.org/10.2478/abcsb-2013-0017.
58. Ahmad H, Ahamed N. Therapeutic properties of meswak chewing sticks: a review. Afr J Biotechnol 2012;11. https://doi.org/10.5897/AJB12.1188.
59. Ahmad H, Rajagopal K. Salvadora persica L. (Meswak) in dental hygiene. Saudi J Dental Res 2014;5: 130–4.
60. Khounganian R, Alwakeel A-A, Albadah A, Almaflehi N. Evaluation of the amount and type of microorganisms in tooth brushes and miswak after immediate brushing. ARC J Dental Sci 2018;3: 15–21.
61. Al-sieni A. The antibacterial activity of traditionally used Salvadora persica L. (Miswak) and commiphora gileadensis (palsam) in Saudi Arabia. Afr J Tradit, Complement Altern Med: AJTCAM/ African Netw Ethnomed 2014;11:23–7.
62. Mukhtar Ii., Leman Z, Ishak MR, Zainudin ES. Sugar palm fibre and its composites: a review of recent developments. Bioresources 2016;11:10756–82.
63. Pickering KL, Efendy MGA, Le TM. A review of recent developments in natural fibre composites and their mechanical performance. Compos Appl Sci Manuf 2016;83:98–112.
64. Kim J-i., Kang PH, Nho YC. Positive temperature coefficient behavior of polymer composites having a high melting temperature. J Appl Polym Sci 2004;92:394–401.
65. Poddar P, Islam MS, Sultana S, Nur HP, Chowdhury AMS. Mechanical and thermal properties of short arecanut leaf sheath fiber reinforced polypropyline composites: TGA, DSC and SEM analysis. J Mater Sci Eng 2016;5. https://doi.org/10.4172/2169-0022.1000270.
66. Dhakal HN, Zhang ZY, Bennett N. Influence of fibre treatment and glass fibre hybridisation on thermal degradation and surface energy characteristics of hemp/unsaturated polyester composites. Compos B Eng 2012;43:2757–61.
67. Le Digabel F, Boquillon N, Dole P, Monties B, Averous L. Properties of thermoplastic composites based on wheat-straw lignocellulosic fillers. J Appl Polym Sci 2004;93:428–36.
68. Abdulkhani A, Hojati Marvast E, Ashori A, Karimi AN. Effects of dissolution of some lignocellulosic materials with ionic liquids as green solvents on mechanical and physical properties of composite films. Carbohydr Polym 2013;95:57–63.
69. Wan Busu WN, Anuar H, Ahmad SH, Rasid R, Jamal NA. The mechanical and physical properties of thermoplastic natural rubber hybrid composites reinforced withHibiscus cannabinus, land short glass fiber. Polym-Plast Technol Eng 2010;49:1315–22.
70. Cho K, Rajan G, Farrar P, Prentice L, Prusty G. Dental resin composites: a review on materials to product realizations. Compos B Eng 2021;230:109495.

71. Khalaf HAR. Effect of mixing silanized poly propylene and siwak fibers on some physical and mechanical properties of heat cure resin denture base. AJPS (Asian J Plant Sci) 2016;16:26–37.
72. Oleiwi JK, Salih SI, Fadhil HS. Effect of siwak and bamboo fibers on tensile properties of self-cure acrylic resin used for denture applications. J Mater Sci Eng 2017;06. https://doi.org/10.4172/2169-0022.1000370.
73. Abdul Qados AMS. Effect of salt stress on plant growth and metabolism of bean plant Vicia faba (L.). J Saudi Soc Agric Sci 2011;10:7–15.
74. Ilbay SG, Güvener S, Alkumru HN. Processing dentures using a microwave technique. J Oral Rehabil 1994;21:103–9.
75. Ohyama A, Imai Y. Differential scanning calorimetric study of acrylic resin powders used in dentistry. Dent Mater J 2000;19:346–51.
76. Viljanen EK, Langer S, Skrifvars M, Vallittu PK. Analysis of residual monomers in dendritic methacrylate copolymers and composites by HPLC and headspace-GC/MS. Dent Mater 2006;22: 845–51.
77. Gul Guven R, Aslan N, Guven K, Matpan Bekler F, Acer O. Purification and characterization of polyphenol oxidase from corn tassel. Cell Mol Biol (Noisy-le-grand) 2016;62:6–11.
78. Simão JA, Carmona VB, Marconcini JM, Mattoso LHC, Barsberg ST, Sanadi AR. Effect of fiber treatment condition and coupling agent on the mechanical and thermal properties in highly filled composites of sugarcane bagasse fiber/PP. Mater Res 2016;19:746–51.

Isah Aliyu, Salit Mohd Sapuan*, Edi Syams Zainudin,
Mohd Zuhri Mohamed Yusoff, Ridwan Yahaya and
Che Nor Aiza Jaafar

11 An overview of mechanical and corrosion properties of aluminium matrix composites reinforced with plant based natural fibres

Abstract: Many researchers have become more interested in utilizing plant based natural fibre as reinforcement for the fabrication of aluminium matrix composites (AMCs) in recent time. The utilization of these environmentally friendly and cost effective plant based natural fibre is necessitated to avoid environmental pollution. The desire for cost-effective and low-cost energy materials in automotive, biomedical, aerospace, marine, and other applications, however, is redefining the research environment in plant based natural fibre metal matrix composite materials. As a result, the goal of this review study is to investigate the impact of agricultural waste-based reinforcements on the mechanical properties and corrosion behaviour of AMCs made using various fabrication routes. Processing settings can be modified to produce homogenous structures with superior AMC characteristics, according to the findings. Plant based natural fibre ash reinforcing materials such as palm kernel shell ash, rice husk ash, sugarcane bagasse, bamboo stem ash, and corn cob ash can reduce AMCs density without sacrificing mechanical qualities. Furthermore, efficient utilization

---

*Corresponding author: Salit Mohd Sapuan,** Advanced Engineering Materials and Composites Research Centre (AEMC), Department of Mechanical and Manufacturing Engineering, Universiti Putra Malaysia, 43400 UPM, Serdang, Selangor, Malaysia; and Laboratory of Biocomposite Technology, Institute of Tropical Forestry and Forest Products (INTROP), Universiti Putra Malaysia, 43400 UPM, Serdang, Selangor, Malaysia, E-mail: sapuan@upm.edu.my
Isah Aliyu, Advanced Engineering Materials and Composites Research Centre (AEMC), Department of Mechanical and Manufacturing Engineering, Universiti Putra Malaysia, 43400 UPM Serdang, Selangor, Malaysia; and Department of Metallurgical Engineering, Waziri Umaru Federal Polytechnic, Brinin Kebbi, Nigeria. https://orcid.org/0000-0003-3564-3500
Edi Syams Zainudin and Mohd Zuhri Mohamed Yusoff, Advanced Engineering Materials and Composites Research Centre (AEMC), Department of Mechanical and Manufacturing Engineering, Universiti Putra Malaysia, 43400 UPM Serdang, Selangor, Malaysia; and Laboratory of Biocomposite Technology, Institute of Tropical Forestry and Forest Products (INTROP), Universiti Putra Malaysia, 43400 UPM, Serdang, Selangor, Malaysia
Ridwan Yahaya, Science and Technology Research Institute for Defence (STRIDE), Kajang, Selangor, Malaysia
Che Nor Aiza Jaafar, Advanced Engineering Materials and Composites Research Centre (AEMC), Department of Mechanical and Manufacturing Engineering, Universiti Putra Malaysia, 43400 UPM Serdang, Selangor, Malaysia

As per De Gruyter's policy this article has previously been published in the journal Physical Sciences Reviews. Please cite as: I. Aliyu, S. M. Sapuan, E. S. Zainudin, M. Z. Mohamed Yusoff, R. Yahaya and C. N. Aiza Jaafar "An overview of mechanical and corrosion properties of aluminium matrix composites reinforced with plant based natural fibres" *Physical Sciences Reviews* [Online] 2022. DOI: 10.1515/psr-2022-0044 | https://doi.org/10.1515/9783110769227-011

of plant based natural fibre reduces manufacturing costs and prevents environmental pollution, making it a sustainable material. Brittle composites, unlike ceramic and synthetic reinforced composites, are not formed by plant based natural fibre reinforcements. As a result of our findings, plant based natural fibre AMCs have a high potential to replace expensive and hazardous ceramic and synthetic reinforced-AMCs, which can be used in a variety of automotive applications requiring lower cost, higher strength-to-weight ratio, and corrosion resistance.

Keywords: aluminium matrix; composite; corrosion behaviour; mechanical properties; plant based natural fibre; reinforcement.

Abbreviations

RHA	rice husk ash
CSA	coconut shell ash
PKSA	palm kernel shell ash
CCA	corn cob ash
GSA	groundnut shell ash
BA	bagasse ash
BLA	bamboo leaf ash
PSA	periwinkle shell ash
FA	fly ash
LBWA	locust bean waste ash
CFA	coconut fibre ash
W	wood
HFA	hemp fibre ash
MSA	melon shell ash
THA	tur husk ash
PLA	pine leaf ash
BPA	bean pod ash
AMCs	alumnium matrix composites
CBA	cow bone ash
ESA	egg shell ash
MSW	maize stalk waste

11.1 Introduction

Reinforcing a metallic material with a non-metallic material has emerged as a major technique in the fabrication of a new material known as composite with improved mechanical properties. Ceramic and synthetic reinforcing materials such as TiC, SiC, Al_2O_3, B_4C, and others are expensive, resulting in high-cost composite materials. The high cost of composite materials made with ceramic and synthetic reinforcements has prompted researchers to look for some alternative ways in reducing cost of producing composites. This resulted in the use of low-cost and readily available plant based natural fibre as a source of reinforcements [1–5].

Most plants are not planted basically for agricultural purpose rather to generate raw materials for industrial application [6]. The plant based natural fibre in different forms are normally waste materials that are disposed without proper procedure thereby creating environmental pollution [7, 8]. The plant based natural fibre include rice husk, sugar palm fibre, coconut shell, groundnut shell, bean pod, corn cob, wool, silk, cotton seed shell, palm sprout shell and teak wood to mention but few [9, 10]. The utilization of plant based natural fibre which include mango seed shell ash, sheanut shell, bamboo leaf, rice husk ash, groundnut shell ash, coconut shell ash, banana peels ash, jute ash and bagasse ash were found to replace the ceramic and synthetic materials as reinforcement in AMCs production [11–22]. Plant based natural fibre ash are product obtained by carbonization and calcination process and has been used since in the early days of civilization as reinforcement materials. The mechanical properties of AMCs are significantly influenced the shape, quantity and size of the plant based natural fibre ash (reinforcement) [23, 24]. Researchers are now devoting their precious time and resources towards the utilization of plant based natural fibre ash as reinforcement materials to produce novel composites. The AMCs exhibit attractive properties such as high strength-to-weight ratio with low density, high hardness, superior wear and corrosion resistance and are gaining recognition in the world of contemporary engineering, particularly in the automotive and aerospace application, to enhance machine performance and minimize fuel consumption [25, 26]. Nowadays, mechanical component manufacturers all around the world are turning to composites to save weight while maintaining the material's properties. Many studies have successfully used the environmentally acceptable and cost-effective plant based natural fibre ash, which has been discovered to be rich in SiO_2 and MgO among others [27].

Aluminium alloy is the most extensively utilized material for making a wide range of light-weight, high-strength products in the globe. These alloys are also reinforced to improve their mechanical properties [28]. AMCs is considered as a potential new generation material which first gained popularity in the 80s as a result of its use in automotive components. Their properties were unproven at the time, and the benefits were sometimes over-sold. As a result, the reputation of AMCs suffered a setback, and as carbon composition has become extensively utilized, engineers and designers have forgotten about AMCs. In the three decades since research and development into the fabrication of those composites is aimed toward weight reduction, AMCs have firmly returned to the limelight [29, 30]. AMCs are used in a variety of applications in our daily lives, including doors, furniture and kitchen utensils [31]. Researchers have conducted research by fabricating low cost AMCs by employing plant based natural fibre as reinforcing materials [32–35]. This type of AMCs are known as high-performance low cost composites because the addition of these plant based natural fibre not only reduced the cost of the composites but also improved their mechanical characteristics and are used in a variety of applications, including biomedical, automotive parts, spacecraft, electrical components, aircraft, and sound insulation [36, 37].

The addition of plant based natural fibre ash of 6 wt% reinforcement combinations of RHA, CSA, egg shell ash and ZnO particles in aluminum AA1100 was studied which lead to enhanced hardness, tensile strength and resistance to wear of AA1100 matrix composite as observed [38].

Taking all of these issues into account, this paper provides an overview of AMCs reinforced with plant bases natural fibre waste produced using various techniques that include stir casting, powder metallurgy, compo casting, and friction stir process. The morphological characteristics, mechanical and corrosion analyses of the produced plant based natural fibre composites were thoroughly reviewed in order to identify the most suitable fabrication routes for the advancement of plant based natural fibre AMCs.

11.1.1 Types of plant based natural fibre

Materials obtained from plant based natural fibre are said to be promising materials that are used as reinforcements material in the fabrication of AMCs, and some are listed in Table 11.1. Those plant based natural fibre have been identified as possible reinforcing materials.

11.1.1.1 Rice husk ash (RHA)

Rice husk is readily available in substantial quantities in many countries that include China, Indonesia, Bangladish, and India [55, 56]. Rice husk (Figure 11.1) is a product obtained from rice mills after the paddy has been milled [57, 58]. They find applications in variety of ways, including production of fertilizer, fuel for the generation of heat and

Table 11.1: Different Al-matrix and plant based natural fibre reinforcements.

Matrix	Plant based natural fibre ash	References
AA 6061	RHA	[39]
Al 1100	RHA	[40]
AA 6063	CCA	[41]
Al 1100	CSA	[42]
Al 1100	CSA	[43]
Al–4.5% Cu	BLA	[44]
Al–Cu–Mg	PKSA	[45]
Al–Mg–Si	PKSA	[46]
AA 6063	GSA + HFA	[47]
Al–Si–Mg	LBWA	[48]
Al 6061	BA	[49]
Pure Al	CFA	[50]
Al–Si–10% Mg	BA	[51]
Al–Si–10% Mg	RHA + FA	[52]
Al 356.2	RHA + FA	[53]
Al–20% Mg–10% Cu	RHA	[54]

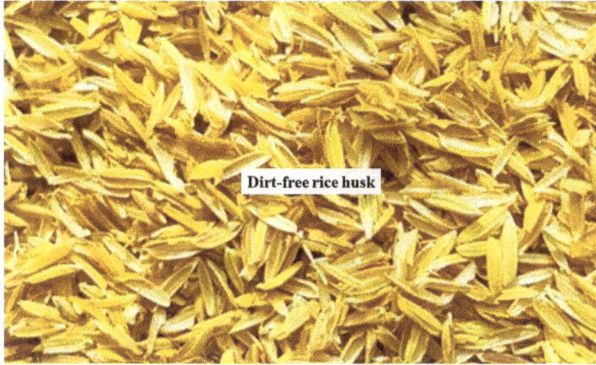

Figure 11.1: Rice hush. Reproduced with permission from ref. [58].

as insulating materials [59–62]. However, discarding it in an open environment has resulted in environmental pollution. As a result, researchers began to use this material as reinforcement in high-performance AMCs since it has a lower density and is more readily available [63, 64]. RHA contain useful percentage of various ceramic particle, around 85–90% silica and having relative low density (2.0–2.65 g/cm^3) [65]. RHA consist of high amount of SiO_2 along with other compounds such as Fe_2O_3, CaO, Al_2O_3 and MgO as depicted in Table 11.2. Marini et al. [66], investigated the effect of RHA addition in AA6061 alloy and found improvement in its wear resistance with the best result was achieved at 6 vol% RHA. Saini et al. [67], carried out extraction of silica from RHA via thermal and chemical treatments and the silica obtained was reinforced in pure-Al thereby resulting in excellent physical, mechanical and thermal properties.

11.1.1.2 Coconut shell ash (CSA)

Coconut shell (Figure 11.2) being among the most well-known plant based natural fibre is readily available in tropical areas in large quantities that result in greenhouse gases when use in generating power, casting, forging industries and are most at times used as fuel in boilers and furnaces [73, 74]. The CSA consists of higher amount of SiO_2 along

Table 11.2: Nominal plant based natural fibre composition used as reinforcement.

Plant based natural fibre	Composition (wt%)							
	CaO	SiO	Fe_2O_3	K_2O	MgO	Al_2O_3	Others	References
GSA	29.72	18.79	17.07	9.81	6.76	5.32	12.53	[68]
RHA	0.91	95.22	0.18	0.24	0.46	2.99	–	[69]
BA	7.87	72.24	4.26	1.26	2.85	9.14	2.38	[70]
CSA	0.57	45.05	12.4	0.52	0.52	16.20	24.74	[71]
PKSA	15.10	46.20	3.20	12.20	3.70	2.30	8.30	[72]
BLA	7.25	76.00	1.40	5.80	1.90	4.20	3.45	[15]

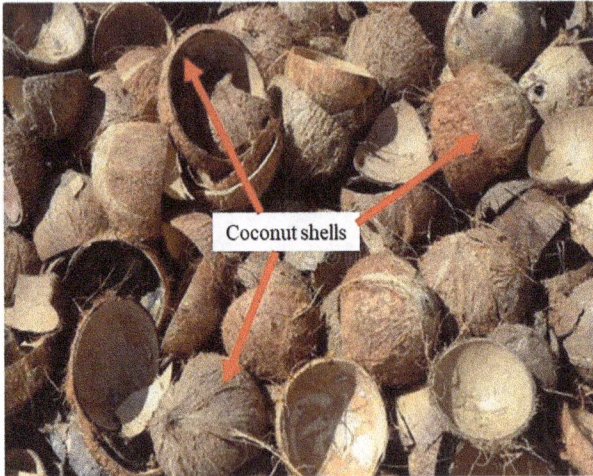

Figure 11.2: Coconut shell. Reproduced with permission from ref. [73].

with other elements like MgO, Al_2O_3 and Fe_2O_3, as depicted in Table 11.2. CSA is an attractive candidate for reinforcement of matrix composites employed in automotive and aerospace applications due to low density, suitable chemical composition and availability [75]. The addition of CSA as particles showed improvement on the properties of Al-composites. Mangalore et al. [76], used a stir casting technique to develop Al 7075 matrix composite reinforced with the addition of coconut shell ash, which resulted in improved mechanical properties.

11.1.1.3 Groundnut shell ash (GSA)

Groundnut shells are the left over after the groundnut seed has been extracted from its pod. Groundnut shell (Figure 11.3) degrades slowly under normal conditions and is used to improve low-cost waste management, reduce pollution, and increase economic

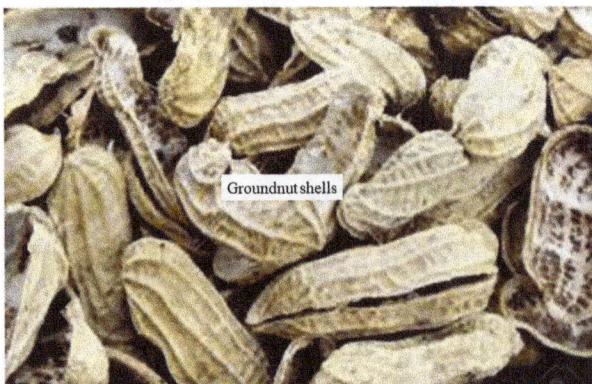

Figure 11.3: Groundnut shell. Reproduced with permission from ref. [77].

value for farmers, all of which encourages increased productivity [77–80]. The burning of the groundnut shell resulted in its ash which constitute of ceramic oxide such as silica, calcium, magnesium, aluminium and iron oxide as shown in Table 11.2; thus, it used as reinforcement in the AMCs [81, 82]. Dwivedi et al. [19], evaluated the use of GSA ash as a reinforcement in an Al-matrix and discovered that the mechanical qualities improved while the cost was reduced.

11.1.1.4 Palm kernel shell ash (PKSA)

Palm kernel shell is a hard biogenic organic waste obtained from shattered oil palm fruit seeds during palm oil production. The PKSs (Figure 11.4) are converted into ash to solve disposal and pollution related problem [83–85]. The main constituents of the ash contain ceramic oxides in the form of SiO_2, Fe_2O_3, Al_2O_3, CaO and K_2O as depicted in Table 11.2. The PKSA physical properties include irregular shape, hard and flaky, calcined at 550 °C, the ash is used in production of AMCs and polymer composites [86, 87]. The addition of PKSA to an Al-metal matrix improved ultimate strength, ductility, elastic modulus with decreased in hardness [88].

11.1.1.5 Bagasse ash (BA)

Bagasse is the leftover of sugarcane after the extraction of the juice as shown in Figure 11.5. Bagasse is typically used as soil amendment, fertilizer composition, animal food and biomass for conversion to fuel [89–91]. Because of the abundance and pozzolanic properties of BA, research in recent years has primarily focused on its use in construction materials [92, 93]. The composition of BA, as shown in Table 11.2, qualifies it to be use as a reinforcement material in manufacturing of Al-matrix composites [94, 95]. Palanivendhan et al. [96], reinforced bagasse ash in Al 6262 and observed improvement in tribological and mechanical properties.

Palm kernel shells

Figure 11.4: Palm kernel shell. Reproduced with permission from ref. [83].

Figure 11.5: Bagasse obtained from sugarcane. Reproduced with permission from ref. [90, 91].

11.1.1.6 Bamboo leaf ash (BLA)

Bamboo is considered as one of the agricultural plants that grows on different continents of the world, with Asia accounting for approximately 65 percent of the total [97, 98]. Bamboo leaf ash has been identified as a promising material for the production of cementitious materials, with the potential to partially replace cement [99]. Bamboo leaf ash contain high amount of SiO_2 with other compounds such as CaO, Al_2O_3, MgO and Fe_2O_3 as depicted in Table 11.2. Kumar & Birru [44], reinforced BLA in Al–4.5% Cu matrix composite and found improvement in strength and hardness of the fabricated composites (Figure 11.6).

11.2 Fabrication techniques of plant based natural fibre Al-matrix composites

The fabrication route employed for the fabrication of Al-matrix composites is generally divided into two types: liquid and solid. Stir casting, friction stir processing and compocasting routes are used in the liquid route, whereas powder metallurgy techniques are used in the solid route [100, 101]. The mechanical, morphological, and corrosion characteristics of plant based natural fibre Al-matrix composite are greatly influenced by the fabrication route used and the mode of plant based natural fibre addition to the Al-matrix [102, 103]. This is a synopsis of these techniques:

11.2.1 Liquid fabrication route

11.2.1.1 Stir casting route

Stir casting among the various techniques is considered to be the simplest with adequate flexibility, most profitable, promising and easy to operate [104–106]. Figure 11.7

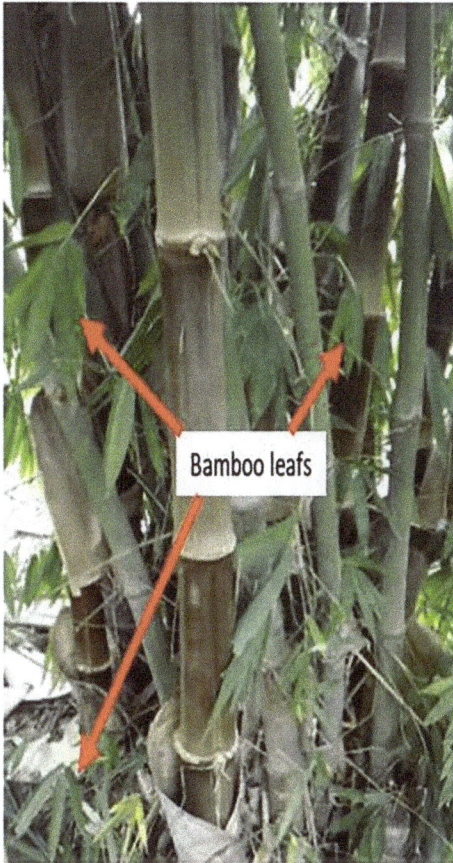

Figure 11.6: Bamboo. Reproduced with permission from ref. [97].

depicts the stir casting technique utilizing agro waste in the manufacture of Al-matrix composites. The lack of wettability between the reinforcement and the metal matrix, on the other hand, is a cause for concern. Some key parameters, such as stirring time, holding time, melt temperature, stirring speed, stirrer positioning and preheating of plant based natural fibre (reinforcement), may be considered prior to composite manufacturing. The quality of plant based natural fibre Al-matrix composite is determined by the appropriate selection of these parameters [107–109].

Many scientists have employed the stir casting technique to produce Al-matrix composites supplemented with plant based natural fibre, as illustrated in Table 11.3. As a result of specific gravity of plant based natural fibre reinforcements that settled at the bottom during manufacturing procedures, appropriate dispersion of the plant based natural fibre particles are frequently problematic in stir casting. Researchers used double stir casting to address the issue of wettability between the plant based natural fibre as reinforcement and Al-matrix, with the goal of improving the composites' characteristics by heating the agro waste before it inclusion into the molten Al-matrix

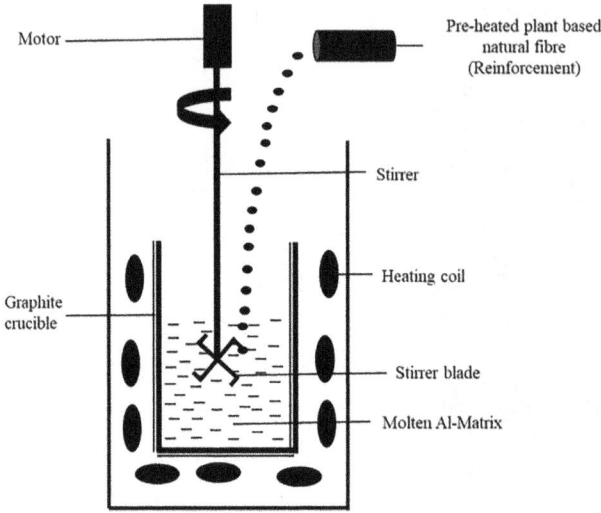

Figure 11.7: Plant based natural fibre Al-matrix stir casting route.

[19, 21]. Wetting agents such as magnesium were also used to reduce wettability between the Al-matrix and plant based natural fibre particles [110, 111]. Furthermore, by employing degassing materials such as hexachloroethene (C_2Cl_6) tablets, porosity that degrades the characteristics of the composites is limited and regulated [20]. The most common method employed in the producing aluminium matrix composites are stir casting, friction stir casting, squeeze casting, and powder technique. Among the various routes investigated, researchers discovered stir casting to be the simplest, convenient, ease process with adequate flexibility, better homogeneity, and the most profitable for the fabrication of natural fibre aluminium matrix composite [105, 112–115]. Several researchers have developed Al-matrix composites with plant based natural fibre reinforcements using stir casting routes, with the resulting composites demonstrating enhanced microstructure and mechanical characteristics, as shown in Table 11.3.

11.2.1.2 Friction stir processing route

Friction stir processing is yet another method for altering and modifying the surface of composites by transforming them into a more uniform microstructure and thus improving their properties. In terms of stiffness and strength, this approach, as demonstrated in Figure 11.8, outperforms powder, stir casting and compo casting techniques while compromising toughness and ductility. Surface agro waste Al-matrix composites with improved surface hardness and wear resistance produced using the friction stir processing method [135, 136]. The distribution of plant based natural fibre

Table 11.3: Fabrication routes and types of plant based natural fibre reinforced AMCs.

Plant based natural fibre	Matrix	Fabrication route	References
CSA	Al 7075	Vortex	[71]
BA	Al 7075	Powder metallurgy	[107]
RHA	Al 7075	Squeeze casting	[116]
BLA	Al–4.5% Cu	Squeeze casting	[117]
CSA	Al 6061	Stir casting	[118]
BA	Al 7075	Friction stir	[39]
BLA	A 356.2	Stir casting	[119]
CH	A 356	Spark plasma sintering	[120]
RHA	Al powder	Stir casting	[121]
BA	Al 7075	Powder metallurgy	[122]
BA	Al 6061	Stir casting	[123]
ESA	Al 6061	Stir casting	[124]
PKSA + PSA	Pure Al	Stir casting	[88]
BLA + Si sand	Al 6063	Stir casting	[125]
CCA	AA 6063	Double stir casting	[41]
FA + RHA	LM 6	Stir casting	[126]
RHA	Al powder	Powder metallurgy	[127]
FA + RHA	A 356	Double stir casting	[128]
CSA	Al cans	Compo casting	[129]
RHA	AA 6061	Friction stir casting	[130]
RHA + G + Al$_2$O$_3$	Al–Si–Mg	Two step stir casting	[131]
BA + FA	LM6	Stir casting	[132]
RHA	Al-can scrap	Stir casting	[133]
RHA + yttrium oxide	Al–Mg	Stir casting	[134]

particles on composite surfaces is improved with this technology. It has been utilized to improve the characteristics of powder metallurgy-fabricated Al-matrix composites as a post-fabrication technique [137].

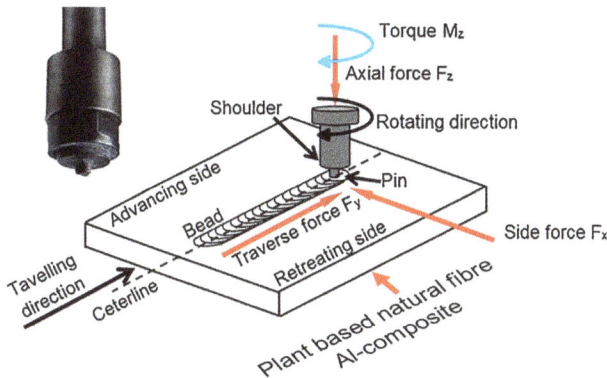

Figure 11.8: Plant based natural fibre Al-matrix composite friction stir casting route. Reproduced with permission from ref. [136].

This route has enabled the fabrication of bulk Al-metal matrix composites, but more research is required before the process can be commercialized [138, 139]. Oghenevweta et al. [140], used the friction stir processing technique to develop MSW reinforced Al-Si-Mg composites. The absence of agglomerations or segregations was revealed by morphological characterization. Furthermore, the developed composite's tensile strength and hardness were shown to have improved. Gupta [141], used a friction stir process to fabricate a PLA-reinforced Al 1120-based composite with a uniform distribution of the PLA and improved hardness and wear values.

11.2.1.3 Compo-casting route

Although the liquid casting method, such as stir casting, is easy and inexpensive, it has poor wettability between the plant based natural fibre and the Al-matrix. To improve wettability, researchers employed wettability agents, preheated, oxidized, and coated plant based natural fibre reinforcements [142–145]. Such procedures, on the other hand, result in higher overall production costs. As a result, when the Al-matrix is semi-solid, the casting temperature is decreased and plant based natural fibre particles are added. As shown in Figure 11.9, this is the most cost-effective way for enhancing wettability. Due to its low cost, convenience of use, near proximity to the finished product, and large-scale production, compo casting is frequently used for the manufacture of Al-matrix composites [146]. Furthermore, this technique has been used to fabricate a variety of plant based natural fibre Al-matrix composites [42]. Several studies have found that compo-casting produces better wettability with even dispersion of the plant based natural fibre particles in Al-matrix composites than stir casting [147]. As a result, compo-casting is widely recognized as the most efficient process for producing

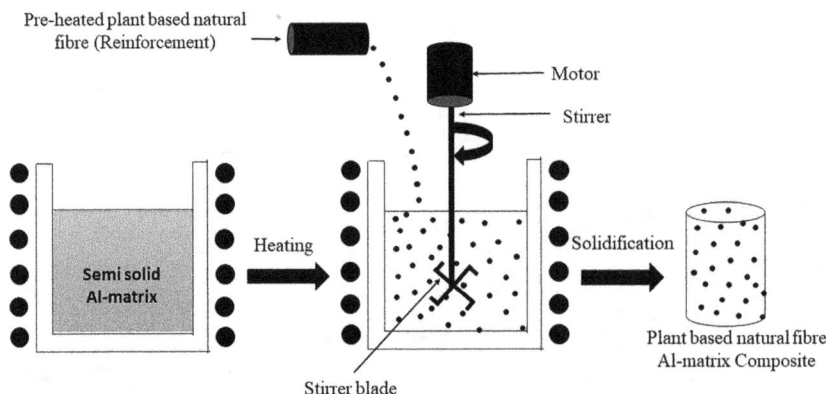

Figure 11.9: Plant based natural fibre Al-matrix flow sheet compo-casting process.

low-cost plant based natural fibre Al-matrix composites. When producing aluminum-based plant based natural fibre composites, several researchers have employed this technology to address the wettability issue faced with the stir casting technique.

11.2.2 Solid fabrication route

11.2.2.1 Powder metallurgy route

The powder metallurgy route is the most widely used solid state process because of its numerous advantages, including the ability to fabricate composites with dimensional accuracy, quality desirable shape, homogeneity of structure, highly efficient use of the initial metal, large productivity, ability to produce porous free materials and low raw material consumption [148].

Al-matrix powders are transformed into materials in this process using a variety of methods such as powder mixing, compacting and sintering, which can be seen in Figure 11.10. This method ensures that plant based natural fibre reinforcements are distributed uniformly throughout the composite substrate. This method is recommended because it prevents agglomeration of particles, low wettability, and the production of hazardous phases that commonly found with the liquid metallurgical method [121]. Sohail et al. [148], studied the mechanical and wear characteristics of THA and alumina reinforced aluminium matrix hybrid composites made by powder metallurgy. The inclusion of the reinforcement improved the mechanical qualities, according to the researchers. Hernandez-Ruiz et al. [94], employed powder metallurgy to create BA-reinforced Al-composites and studied the influence of the ash content (wt%) on hardness. The BA content was discovered to be a strong determinant of hardness improvement.

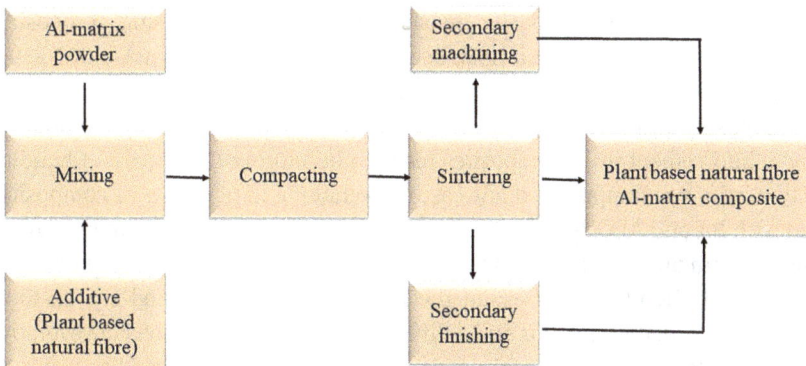

Figure 11.10: Agro waste Al-matrix powder metallurgy route.

Powder metallurgy, compo-casting and friction stir casting routes have all been used to fabricate aluminum-based composites. Due to high operating cost and complexities associated with those processes, researchers are turning to stir casting technique. Stir casting accounts for roughly 67% of all Metal Matrix Composites (MMCs) produced. Stir casting is a proven and promising process for producing AMCs due to its simplicity, adequate flexibility, cost-effectiveness, lower production cost, and mass production capability [112–115, 121, 135–150].

11.3 Properties of plant based natural fibre reinforced AMCs

Table 11.3 shows the different plant fibres, matrix and fabrication route of various plant based natural fibre reinforced Al-matrix composites such as palm kernel shell ash, coconut shell ash, rice husk ash, bamboo leaf ash, corn cob ash and bagasse ash. The effects of various plant based natural fibre reinforcements on the mechanical and corrosion properties of AMCs are investigated in the sub-section below.

11.3.1 Mechanical properties of plant based natural fibre reinforced AMCs

A number of researchers have looked at the impact of these plant based natural fibre used as reinforcements on the mechanical characteristics of Al-based composites. Satheesh & Pugazhvadivu [151], investigated and found that the densities of reinforced GSA and fixed proportion silica carbide Al composites manufactured using the double stir casting technique decreased, while the composite's tensile strength and hardness rose, as shown in Figure 11.11. It has been discovered that plant based natural fibre reinforcements can reduce composite densities without compromising composite mechanical properties. Udoye, et al. [152], used stir casting to create an AA6063-RHA metal matrix composite with increased hardness and tensile strength. The researchers discovered that substituting plant based natural fibre ash particles for minor amounts of ceramic reinforcement in Al composites resulted in improved properties. Chandla et al. [123], used a vacuum-assisted stir casting technique to make an Al composite reinforced with BA and Al_2O_3 and found that increasing the weight fraction improved the tensile strength and hardness, but with decreased densities and increased porosities, as shown in Figure 11.12. Despite the fact that vacuum-assisted stir casting techniques improve wettability, it is still a difficult obstacle to overcome. However, by employing the vortex method in the fabrication of AMCs, this problem can be resolved. Kumar et al. [70], fabricated and compared the properties of bagasse ash and titanium nitride reinforced hybrid composites. They reached a conclusion that the properties

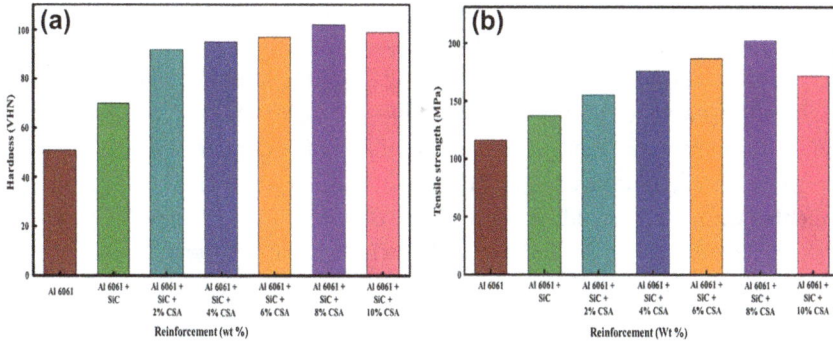

Figure 11.11: Variation of SiC and CSA reinforcements on (a) hardness and (b) tensile strength of Al-matrix composite. Reproduced with permission from ref [134].

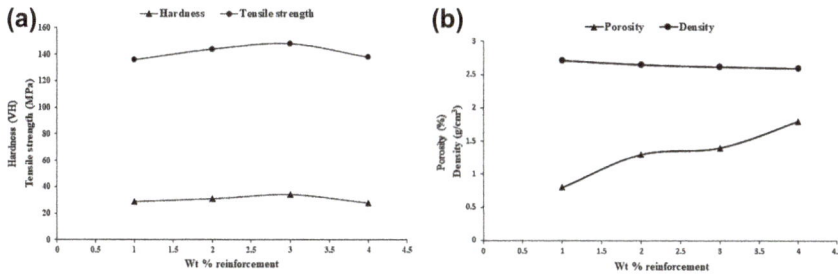

Figure 11.12: Variation of (a) tensile strength and hardness (b) porosity and density with wt% Al$_2$O$_3$ and BA in Al matrix composite. Reproduced with permission from ref [123].

improved more with the addition of bagasse ash than titanium nitride but titanium is a major contributor in influencing the mechanical properties. Kumar & Kumar [153], fabricated Al composites reinforced with RHA and B$_4$C using a high vacuum casting machine and noticed that increasing the weight percentages of RHA and Boron Carbide (B$_4$C) lead to improvement in the hardness, tensile, compressive and flexural strength of the composite.

Kumar [116], investigated the mechanical properties of RHA and Mica reinforced Al7075 composites and discovered significant improvements in the composite's hardness, brittleness and tensile strength. Similarly, Udoye, et al. [152], studied the mechanical properties of RHA reinforced in AA6061 matrix composites and discovered that the inclusion of RHA significantly improved the composite's tensile strength, hardness, and elongation behaviour. Babaremu et al. [154], used stir casting to explore the mechanical properties of cow horn and CCA reinforced Al 8011 composites, discovering that increasing the percentage weight of cow horn and CCA increased the composite's hardness, ultimate tensile, and yield strength. Apasi et al. [155], employed double stir casting to analyze Al-Si-Fe matrix reinforced CSA particles as reinforcement and discovered that as the wt% of CSA increased, the hardness, tensile, and yield strength increased. Even while double stir casting procedures enhance wettability, it is still a difficult obstacle to overcome. This challenge, however, can be overcome by employing the vortex approach in the synthesis of Al-matrix composites. Furthermore, Yashpal et al. [156], employed stir casting to manufacture Al 6061 composites supplemented with alumina and BA particles and discovered that, despite being higher than the Al 6061 alloy, the inclusion of BA lowered hardness, tensile, and impact strength. The impact of plant based natural fibre reinforcements on the mechanical characteristics of Al-matrix composites is shown in Table 11.4. The plant based natural fibre component was found to boost hardness, tensile, impact, and compressive characteristics to a certain level, but only to that extent.

11.3.2 Corrosion characteristics of plant based natural fibre Al-matrix composite

Several researchers studied the corrosion behaviour of Al-matrix composite reinforced with plant based natural fibre in corrosive environments such as H_2SO_4, HCl, NaCl, and others [165, 166]. Corrosive environments and plant based natural fibre used to study the corrosion characteristics of the developed plant based natural fibre Al-matrix composites are shown in Table 11.5. In seawater, Suriani et al. [167], investigated the corrosion behaviour of CSA and silica carbide reinforced in aluminium using potentio-dynamic polarization, electrochemical impendence spectroscopy, and weight loss measurement methods. With potentio-dynamic polarization, electrochemical impendence spectroscopy, and weight loss measurement methods, the addition of 5 wt% of CSA resulted in better corrosion resistance than additions of ash greater than 5 wt%. Udoye et al. [168], investigated the corrosion characteristics of A6061 incorporated with RHA in 0.75 M H_2SO_4 and found that the corrosion performance improved. The formation of non-energetic sites as a result of RHA addition is most likely the cause of a decrease in current exchange between RHA and A6061, lowering the corrosion rate. Rao et al. [169], investigated the corrosion characteristics of Al–4.5% Cu with BLA as reinforcement immersed in 0.5 M NaOH solution buffered with NaCl. The corroded composites were analyzed using weight loss and surface analysis. When compared to the matrix alloy, the weight loss on the fabricated composite was minimal, while surface morphology revealed significantly higher pits on the matrix alloy and lower pits on the composites. When compared to the Al-matrix, the incorporation of non-metallic plant based natural fibre in Al-matrix composite does not form an electrolytic cell of anode and cathode, resulting in a reduction in corrosion rate.

Table 11.4: Types nature fibres and matrix combination with their mechanical properties.

Compositions	Tensile strength	Hardness	Elongation	References
Al 6061 + 10%SiC + 10% FA	173.00 MPa	58.70 BHN	12.80	
Al 6061 + 10% SiC + 12.5% FA + 2.5% CSA	187.00 MPa	64.30 BHN	10.70	
Al 6061 + 10% SiC + 15% FA + 5% CSA	194.00 MPa	70.50 BHN	8.20	[157]
Al 6061 + 10% SiC + 20% FA + 10% CSA	213.00 MPa	81.10 BHN	7.40	
Al 6061 + 5% RHA	171.30 MPa	48.40 BHN	4.94	
Al 6061 + 1% RHA + 4% SiC	113.30 MPa	43.70 BHN	0.95	[158]
Al 6061 + 2% RHA + 3% SiC	108.60 MPa	40.20 BHN	0.97	
ADC 12 + 5% B_4C	164.36 MPa	88.00 VHN	2.76	
ADC 12 + 5% B_4C + 3% RHA	168.45 MPa	92.00 VHN	2.55	
ADC 12 + 5% B_4C + 6% RHA	174.88 MPa	95.00 VHN	2.45	[159]
ADC 12 + 5% B_4C + 9% RHA	177.88 MPa	97.00 VHN	2.20	
ADC 12 + 5% B_4C + 12% RHA	178.51 MPa	99.00 VHN	2.18	
Al	90.80 N/mm^2	58.26 VHN	29.70	
Al + 5% CSA	101.60 N/mm^2	62.26 VHN	23.30	[160]
Al + 10% CSA	127.00 N/mm^2	79.41 VHN	18.30	
Al + 15% CSA	143.00 N/mm^2	83.73 VHN	16.00	
Al-2% Mg	40.19 MPa	52.33 BHN	–	
Al-2% Mg + 5% W	44.21 MPa	56.00 BHN	–	
Al-2% Mg + 10% W	59.13 MPa	62.00 BHN	–	[161]
Al-2% Mg + 15% W	96.80 MPa	58.00 BHN	–	
Al-2% Mg + 20% W	97.69 MPa	57.00 BHN	–	
AA 7075	200.10 MPa	70.08 VHN	–	
AA 7075 + 5% B_4C	260.56 MPa	105.15 VHN	–	[162]
AA 7075 + 5% B_4C + 3% RHA	235.28 MPa	115.18 VHN	–	
AA 7075 + 5% B_4C + 5% RHA	220.08 MPa	121.42 VHN	–	
AA 6061	6000 KPa	152.00 KN	2.00	
AA 6061 + 2% RHA	6030 KPa	170.00 KN	2.40	
AA 6061 + 4% RHA	6132 KPa	175.00 KN	2.50	[163]
AA 6061 + 6% RHA	6225 KPa	178.00 KN	2.70	
AA 6061 + 8% RHA	6339 KPa	188.00 KN	2.90	
AA 6061	1.46 N/mm^2	7.80 BHN	–	
AA 6061 + 2% Al_2O_3 + 2% CSA	1.95 N/mm^2	14.60 BHN	–	
AA 6061 + 4% Al_2O_3 + 4% CSA	2.09 N/mm^2	19.00 BHN	–	[118]
AA 6351 + 8% SiC	186 MPa	72.50 VHN	–	
AA 6351 + 6% SiC + 2% RHA	182 MPa	70.06 VHN	–	
AA 6351 + 4% SiC + 4% RHA	177 MPa	66.97 VHN	–	[164]
AA 6351 + 2% SiC + 6% RHA	173 MPa	64.36 VHN	–	

11.3.3 Microstructure of plant based natural fibre reinforced AMCs

The introduction of plant based natural fibres in Al-matrix has a great effect on the morphological characteristics which directly affect the mechanical and corrosion

Table 11.5: Plant based natural fibre Al-matrix composite and corrosive media used for corrosion study.

Plant based natural fibre	Al-matrix	Corrosive media	References
GSA	Al–Mg–Si	NaCl and H_2SO_4	[170]
B_4C + RHA	Al 6061	$AlCl_3$	[171]
CSA	AA 6063	H_2SO_4	[172]
RHA + Clay	AA 6061	H_2SO_4	[161]
Fly Ash	AA 7075	NaCl	[173]
CSA + CBA	AA 7075	NaCl	[174]
RHA + ESA	Al 6063	NaCl	[175]

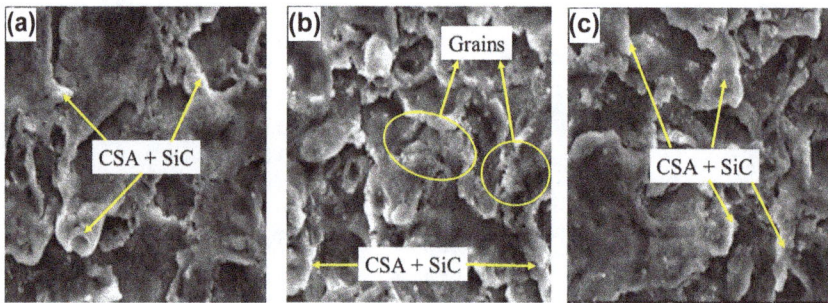

Figure 11.13: SEM image of (a) Al 6061 + SiC + CSA, (b) Al 6061 + SiC + CSA and (c) Al 6061 + SiC + CSA. Reproduced with permission from ref. [151].

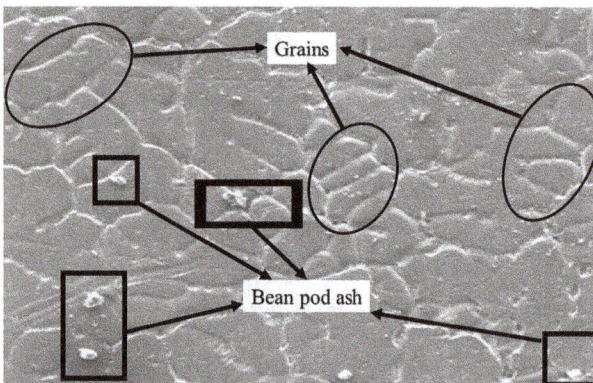

Figure 11.14: SEM image of aluminium alloy with uniform distribution of BPA with fine grains. Reproduced with the author's permission [25].

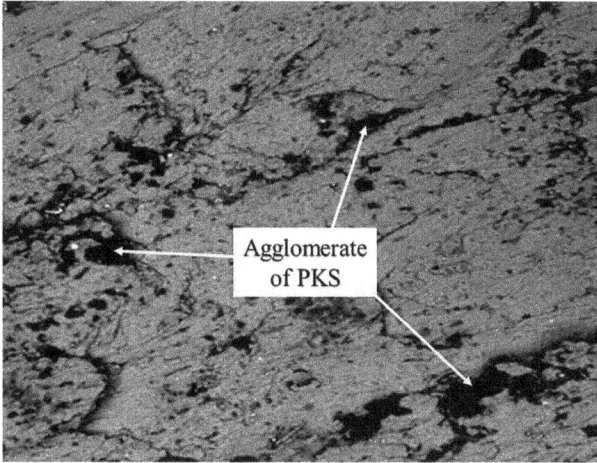

Figure 11.15: SEM morphology of AA6063 with concentrated agglomerate of PKS. Reproduced with permission from ref. [177].

behaviour of AMCs. The microstructure, corrosion and stress analysis of RHA reinforced Al–Cu–Mg composite manufactured via stir casting were explored by [176]. They discovered that the presence of RHA particles caused the master phase of the alloy to be shifted from its initial position, resulting in a reduced grain size via modifying the crystallographic arrangement. The insertion of RHA-particle increased the tensile strength while lowering the corrosion rate due to the production of hard phases of SiO_2, $CuMgO_5$, and C. Satheesh & Pugazhvadivu [151], observed uniform distribution of SiC and CSA in Al 6061 which lead to improve hardness, tensile strength and wear characteristics of the composite. The homogeneous distribution of reinforced CSA and SiC in the Al 6061 matrix could be responsible for grain refinement is depicted in Figure 11.13. The improvement on tensile strength and hardness noticed on Al-Cu-Mg matrix strengthened with bean pod ash produced via double stir casting could due to uniform distribution of bean pod ash in the matrix. The even distribution of BPA in the matrix resulted in the formation of fine grains as shown in Figure 11.14 [25]. The presence of dark portions of PKS as reinforcement in AA6063 alloy in the SEM morphology depicted in Figure 11.15 forms agglomerate and is evenly distributed which could lower the mechanical properties such as hardness, impact and tensile strength [177].

11.4 Conclusions

The impact of various plant based natural fibre reinforcements on mechanical, morphological and corrosion behaviour as well as fabrication techniques of those natural plant fibres are thoroughly explored in this reviewed in order to identify the most suitable fabrication routes for the advancement of plant based natural fibre in AMCs. The following are some of the significant findings:

–   The chemical compositions that makeup the plant based natural fibre ash varies, with SiO_2 being the dominant component.
–   Grain formation and homogeneous distribution of plant based natural fibre particles were revealed in the morphologies of the Al-matrix composites, which are crucial in defining the mechanical properties of the Al-matrix composite.
–   The use of plant based natural fibre as reinforcement in AMCs, such as rice husk ash, coconut shell ash, bamboo leaf ash, and so on, reduces AMCs density.
–   Various plant based natural fibre combination as reinforcements are added to Al-matrix to obtain a composites with desirable properties that are difficult to be achieved with ceramic and synthetic reinforcements.
–   The mechanical properties of AMCs can be adjusted by varying the amount and sizes of plant based natural fibre ash used as reinforcement.
–   Among the fabrication techniques, stir casting has been found as the most widely and cost effective fabrication route for the advancement of plant based natural fibre AMCs.
–   These plant based natural fibre is said to have the potential to replace ceramic and synthetic reinforcements.
–   The corrosion rate of plant based natural fibre Al-matrix composites is found to decrease as the weight percent of plant based natural fibre in the composite increases.

References

1. Gayathri J, Elansezhian R. Materials today: proceedings enhancement of mechanical properties of aluminium metal matrix composite by reinforcing waste alumina catalyst and nano Al_2O_3. Mater Today Proc 2021;45:462–6.
2. Gowrishankar MC, Shivaprakash YM, Kumar D, Prafful AJS, Prasad Y. Production and mechanical testing of aluminium alloy based hybrid metal matrix composite. Mater Today Proc 2018;5: 23872–80.
3. Edoziuno FO, Nwaeju CC, Adediran AA, Odoni BU, Arun VR. Mechanical and microstructural characteristics of aluminium 6063 alloy/palm kernel shell composites for lightweight applications. Sci African 2021;12:e00781.
4. Kumar KR, Pridhar T, Balaji VSS. Mechanical properties and characterization of zirconium oxide (ZrO 2) and coconut shell ash (CSA) reinforced aluminium (Al 6082) matrix hybrid composite. J Alloys Compd 2018;765:171–9.
5. Arunkumar S, Sundaram MS, Suketh KM, Vigneshwara S. Materials today: proceedings a review on aluminium matrix composite with various reinforcement particles and their behaviour. Mater Today Proc 2020;33:484–90.
6. Kerni L, Singh S, Patnaik A, Kumar N. A review on natural fiber reinforced composites. Mater Today Proc 2020;28:1616–21.

7. Yadav R, Kumar VD, Prakash SD. Eggshell and rice husk ash utilization as reinforcement in development of composite material: a review. Mater Today Proc 2021;43:426–33.
8. Dayma S, Patel C. Development of aluminium-based metal matrix composites (MMCs) using industrial and agro wastes as reinforcement-A review various engineering applications as reinforcement enhances mechanical properties like hardness , due to higher reinforcement cost. In: International research conference on innovations, startup and investments (ICOSTART-2019); 2019:55–61 pp.
9. Babu RBMR. Effect of mechanical properties on palm sprout shell ash reinforced with Al-6061 alloy metal matrix composites. Int J Sci Res Eng Trends 2019;5:2395–566.
10. Pravinkumar M. Characterization of aluminium hybrid composite reinforcement with teak wood ash and bamboo ash by using stir casting process. Int J Eng Res Technol 2018;7:423–30.
11. Ochuokpa EO, Yawas DS, Okorie PU, Sumaila M. Evaluation of mechanical and metallurgical properties Al-Si-Mg/mangiferaindica seed shell ash (MSSA) particulate composite for production of motorcycle hub. J Sci Educ Technol 2021;9:221–38.
12. Zubairu PT, Hassan AB, Muriana RA, Musa NA. Effect of shea nutshell as on the mechanical properties of cast Al6061. J Arid Environ 2021;10:95–102.
13. Kanthasamy S, Ravikumar TS, Tamilanban T. Influence of coconut shell ash particle (CSAp) as reinforcement on mechanical and wear behavior of AZ31 magnesium alloy. Mater Today Proc 2020. https://doi.org/10.1016/j.matpr.2020.11.832.
14. Babu PM, Rajamuneeswaran S, Pritima D, Marichamy S, Vairamuthu J. Spark erosion machining behaviour of coconut shell ash reinforced silicon metal matrix. Mater Today Proc 2020;33: 4602–4.
15. Olaniran O, Uwaifo O, Bamidele E, Olaniran B. An investigation of the mechanical properties of organic silica, bamboo leaf ash and rice husk reinforced aluminium hybrid composite. Int J Mater Sci 2019;3:129–34.
16. Butola R, Pratap C, Shukla A, Walia RS. Effect on the mechanical properties of aluminum-based hybrid metal matrix composite using stir casting method. Mater Sci Forum 2019;969:253–9.
17. Butola R, Kanwar S, Tyagi L, Singari RM, Tyagi M. Optimizing the machining variables in CNC turning of aluminum based hybrid metal matrix composites. SN Appl Sci 2020;2:1–9.
18. Joseph OO, Babaremu KO. Agricultural waste as a reinforcement particulate for aluminum metal matrix composite (AMMCs): a review. Fibers 2019;4:1–9.
19. Krushna MG, Shekhar PS, Kumar SA. Effect of hot forging on high temperature tribilogical properties of aluminium composite reinforced with agro and industrial waste. Int J Eng Adv Technol 2019;8:1607–12.
20. Panda SS, Senapati AK, Rao PS. Effect of particle size on properties of industrial and agro waste-reinforced aluminum-matrix composite. JOM 2021;73:2096–103.
21. Dwivedi SP, Maurya M, Maurya NK, Srivastava AK, Sharma S, Saxena A. Utilization of groundnut shell as reinforcement in development of aluminum based composite to reduce environment pollution: a review. Evergr 2020;7:15–25.
22. Kanthasamy S, Ravikumar TS, Tamilanban T. Mechanical and corrosion behavior of groundnut shell ash particle (GSAp) reinforced AZ31 magnesium composite. Mater Today Proc 2020:1–5. https://doi.org/10.1016/j.matpr.2020.11.834.
23. Nassar MMA, Arunachalam R, Alzebdeh KI. Machinability of natural fiber reinforced composites: a review. Int J Adv Manuf Technol 2017;88:2985–3004.
24. Udoye NE, Inegbenebor AO. The study on improvement of aluminium alloy for engineering application: a review. Int J Mech Eng Technol 2019;10:380–5.
25. Aigbodion VS. Bean pod ash nanoparticles a promising reinforcement for aluminium matrix biocomposites. J Mater Res Technol 2019;8:6011–20.

26. Srivyas PD, Charoo MS. Aluminum metal matrix composites a review of reinforcement; mechanical and tribological behavior. Int J Eng Technol 2018;7:117–22.
27. Subramaniam B, Purusothaman VR, Karuppusamy SM, Ganesh SH, Markandan RK. Review on properties of aluminium metal matrix composites. J Mech Energy Eng 2020;4:57–66.
28. Sathish T, Karthick S. Wear behaviour analysis on aluminium alloy 7050 with reinforced SiC through taguchi approach. J Mater Res Technol 2020;9:3481–7.
29. Moona G, Walia RS, Rastogi V, Sharma R. Aluminium metal matrix composites: a retrospective investigation. Indian J Pure Appl Phys 2018;56:164–75.
30. Thompson R. Aluminium matrix composites: a sustainable solution. Reinf Plast 2021;65:10–3.
31. Yadav R, Gupta RK, Goyal A. Study of tribological behaviour of hybrid metal matrix composites prepared by stir casting method. Mater Today Proc 2020;28:2218–22.
32. Fayomi OSI, Joseph OO, Akande IG, Ohiri CK, Enechi KO, Udoye NE. Effect of CCBP doping on the multifunctional Al-0.5 Mg-15CCBP superalloy using liquid metallurgy process for advanced application. J Alloys Compd 2019;783:246–55.
33. Olusesi OS, Udoye NE. Development and characterization of AA6061 aluminium alloy/clay and rice husk ash composite. Manuf Lett 2021;29:34–41.
34. Ezema IC, Aigbodion VS, Okonkwo EG, Obayi CS. Fatigue properties of value-added composite from Al-Si-Mg/palm kernel shell ash nanoparticles. Int J Adv Manuf Technol 2020;107:3247–57.
35. Varalakshmi K, Ch Kishore Kumar K, Ravindra Babu P, Ch Sastry MR. Characterization of Al 6061-coconut Shell ash metal matrix composites using stir casting. Int J Eng Sci 2019;2:41–9.
36. Nakamura M. Organosilica nanoparticles and medical imaging. In: The Enzymes, 1st ed. Cambridge, Massachusetts: Academic Press; 2018, vol. 44:137–73 pp.
37. Croissant JG, Fatieiev Y, Almalik A, Khashab NM. Mesoporous silica and organosilica nanoparticles: physical chemistry, biosafety, delivery strategies, and biomedical applications. Adv Healthc Mater 2018;7:1–75.
38. Nagaraj A. A study on mechanical and tribological properties of aluminium 1100 alloys 6% of RHAp, BAp, CSAp, ZnOp and egg shellp composites by ANN. Silicon 2020;10:3367–76.
39. Fatchurrohman N, Farhana N, Marini CD. Investigation on the effect of friction stir processing parameters on micro-structure and micro-hardness of rice husk ash reinforced Al6061 metal matrix composites. IOP Conf Ser Mater Sci Eng 2018;319:012032.
40. Ghanaraja S, Gireesha BL, Ravikumar KS, Likith P. Effect of forging on mechanical properties of rice husk ash-silicon carbide reinforced Al1100 hybrid composites. In: AIP Conference Proceedings, American Institute of Physics. College Park, Maryland; 2018, 020050:020050 p.
41. Odoni BU, Odikpo F, Chinasa NC, Akaluzia RO. Experimental analysis, predictive modelling and optimization of some physical and mechanical properties of aluminium 6063 alloy based composites reinforced with corn cob ash. Eng Struct 2020;7:451–65.
42. Sankara RSR, Panigrahi MK, Ganguly RI, Srinivasa GR. Optimization of tribological behaviour on Al-coconut shell ash composite at elevated temperature optimization of tribological behaviour on Al-coconut shell ash composite at elevated temperature. IOP Conference Series 2018;18:5332–9.
43. Raju SS, Rao SG, Samantra C. Wear behavioral assessment of Al-CSAp-MMCs using grey-fuzzy approach. J Int Meas Confed 2019;140:254–68.
44. Kumar BP, Birru AK. Microstructure and mechanical properties of aluminium metal matrix composites with addition of bamboo leaf ash by stir casting method. T Nonferr Metal Soc 2017;27:2555–72.
45. Victor GA. Abrasive wear behaviour of Al–Cu–Mg/palm kernel shell ash particulate composite. Leonardo El J Pract Technol 2017;31:77–92.
46. Oyedeji OE, Dauada M, Yaro SA, Abdulwahab M. Characterization of Al–Mg–Si alloy reinforced with optimum palm kernel shell ash (PKSA) particle and its consequence on the dynamic properties for aerospace application. Res Sq 2021:1–14.

47. Palanivendhan M, Chandradass J. Experimental investigation on mechanical and wear behavior of agro waste ash based metal matrix composite. Mater Today Proc 2021;45:6580–9.

48. Usman Y, Dauda ET, Abdulwahab M, Dodo RM. Effect of mechanical properties and wear nehaviour onlocust bean waste ash (LBWA) particle reinforced aluminium alloy (A356 alloy) composites. Fudma J Sci 2002;4:416–21.

49. Virkunwar AK, Ghosh S, Basak R. Study of mechanical and tribological characteristics of aluminium alloy reinforced with sugarcane bagasse ash. In: An International Conference on Tribology. Mumbai; 2018;1–5 pp.

50. Jain S, Aggarwal V, Tyagi M, Walia RS, Rana R. Development of aluminium matrix composite using coconut husk ash reinforcement. In: International conference on latest developments in materials, manufacturing and quality control (MMQC-2016); 2016:252–9 pp.

51. Shankar S, Balaji A, Kawin N. Investigations on mechanical and tribological properties of Al-si10-mg alloy/sugarcane bagasse ash particulate composites investigations on mechanical and tribological properties of Al-Si10-Mg alloy/sugarcane bagasse ash particulate composites. Part Sci Technol An Int J 2017;6351:0–32.

52. Subrahmanyam APSVR, Narsaraju G, Rao BS. Effect of rice husk ash and fly ash reinforcements on microstructure and mechanical properties of aluminium alloy (AlSi10Mg) matrix composites effect of rice husk ash and fly ash reinforcements on microstructure and mechanical properties of aluminium al. Int J Adv Sci Technol 2015;76:1–8.

53. Subrahmanyam APSVR, Madhukiran J, Naresh G, Madhusudhan S. Fabrication and characterization of Al356.2, rice husk ash and fly ash reinforced hybrid metal matrix composite. Int J Adv Sci Technol 2016;94:49–56.

54. Manoj Kumar D, Guru Dattatreya GS, Krishnarjun Rao N. Synthesis characterization and mechanical behavior of rice husk ash reinforced al-20mg 10 cu alloy matrix hybrid composites. Int J Mech Prod Eng Res Dev 2018;8:111–8.

55. Singh B. 13. Rice husk ash. Waste and supplementary cementitious material in concrete. Bristol, England: Woodhead publishing; 2018:417–60 pp.

56. Zou Y, Yang T. Rice husk, rice husk ash and their applications. Rice bran and rice bran oil. College Park, Maryland: AOCS Press; 2019:207–46 pp.

57. Irawan A, Latifat Upe S, Meity Dwi IPI. Effect of torrefaction process on the coconut shell energy content for solid fuel effect of torrefaction process on the coconut shell energy content for solid fuel. AIP Conf Proc 2020;020010:1–7.

58. Lubwama M, Yiga VA. Characteristics of briquettes developed from rice and coffee husks for domestic cooking applications in Uganda. Renew Energy 2018;118:43–55.

59. Oladele SO, Adeyemo AJ, Awodun MA. Geoderma influence of rice husk biochar and inorganic fertilizer on soil nutrients availability and rain-fed rice yield in two contrasting soils. Geoderma 2019;336:1–11.

60. Ríos-badran M, Luzardo-Ocampo I, Garcia-Trejo JF, Santos-Crus J, Gutierrez-Antonio C. Production and characterization of fuel pellets from rice husk and wheat straw. Renew Energy 2020;145: 500–7.

61. Krishnadevi K, Devaraju S, Sriharshitha S, Alagar M, Priya YK. Environmentally sustainable rice husk ash reinforced cardanol based polybenzoxazine bio – composites for insulation applications. Polym Bull 2020;77:2501–20.

62. Tiwari S, Pradhan MK. Science direct effect of rice husk ash on properties of aluminium alloys: a review. Mater Today Proc 2017;4:486–95.

63. Saini S, Gupta A, Mehta AJ, Pramanik S. Rice husk-extracted silica reinforced graphite/aluminium matrix hybrid composite. J Therm Anal Calorim 2020:147:0123456789.

64. Garud V, Bhoite S, Patil S, Ghadage S, Gaikwad N. Performance and combustion characteristics of thermal barrier coated (YSZ) low heat rejection diesel engine. Mater Today Proc 2017;4:188–94.
65. Kandpal BC, Johri N, Kumar N, Srivastava A. Effect of industrial/agricultural waste materials as reinforcement on properties of metal matrix composites. Mater Today Proc 2021;46:1–5.
66. Marini CD, Fatchurrohman N, Zulkfli Z. Proceedings morphological study of friction stir processed aluminium metal matrix composites. Mater Today Proc 2021;46:1745–8.
67. Saini S, Singh S, Singh K, Singh A. Some studies into weldability of rice husk ash aluminium matrix composites using TIG welding. Mater Today Proc 2020;24:298–307.
68. Venkatesh L, Arjunan TV, Ravikumar K. Microstructural characteristics and mechanical behaviour of aluminium hybrid composites reinforced with groundnut shell ash and B4C. J Braz Soc Mech Sci Eng 2019;41:1–13.
69. Bhasha CA, Balamurugan K. Fabrication and property evaluation of Al 6061 + x% (RHA + TiC) hybrid metal matrix composite. SN Appl Sci 2019;1:1–9.
70. Kumar KR, Pridhar T, Vettivel SC. Influence on mechanical behaviour and characterization of a6063/bagasse and titanium nitride hybrid composites. Rev Cuba Fis 2021;74:473–86.
71. Kasagani S, Bellamkonda PN, Sudabathula S. Wear and friction behaviour of coconut shell ash reinforced aa-7075 metal. Int J Res 2020;VIII:477–85.
72. Ikubanni PP, Oki M, Adeleke AA, Adediran AA, Adesina OS. Influence of temperature on the chemical compositions and microstructural changes of ash formed from palm kernel shell. Results Eng 2020;8:100173.
73. Prakash R, Thenmozhi R, Raman SN, Subramanian C, Divyah N. Mechanical characterisation of sustainable fibre - reinforced lightweight concrete incorporating waste coconut shell as coarse aggregate and sisal fibre. Int J Environ Sci Technol 2021;18:1579–90.
74. Tanko J, Ahmadu U, Sadiq U, Muazu A. Characterization of rice husk and coconut shell briquette as an alternative solid fuel. Adv Energy Convers 2020;2:1–12.
75. Panda B, Niranjan CA, Vishwanatha AD, Harisha P, Chandan KR, Kumar R. Development of novel stir cast aluminium composite with modified coconut shell ash filler. Mater Today Proc 2019;22: 2715–24.
76. Mangalore P, Ulvekar A, Abhiram A, Sanjay J, Advaith. Mechanical properties of coconut shell ash reinforced aluminium metal matrix composites mechanical properties of coconut shell ash reinforced aluminium metal matrix composites. AIP Conf Proc 2019;020014:1–6.
77. Duc PA, Dharanipriya P, Velmurugan BK, Shanmugavadivu M. Groundnut shell- a beneficial bio-waste. Biocatal Agric Biotechnol 2019;20:1–5.
78. Amu O, Adetayo O, Faluyi F, Akinyele E. Experimental study of improving the properties of lime-stabilized structural lateritic soil for highway structural works using groundnut shell ash. Walailak J 2021;18:2–17.
79. Nalluri N, Karri VR. Use of groundnut shell compost as a natural fertilizer for the cultivation of vegetable plants. Int J Adv Res Sci Eng 2018;07:97–104.
80. Dzomeku IK, Illiasu O. Effects of groundnut shell, rice husk and rice straw on the productivity of maize (Zea mays L) and soil fertility in the Guinea savannah zone of Ghana. ACTA Sci Agric 2018;2: 29–35.
81. Alaneme KK, Bodunrin MO, Awe AA. Microstructure, mechanical and fracture properties of groundnut shell ash and silicon carbide dispersion strengthened aluminium matrix composites. J King Saud Univ - Eng Sci 2018;30:96–103.
82. Jadhav S, Aradhye A, Kulkarni S, Shinde Y, Vaishampayan V. Effect of hybrid ash reinforcement on microstructure of A356 alloy matrix composite. AIP Conf Proc 2019;2105:020010.
83. Fanijo E, Babafemi AJ, Arowojolu O. Performance of laterized concrete made with palm kernel shell as replacement for coarse aggregate. Constr Build Mater 2020;250:118829.

84. Imoisili PE, Ukoba KO, Jen TC. Synthesis and characterization of amorphous mesoporous silica from palm kernel shell ash. Bol Soc Esp Ceram Vidr 2020;59:159–64.
85. Patrick DO, Shahbaz M. Thermogravimetric kinetics of catalytic and non-catalytic pyrolytic conversion of palm kernel shell with acid-treated coal bottom ash. Bionergy Res 2020;13:452–62.
86. Iyasele EO. Comparative analysis on the mechanical properties of a metal-matrix composite (MMC) reinforced with palm kernel/periwinkle shell ash. Glob Sci J 2018;6:1–24.
87. Baffour-Awuah E, Akinlabi SA, Jen TC, Hassan S, Okokpujie IP, Ishola F. Characteristics of palm kernel shell and palm kernel shell-polymer composites: a review. IOP Conf Ser Mater Sci Eng 2021;1107:012090.
88. Omondiale EI. Comparative analysis on the mechanical properties of a metal-matrix composite (mmc) reinforced with palm kernel/periwinkle shell ash. Glob Sci J 2018;6:1–24.
89. Paya J, Monzo J, Borrachero MV, Tashima MM, Soriano L. Bagasse ash. In: Waste and supplementary cementitious material in concrete. Amstersam, the Netherlands: Woodhead publising; 2018;559–98 pp.
90. Torgbo S, Quan MV, Sukyai P. Cellulosic value-added products from sugarcane bagasse. Cellulose 2021;28:5219–40.
91. Narisetty V, Castro E, Durgapal S, Coulon F, Jacob S, Kumar D, et al. High level xylitol production by pichia fermentans using non-detoxified xylose-rich sugarcane bagasse and olive pits hydrolysates. Bioresour Technol 2021;342:126005.
92. Hernández-Olivares F, Medina-alvarado RE, Burneo-valdivieso XE, Zúñiga-suárez AR. Short sugarcane bagasse fibers cementitious composites for building construction. Constr Build Mater 2020;247:8–10.
93. Murugesan T, Vidjeapriya R, Bahurudeen A. Development of sustainable alkali activated binder for construction using sugarcane bagasse ash and marble waste. Sugar Tech 2020;22:885–95.
94. Dada A, Ajibola W. Experimental analysis of metal matrix composite. Int J Mod Trends Eng Res 2019;6:13–28.
95. Hernandez-Ruiz JE, Pino-Rivero L, Villar-Cocina E. Aluminium matrix composite with sugarcane bagasse ash as reinforcement material. Construct Build Mater 2019;36:55–9.
96. Palanivendhan M, Chandaradass J, Philip J. Fabrication and mechanical properties of aluminium alloy/bagasse ash composite by stir casting method. Mater Today Proc 2021;45:6547–52.
97. Akinlabi ET, Anane FK, Akwada DR. Properties of bamboo. Amstersam, the Netherlands: Springer International Publishing; 2017;87–147 pp.
98. Khairul M, Mohd H, Sapawe N. Effective performance of silica nanoparticles extracted from bamboo leaves ash for removal of phenol. Mater Today Proc 2020;31:27–32.
99. Silva LHP, Tamashiro JR, Guedes de Paiva FF, Fernando do Santo L, Teixeira SR, Kinoshita A, et al. Bamboo leaf ash for use as mineral addition with Portland cement. J Build Eng 2021;42:1–9.
100. Patel NS, Patel AD. Studies on properties of Al-sic MMCs for making valves for automobile components – a technical review. Trans Indian Inst Met 2017;2:305–8.
101. Parveez B, Maleque MA, Jamal NA. Influence of agro-based reinforcements on the properties of aluminum matrix composites: a systematic review. J Mater Sci 2021;56:16195–222.
102. Akinwamide SO, Abe BT, Akinribide OJ, Obadele BA, Olubambi PA. Characterization of microstructure, mechanical properties and corrosion response of aluminium-based composites fabricated via casting—a review. Int J Adv Manuf Technol 2020;109:975–91.
103. Bodunrin MO, Alaneme KK, Chown LH. Aluminium matrix hybrid composites: a review of reinforcement philosophies; Mechanical, corrosion and tribological characteristics. J Mater Res Technol 2015;4:434–45.

104. Hynes JNR, Sankaranarayanan R, Tharmaraj R, Pruncu CI, Dispinar D. A comparative study of the mechanical and tribological behaviours of different aluminium matrix – ceramic composites. J Braz Soc Mech Sci Eng 2019;41:1–12.
105. Manikandan R, Arjunan TV, Akhil R, Nath OP. Studies on micro structural characteristics, mechanical and tribological behaviour of boron carbide and cow dung ash reinforced aluminium (Al 7075) hybrid metal matrix composite. Compos Part B 2020;183:107668.
106. Kereem A, Qudeiri JA, Abdudeen A, Ahammed T, Ziout A. A review on AA 6061 metal matrix composites produced by stir casting. Materials 2021;14:1–22.
107. Jayabalaji G, Manimaran P, Arun Shankar VV. Investigation on influence of process parameters in WEDM of hybrid aluminiun composite processed through squeeze casting. J Crit Rev 2020;7: 197–9.
108. AbuShanab WS, Moustafa EB, Ghandourah E, Taha MA. A comprehensive study of Al–Cu–Mg system reinforced with nano-ZrO 2 particles synthesized by powder metallurgy technique Waheed S. AbuShanab; 2021:1–34 pp.
109. Bhoi NK, Singh H, Pratap S. Developments in the aluminum metal matrix composites reinforced by micro/nano particles – a review. J Compos Mater 2019;64:1–21.
110. Muni RN, Singh J, Kumar V, Sharma S. Influence of rice husk ash, cu, mg on the mechanical behaviour of aluminium matrix hybrid composites. Int J Appl Eng Res 2019;14:1828–34.
111. Vanam JP, Rao KN. Characterization and tribological behaviour of aluminium metal matrix composite. J Eng Res Appl 2018;8:6–11.
112. Ramamoorthi R, Hillary JJM, Sundaramoorthy R, Jim Joseph JD, Kalidas K, Manickaraj K. Influence of stir casting route process parameters in fabrication of aluminium matrix composites – a review. Mater Today Proc 2021;45:6660–4.
113. Panwar N, Chauhan A. Fabrication methods of particulate reinforced aluminium metal matrix composite–a review. Mater Today Proc 2018;5:5933–9.
114. Sundriyal P, Sah PL. Enhancement of mechanical properties of graphite particulate aluminum metal matrix composites by magnesium addition. Mater Today Proc 2017;4:9481–6.
115. Yashpal S, Jawalkar CS, Verma AS, Suri NM. Fabrication of aluminium metal matrix composites with particulate reinforcement: a review. Mater Today Proc 2017;4:2927–36.
116. Kumar NS. Fabrication and characterization of Al7075/RHA/Mica composite by squeeze casting. Mater Today Proc 2020;37:750–3.
117. Bannaravuri PK, Birru AK. Strengthening of Al-4.5% Cu alloy with the addition of silicon carbide and bamboo leaf ash. Int J Struct Integr 2019;10:149–61.
118. Kaladgi ARR, Fazlur Rehman K, Afzal A, Baig MA, Soudagar MEM, Bhattacharyya S. Fabrication characteristics and mechanical behaviour of aluminium alloy reinforced with Al 2 O 3 and coconut shell particles synthesized by stir casting. IOP Conf Ser Mater Sci Eng 2021;1057:012017.
119. Ebenezer NS, Vinod B, Jagadesh HS. Corrosion behaviour of bamboo leaf ash-reinforced nickel surface-deposited aluminium metal matrix composites. J Bio Tribo Corros 2021;7:1–7.
120. Okoye CNC, Ochieze BQ. Age hardening process modeling and optimization of aluminum alloy A356/Cow horn particulate composite for brake drum application using RSM, ANN and simulated annealing. Def Technol 2018;14:336–45.
121. Shaikh MBN, Arif S, Aziz T, Waseem A, Shaikh MAN, Ali M. Microstructural, mechanical and tribological behaviour of powder metallurgy processed SiC and RHA reinforced Al-based composites. Surface Interfac 2019;15:166–79.
122. Sathish Kumar N, Harshavardhan KP, Arul Murugan R. Characterization of aluminium hybrid composite reinforced with baggase ash and graphite processed through squeeze casting. J Crit Rev 2020;7:176–8.

123. Chandla NK, Yashpal, Kant S, Goud MM, Jawalkar CS. Experimental analysis and mechanical characterization of Al 6061/alumina/bagasse ash hybrid reinforced metal matrix composite using vacuum-assisted stir casting method. J Compos Mater 2020;54:4283–97.
124. Kesarwani S, Niranjan MS, Singh V. To study the effect of different reinforcements on various parameters in aluminium matrix composite during CNC turning. Compos Commun 2020;22: 100504.
125. Bodunrin M, Oladijo P, Daramola OO, Alaneme KK. Porosity measurement and wear performance of aluminium hybrid composites reinforced. Ann Fac Eng Hunedoara-Int J Eng 2016;14:231–8.
126. Senapati AK, Manas VS, Singh A, Dash S. A comparative investigation on physical and mechanical properties of MMC reinforced with waste materials. Int J Res Eng Technol 2016;05:172–8.
127. Cheng L, Yu D, Hu E, Tang Y, Hu K, Dearn KD, et al. Surface modified rice husk ceramic particles as a functional additive: improving the tribological behaviour of aluminium matrix composites. Carbon Lett 2018;26:51–60.
128. Ramanathan S, Vinod B, Narayanasamy M, Anandajothi P. Dry sliding wear mechanism maps of Al – 7Si – 0. 3Mg hybrid composite: novel approach of agro-industrial waste particles to reduce cost of material. J Bio Tribo Corros 2019;5:1–19.
129. Bello AS, Raheem AI, Raji KN. Study of tensile properties, fractography and morphology of aluminium (1xxx)/coconut shell micro particle composites. J King Saud Univ – Eng Sci 2017;29: 269–77.
130. Dinaharan I, Kalaiselvan K, Murugan N. Influence of rice husk ash particles on microstructure and tensile behavior of AA6061 aluminum matrix composites produced using friction stir processing. Compos Commun 2017;3:42–6.
131. Alaneme KK, Sanusi KO. Microstructural characteristics, mechanical and wear behaviour of aluminium matrix hybrid composites reinforced with alumina, rice husk ash and graphite. Eng Sci Technol An Int J 2015;18:416–22.
132. Anas M, Khan MZ. Comparison of hardness and strength of fly ash and bagasse ash Al-MMCs. Int J Latest Res Sci Technol 2015;1:88–93.
133. Durowoju MO, Agunsoye JO, Mudashiru LO, Yekinni AA, Bello SK, Rabiu TO. Optimization of stir casting process parameters to improve the hardness property of Al/RHA matrix composites. Eur J Eng Res Sci 2017;2:5.
134. Moosa A, Awad AY. Influence of rice husk ash-yttrium oxide addition on the mechanical properties behavior of aluminum alloy matrix hybrid composites. Int J Curr Eng Technol 2016;6:804–12.
135. Rana H, Badheka V. Influence of friction stir processing conditions on the manufacturing of Al–Mg–Zn–Cu alloy/boron carbide surface composite. J Mater Process Tech. 2018;255:795–807.
136. Weglowski MS. Friction stir processing – state of the art. Arcives Civ Mech Eng 2017;8:114–29.
137. Banoth R, Prasad KR. A comparative based review on aluminium based metal matrix composites by friction stir processing. Int J Recent Technol Eng 2018;8:476–86.
138. Boopathi S, Thillaivanan A, Pandian M, Subbiah R, Shanmugam P. Friction stir processing of boron carbide reinforced aluminium surface (Al-B4C) composite: mechanical characteristics analysis. Mater Today Proc 2021;50:2430–35.
139. Shaik D, Sudhakar I, Bharat GCSG, Varshini V, Vikas S. Tribological behavior of friction stir processed AA6061 aluminium alloy. Mater Today Proc 2021;44:860–4.
140. Oghenevweta JE, Aigbodion VS, Nyior GB, Asuke F. Mechanical properties and microstructural analysis of Al–Si–Mg/carbonized maize stalk waste particulate composites. J Eng Sci King Saud Univ 2016;28:222–9.
141. Gupta MK. Effects of tool pin profile and feed rate on wear performance of pine leaf ash/Al composite prepared by friction stir processing. J Adhes Sci Technol 2021;35:256–68.

142. Van Trinh P, Lee J, Ngoc PM, Dinh DP, Hyung SH. Effect of oxidation of SiC particles on mechanical properties and wear behavior of SiC p/Al6061 composites. J Alloys Compd 2018;769:282–92.

143. Guo B, Zhang X, Cen X, Chen B, Wang X, Song M, et al. Enhanced mechanical properties of aluminum based composites reinforced by chemically oxidized carbon nanotubes. Carbon N Y 2018;139:459–71.

144. Chen G, Yang W, Xin L, Wang P, Lui S, Qiao J, et al. Mechanical properties of Al matrix composite reinforced with diamond particles with W coatings prepared by the magnetron sputtering method. J Alloys Compd 2018;735:777–86.

145. Azadi M, Zolfaghari M, Rezanezhad S, Azadi M. Preparation of various aluminium matrix composites reinforcing by nano-particles with different dispersion methods preparation of various aluminium matrix composites reinforcing by nano-particles with different dispersion methods. In: Proc Iran Int Alu Conf. Bangalore. India; 2018.

146. Mussatto A, Ahad IUI, Mousavian RT, Delaure Y, Brabazon D. Advanced production routes for metal matrix composites. Eng Reports 2020;5:1634–43.

147. Jose J, Peter EP, Feby JA, George AJ, Joseph J, Chandra GR, et al. Manufacture and characterization of a novel agro-waste based low cost metal matrix composite (MMC) by compocasting. Mater Res Express 2018;5:066530.

148. Sohail MA, Motgi BS, Purohit KG. Fabrication of Al 7068 reinforced with tur husk ash (THA) and alumina hybrid metal matrix composite by powder metallurgy and evaluating its microstructure and mechanical properties. Int J Sci Res Eng Trends 2019;5:1634–43.

149. Panwar N, Chauhan A. Fabrication methods of particulate reinforced aluminium metal matrix composite-a review. Mater Today 2018;5:5933–9.

150. Ramanathan A, Krishnan PK, Muraliraja R. A review on the production of metal matrix composites through stir casting – furnace design, properties, challenges, and research opportunities. J Manuf Process 2019;42:213–45.

151. Satheesh M, Pugazhvadivu M. Investigation on physical and mechanical properties of Al6061-silicon carbide (SiC)/coconut shell ash (CSA) hybrid composites. Phys B Condens Matter 2019;572:70–5.

152. Udoye NE, Nnamba OJ, Fayomi OSI, Inegbenebor AO, Jolayemi KJ. Analysis on mechanical properties of AA6061/Rice husk ash composites produced through stir casting technique. Mater Today Proc 2020;43:1415–20.

153. Kumar A, Kumar M. Mechanical and dry sliding wear behaviour of B4C and rice husk ash reinfroced Al 7075 alloy hybrid composite for armors application by using taguchi techniques. Mater Today Proc 2019;27:2617–25.

154. Babaremu KO, Joseph OO, Akinlabi ET, Jen TC, Oladijo OP. Morphological investigation and mechanical behaviour of agrowaste reinforced aluminium alloy 8011 for service life improvement. Heliyon 2020;6:e05506.

155. Apasi A, Yawas DS, Abdulkareem S, Kolawole MY. Improving mechanical properties of aluminium alloy through addition of coconut shell-ash. J Sci Technol 2016;36:34–43.

156. Yashpal CS, Jawalkar S, Kant N, Panwar M, Sharma D, Pali HS. Effect of particle size variation of bagasse ash on mechanical properties of aluminium hybrid metal matrix composites. Mater Today Proc 2020;21:2024–9.

157. Devanathan R, Ravikumar J, Boopathi S, Christopher Selvam D, Anicia SA. Influence in mechanical properties of stir cast aluminium (AA6061) hybrid metal matrix composite (HMMC) with silicon carbide, fly ash and coconut coir ash reinforcement. Mater Today Proc 2019;22:3136–44.

158. Sarkar S, Bhirangi A, Mathew J, Oyyaravelu R, Kuppan P, Balan ASS. Fabrication characteristics and mechanical behavior of rice husk ash-silicon carbide reinforced Al-6061 alloy matrix hybrid composite. Mater Today Proc 2018;5:12706–18.

159. Muralimohan R, Kempaiah UN, Seenappa. Influence of rice husk ash and B4C on mechanical properties of ADC 12 alloy hybrid composites. Mater Today Proc 2018;5:25562–9.

160. Thimothy P, Sankara S, Ratnam C. Development and accretion of tribological performance on Al-CSA composites using orthogonal array. Mater Today Proc 2019;18:5332–9.

161. Omoniyi P, Adekunle A, Ibitoye S, Olorunpomi O, Abolusoro O. Mechanical and microstructural evaluation of aluminium matrix composite reinforced with wood particles. J King Saud Univ – Eng Sci 2021;34:445–50.

162. Verma N, Vettivel SC. Characterization and experimental analysis of boron carbide and rice husk ash reinforced AA7075 aluminium alloy hybrid composite. J Alloys Compd 2018;741:981–98.

163. Udoye NE, Inegbenebor AO, Fayomi OSI. Corrosion performance and wear behaviour of AA6061 reinforced hybrid: nano – rice husk ash/clay particulate for cooling tower fan blade in 0 1 75 M H. J Bio Tribo Corros 2020;6:1–9.

164. Arora G, Sharma S. Effects of rice husk ash and silicon carbide addition on AA6351 hybrid green composites. Emerg Mater Res 2020;9:141–6.

165. Alaneme KK, Ekperusi JO, Oke SR. Corrosion behaviour of thermal cycled aluminium hybrid composites reinforced with rice husk ash and silicon carbide. J King Saud Univ – Eng Sci 2018;30: 391–7.

166. Ikumapayi OM, Akinlabi ET, Majumdar JD, Fayomi OSI, Akinlabi A. Corrosion study and quantitative measurement of crystallite size of high strength aluminum hybrid composite developed via friction stir processing korrosionsuntersuchung und quantitative messung der. Materwiss Werksttech 2020;51:732–9.

167. Suriani MJ, Zulkifli F, Khairul MA, Azaman MD, Jorkoni MNK. Corrosion behavior of hybrid coconut shell powder and silica carbide reinforced aluminium composite towards aggressive biofilm attack corrosion behavior of hybrid coconut shell powder and silica carbide reinforced aluminium composite towards aggressive bi. IOP Conf Ser Mater Sci Eng 2019;670:012006.

168. Udoye NE, Nnamba OJFOSI, Inegbennebor AO. Corrosion performance of AA6061/rice husk ash composite for engineering application. In: Int conf eng sustain world (ICESW 2020); 2021.

169. Rao GB, Bannaravuri PK, Birru AK. Effect on corrosion behaviour of the surface of aluminium 4. 5 Cu with bamboo leaf ash composites by laser treatment. Mater Res Express 2020;7:016594.

170. Alaneme KK, Eze HI, Bodunrin MO. Corrosion behaviour of groundnut shell ash and silicon carbide hybrid reinforced Al-Mg-Si alloy matrix composites in 3.5% NaCl and 0.3M H 2SO4 solutions. Leonardo Electron J Pract Technol 2015;14:141–58.

171. Nithyanandhan T, Sivaraman P, Ramamoorthi R, Kumar PS, Kannakumar R. Enhancement of corrosion behaviour of AL6061-B 4 C-RHA reinforced hybrid composite. Mater Today Proc 2020: 2–7. https://doi.org/10.1016/j.matpr.2020.04.167.

172. Daramola OO, Adediran AA. Evaluation of the mechanical properties and corrosion behaviour of coconut shell ash reinforced aluminium (6063) alloy composites. Leonardo Electron J Pract Technol 2015;27:107–19.

173. Ikumapayi OM, Akinlabi E. A comparative assessment of tensile strength and corrosion potection in friction stir processed AA7075-T651 matrix composites using fly ashes nanoparticles as reinforcement inhibi … protection in friction stir processed AA7075-T651 matrix. Int J Mech Prod Eng Res Dev 2019;9:839–54.

174. Ikumapayi OM, Akinlabi ET, Abegunde OO, Fayomi OSI. Electrochemical investigation of calcined agrowastes powders on friction stir processing of aluminium-based matrix composites. Mater Today Proc 2019;26:3238–45.

175. Idusuyi N, Oviroh PO, Adekoya AH. A study on the corrosion and mechanical properties of an Al6063 reinforced with egg shell ash and rice husk ash. In: International mechanical engineering congress and exposition (IMECE2018), American Society of Mechanical Engineers. New York, NY:

International Mechanical Envgineering Congress and Exposition (IMECE); 2018, 52170: V012T11A016 p.

176. Aigbodion VS, Ozor PO, Eke MN, Nwoji CU. Explicit microstructure, corrosion, and stress analysis of value-added Al-3.7% Cu-1.4%Mg/1.5% rice husk ash nanoparticles for pump impeller application. Chem Data Collect 2021;33:100675.

177. Edoziuno FO, Adediran AA, Odoni BU, Utu OG, Olayanju A. Physico-chemical and morphological evaluation of palm kernel shell particulate reinforced aluminium matrix composites. Mater Today Proc 2019;38:652–7.

I. Palle*, W. L. Zen, A. A. Mohd Yunus, M. S. Gilbert and
M. A. Abd Ghani

12 Physical and mechanical properties of *Acacia mangium* plywood after sanding treatment

Abstract: The purpose of this study was to evaluate the effect of sanding treatment on *Acacia mangium* plywood panel properties. Three layered *A. mangium* plywood panels were produced from sanded veneer using three different grit sizes: S180, S240 and S320, from medium to ultrafine respectively, with one non-sanded panel as control. Melamine urea formaldehyde resin was unvaryingly used as binder. The outturn of the sanding treatment was observed through a set of evaluations which includes the contact angle test for surface properties, shear strength and static bending for the mechanical properties, and thickness swelling and water absorption rate for the physical properties and dimensional stability. The study shows that almost all of the properties exhibit significant differences at $p \leq 0.05$ except for the density and shear test. From the results, the modulus of elasticity of plywood decreases with the increase of grit sizes but shows no significant difference with the control and S240 samples. The S180 samples notably shows an impressive mechanical property with the highest modulus of rupture, albeit the lowest dimensional stability. Nonetheless, the result of this study indicated that the sanding treatment shows a notable increase in the physical, mechanical and dimensional stability of the panel, with the S180 medium grit proving to be adequate and best in improving the plywood panels. As excessive smoothness may negatively affect the wettability itself due to the production of uneven surface and the development of loose fiber that compromises the adherence of the MUF and veneer, the higher grits negatively impacted its treatment on the surface.

Keywords: contact angle; mechanical strength; resin penetration; sanded veneer; wettability.

---

*Corresponding author: I. Palle, Faculty of Tropical Forestry, University Malaysia Sabah, Jalan UMS, 88400 Kota Kinabalu, Sabah, Malaysia, E-mail: isspalle@ums.edu.my
W. L. Zen, A. A. Mohd Yunus, M. S. Gilbert and M. A. Abd Ghani, Faculty of Tropical Forestry, University Malaysia Sabah, Jalan UMS, 88400 Kota Kinabalu, Sabah, Malaysia

As per De Gruyter's policy this article has previously been published in the journal Physical Sciences Reviews. Please cite as: I. Palle, W. L. Zen, A. A. Mohd Yunus, M. S. Gilbert and M. A. Abd Ghani "Physical and mechanical properties of *Acacia mangium* plywood after sanding treatment" *Physical Sciences Reviews* [Online] 2022. DOI: 10.1515/psr-2022-0045 | https://doi.org/10.1515/9783110769227-012

12.1 Introduction

Plywood manufacturing such as cutting, drying and laminating can affect the physical and chemical surface properties of veneer [1]. It is reported that wood surfaces can undergo surface inactivation due to high temperatures experienced resulting in a loss of bonding ability. Rough wood surfaces that can prevent surface contact can contribute to a poor finish. In order to improve bonding capability, surface modifications can be applied to the surface properties panel. According to Sulaiman et al. [2], the roughness of wood surface can be improved to a certain extent by sanding. Sanding is one of the mechanical pre-treatments for wood, in which produces a smooth and even surface that creates a low contact angle for the adhesive through the leveling of the surface. Through the smoothing of the surface, jagged surface is removed, allowing better contact and wettability of adhesive to the wood surface [3, 4]. A study by Aydin [1] shows that panels produced from sanded veneers have higher shear and bending strength values than non-sanded veneers.

Surface roughness is affected by the sanding process in relation to grit size [5]. Karliati et al. [6] studied the effect of different grit sizes of sandpaper on the roughness of laminae. The treatments were performed using KAG 80, 100, 150, 220, 300 and 400 grit sizes. The study found that the laminae surface roughness decreased in the order of grit size. Similar study has been reported by Sulaiman et al. [2], where the sanded surface roughness decreased in the order of grit size 120, 150 and 180. Low grit sizes increased surface roughness because capillary forces could suck in more liquid.

Alamsyah et al. [7] reported that the shear bond strength of *Acacia mangium* laminated board adhered with resorcinol formaldehyde resin is weak. This issue was related to *A. mangium's* poor wettability, which prevented any liquid from sufficiently penetrating the surface of the wood. In order to increase *A. mangium's* surface wettability, pretreatment is essential. This study aimed to evaluate the effect of sanding treatment by using different grit sizes of sandpaper (180, 240 and 320 grit size) on the physical and mechanical properties of *A. mangium* plywood panel bonded by melamine urea formaldehyde adhesive. The wettability test of *A. mangium* veneer before and after sanding treatment was also determined through contact angle test.

12.2 Materials and methods

12.2.1 *A. mangium* log

A. mangium log with 45 cm diameter was obtained from the Sabah Forestry Development Authority (SAFODA), Kinarut, Sabah. The logs were debarked manually before undergoing the peeling process using the Hydraulic Spindleless Peeling machine to obtain the veneers with an average thickness of 1.2 mm. The veneer obtained were air-dried for 24 h before undergoing oven drying at 70 °C for 2 h to get 8–12% MC.

12.2.2 Sanding treatment of veneer surface

The veneer surfaces were sanded at three different grit sizes: S180, S240 and S320 which ranges from medium to ultra-fine grit. The surface is the cleaned of any dust to allow the subsequent testing such as the contact angle test and adhesive application.

12.2.3 *A. mangium* plywood

Plywood panel with three plies were manufactured from the sanded and non-sanded veneers using melamine urea formaldehyde (MUF) as binder. Approximately 160 g/m^2 of adhesive was manually spread on a single surface of veneer. The plywood then underwent hot pressing at the pressure of 12 kg/m^2 and temperature of 110 °C for 5 min to activate and cure the MUF. Eight replicates of plywood panels were fabricated for each parameter.

12.2.4 Testing

12.2.4.1 Surface properties

The surface property was measured based on the contact angle to identify the wettability of the veneers. The contact angle analysis was carried through the measuring of the contact angles of different solutions such as distilled water, MUF resin, 0.05 M of hydrochloric acid (HCl), and 0.05 M of sodium hydroxide (NaOH) when being dropped on the veneer surfaces. The contact angle measurement was taken in 5 s after the liquid deposited on the veneers. The value of contact angle was measured using a light microscope.

12.2.4.2 Mechanical properties

Shear strength and static bending tests of plywood panels were analyzed following British Standard (BS) 6566: Part 8: 1985 and ASTM D1037-06a, respectively. For shear strength test, samples were submerged in warm water at 67 ± 2 °C for 3 h ± 10 min, followed by a cooling process in water at 15 ± 5 °C for 1 h before the test were done. A total of 32 samples were used for mechanical test.

12.2.4.3 Physical properties and dimensional stability

The samples physical stability in water was measured through the thickness welling test and water absorption rate. Both tests were carried out according to ASTM D

1037-06a. A total of 40 samples were used for the physical and dimensional stability test for each parameter.

12.2.4.4 Data analysis

The results were analyzed by using Statistical Analysis System (SAS). Analysis of variance (ANOVA) and mean separation using least significant difference (LSD) were carried out to evaluate the effect of different sanding treatment on plywood panel properties. For the ease of identification, the samples are formed into 4 groups as shown in Table 12.1.

12.3 Results and discussion

12.3.1 Wettability of veneer on surface properties

A smaller contact angle means that the surface is more wettable and more hydrophilic [2, 8]. The contact angle values of veneer samples are shown in Figure 12.1. Evidently, sanded panel had lower contact angle value compared to non-sanded veneer. This result is in agreement with results obtained by [1, 2, 8]. Sanding treatment pull more surface material and produced higher roughness value. According to Sulaiman et al. [2], rough surfaces arguably promotes excess penetration because of the higher surface area to volume ratio due to the unpolished surface and increasing the tendency in absorbing more liquids, however negatively resulting in high contact angle of the surface. In terms of the impact of various sanding grit sizes, the contact angle value legibly decreased as grit sizes increased. However, it was found that there were no significant differences between the 180, 240 and 320-grit sizes.

pH is another factor that can affect the glue bond quality other than the surface condition. Comparison of the contact angle were also made between different types of pH solutions (Figure 12.1). Distinctly, the wettability of treated *A.mangium* veneers were enhanced concurrently with the increasing of pH solution. The acidity of *A. mangium* wood with a pH range between 5.21 and 5.56 resulted in a greater degree of interaction

Table 12.1: Test groups based on plywood made from different sandpaper grit sizes.

Code	Description
C	Non sanded (control)
S180	Plywood made from veneer sanded with 180 grit sandpaper
S240	Plywood made from veneer sanded with 240 grit sandpaper
S320	Plywood made from veneer sanded with 320 grit sandpaper

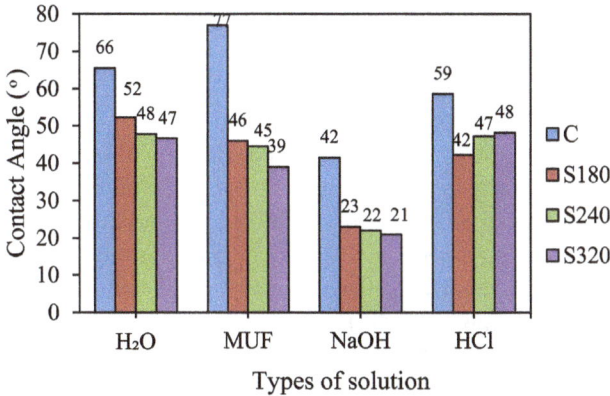

Figure 12.1: Wettability of sanded and unsanded *Acacia mangium* veneer using different solvents.

with the alkaline solution, as the acidic cellulose has a strong electron-accepting ability [9] which signifies that its surfaces were dominated by the electron donating sites. In addition, Redzuan et al. [5] reported that NaOH can easily dissolve some of the components of extractives, which removes the barrier, thus promoting better penetration. By changing the solution to acidic (HCl), the contact angle notably increased and presumably creates a rather hydrophobic and polar wood surface.

12.4 Physical and mechanical properties of plywood

The moisture content (MC) of the plywood ranged between 12.86 and 14.10% while the density ranged between 0.77 and 0.80 g/mm^2 (Table 12.2). The MC of plywood increased with the increasing of sandpaper grit size. The non-sanded plywood showed the lowest MC which is 12.86% while the highest was S320, 14.1%. The ANOVA results revealed that different grit sizes did not affect the density of plywood at $p > 0.05$. Nevertheless, there is a significant effect of MC between un-sanded and sanded panels

Table 12.2: The physical and dimensional stability properties of *Acacia mangium* plywood.

Types of board	Moisture content (%)	Density (g/cm^3)	Water absorption (%)	Thickness swelling (%)
C	12.86[b] (0.49)	0.78[a] (0.05)	27.03[a] (1.62)	7.14[a] (1.11)
S180	13.83[a] (0.42)	0.80[a] (0.02)	23.54[b] (3.18)	4.16[b] (0.55)
S240	13.97[a] (0.22)	0.78[a] (0.05)	24.51[b] (3.14)	4.29[b] (0.46)
S320	14.10[a] (0.34)	0.77[a] (0.02)	27.70[a] (2.19)	6.94[a] (1.01)

Means with the same letters [a,b,c,d] in the same column were not significantly different ($p > 0.05$). ( ) Values in parentheses indicate the standard deviation.

at p ≤ 0.05. Increasing of grit size corresponds to the increment of wettability properties consequently raising the moisture uptake [3].

The water absorption and thickness swelling of plywood panels after 24 h soaking period is shown in Figure 12.2. Obviously, un-sanded panel showed the highest swelling and water absorption followed by S320, S240 and S180, respectively. S180 showed less prone to swell when expose to water, might due to high bonding ability which typically can increase the resistance towards shrinking and swelling [3]. A significant increase in the rate of thickness swelling took place after increasing grit size from S240 to S320. This is in line with the results on contact angle that showing S240 and S320 have lower contact angle value (Figure 12.1) implied the liquid easier to penetrate into cell layer. Over penetration of adhesive into cell walls produce weak bond line that fail to withstand the wood expansion during soaking condition [10] hence increase tendency in capturing high amount of water.

Bending strength values of plywood made from sanded veneer (S180) was found to be higher than non-sanded veneer (Table 12.3). This phenomenon is related to the contact angle which typically used to estimate the wettability of materials [7]. As shown in Figure 12.1, the sanded veneer surfaces were certainly more wettable and are easier to spread. Sanding treatment had improved the surface roughness which then enhanced the resin penetration and led to good bonding strength [2, 11]. This is confirming through resin penetration observation where the untreated plywood shows no sign of penetration while the treated plywood has significant penetration of resin (Figure 12.3). The wettability also affected by the resulting surface roughness associated with sandpaper grit size. According to Ayrilmis et al. [8], the size of the aluminium oxide on the sandpaper gets smaller with the increasing of grit size. It is apparently clear that the applications of higher grit of sanding (>S180) reduced the surface roughness of *A. mangium* plywood. Surface roughness was proposed to enhance intrinsic adhesion.

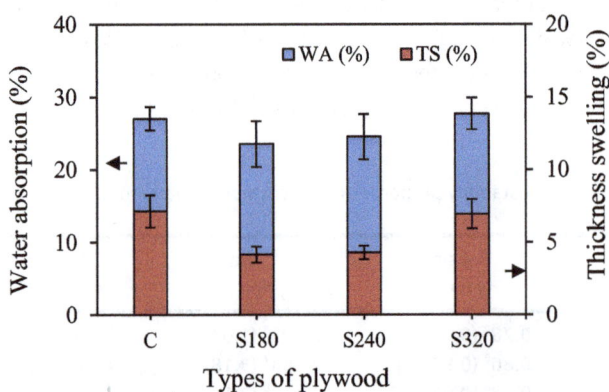

Figure 12.2: Thickness swelling and water absorption properties of plywood.

Table 12.3: The mechanical properties of *Acacia mangium* plywood.

Types of board	Shear strength (N/mm²)	Modulus of elasticity (N/mm²)	Modulus of rupture (N/mm²)
C	3.70[a] (0.518)	8077.6[a] (270.92)	45.78[ab] (5.567)
S180	3.33[a] (0.504)	8255.0[a] (285.55)	49.94[a] (2.302)
S240	3.48[a] (0.397)	8217.7[a] (538.44)	45.45[b] (6.907)
S320	3.66[a] (0.383)	7573.7[b] (139.36)	48.60[ab] (4.503)

means with the same letters [a,b,c,d] in the same column were not significantly different (p > 0.05). ( ) Values in parentheses indicate the standard deviation.

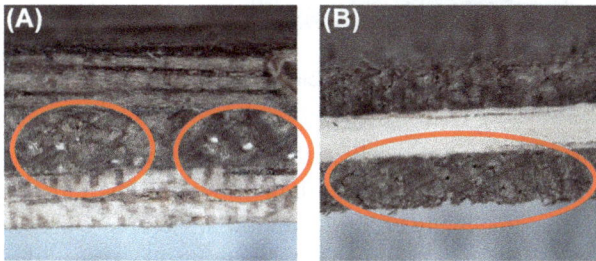

Figure 12.3: Resin penetration in vessel of treated (A) and untreated (B) veneer (red circled area).

The adhesion mechanism increased by filling up the wood pores through rough surfaces to create mechanical interlock [1, 3, 10]. This scenario might correspond to the increment of bending strength of plywood after sanding using S180 grit size. In addition, higher panel density of S180 had increased the stiffness of the panel [2] resulted in high MOE and MOR values of plywood panel. However, by increasing grit sizes, the bending strength of plywood reduced significantly. Too much sanding can alter the surface condition from a flat surface to an uneven surface, resulted excessive loose fiber that would interfere with adhesion [12]. This scenario might be the reason for the decrement of bending properties of plywood after sanding using more that 180-grit size.

The shear strength values of treated plywood subjected to the wet test were increased with increasing grit sizes but slightly lower than control. Statistical analysis shows there is no significant difference between parameters studied. Karliati et al. [4] reported that the shear strength of laminated wood decreased with increasing sandpaper grit sizes in the wet test condition. The weakening adhesive bonds as stress which developed from dimensional changes during soaking leads to the decrement of strength. Figure 12.4 shows the types of bond failure of the specimens after testing. The S180 sample shows the failure occurs entirely within the core veneer which indicates the mixed mode failure (Figure 12.3A). This shows that the bondline is sufficiently strong

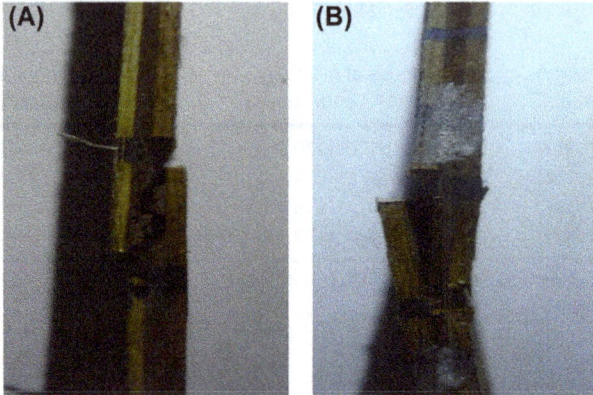

Figure 12.4: Types of bond failure on plywood panel, mix-mode failure (A) glue line failure (B).

where the fracture changes from the bondline to within the wood. Meanwhile, control group shows the failure occurs at the interface between the adhesive and the adherent (Figure 12.3B). This kind of failure frequently occur due to low strength of adhesive bonding [13], where the presence of surface contaminants during the preparation process can degrade the interfacial strength due to the two adherents are failure to generate a chemically active surface [14].

12.5 Conclusions

The results of this study indicated that sanding treatment does increase the physical, mechanical and dimensional stability properties of plywood. Nonetheless, plywood surfaces sanded with lower grit sizes (S180) were more wettable than those with higher grit sizes. The rougher surface was shown to be more wettable than a smooth one as too much sanding alters the condition of the surface, resulting in an excess of loose fibre that is not ideal in the adhesive-substrate adhesion mechanism. S180 shows the more superior dimensional properties and the highest mechanical properties, where the MOR value is beyond the ASTM standard which is 49.94 N/mm^2. Thus, the best grit size of sandpaper used in this study is 180.

References

1. Aydin I. Activation of wood surfaces for glue bonds by mechanical pre-treatment and its effects on some properties of veneer surfaces and plywood panels. Appl Surf Sci 2004;233:268–74.
2. Sulaiman O, Hashim R, Subari K, Liang CK. Effect of sanding on surface roughness of rubberwood. J Mater Process Technol 2009;2019:3949–55.

3. Frihart CR, Hunt CG. Adhesives with wood materials: bond formation and performance. In: Forest products laboratory. Wood handbook—wood as an engineering material. General technical report FPL-GTR-190. Madison, WI: U.S. Department of Agriculture Forest Service, Forest Products Laboratory; 2010.

4. Frihart CR. Wood adhesion and adhesives. In: Rowel RM, editor. Handbook of wood chemistry and wood composite. New York: CRC Press; 2005:215–278 pp.

5. Mitchell P, Lemaster R. Investigation of machine parameters on the surface quality in routing soft maple. For Prod J 2002;52:85–90.

6. Karliati T, Febrianto F, Syafii W, Wahyudi I, Sumardi I, Lee SH, et al. Properties of laminated wood bonded with modified Gutta-Percha adhesive at various surface roughness profile of laminae. Bioresources 2019;12:8241–9.

7. Alamsyah EM, Yamada M, Taki K. Bondability of tropical fast-growing tree species. III: curing behavior of resorcinol formaldehyde resin adhesive at room temperature and effects of extractives of *Acacia mangium*, wood on bonding. J Wood Sci 2008;54:208–13.

8. Ayrilmis N, Candan Z, Akbulut T, Balkiz OD. Effect of sanding on surface properties of medium density fiberboard. Drv Ind 2010;61:175–81.

9. Nawawi DS. The acidity of five tropical woods and its influence on metal corrosion. J For Prod Technol 2002;XV:2.

10. Serrano E, Kallander B. Building and construction – timber. In: Adhesive bonding – science, technology and applications. England: CRC Press LLC; 2005.

11. Cumble RP, Dareka DH. Influence of surface roughness of adherend on strength of adhesive joint. IJEDR 2017;5:100–6.

12. Vick CB. Adhesive bonding of wood materials. In: Wood handbook – wood as an engineering material. Gen. Tech. Rep. FPL-GTR-113. Madison. WI: U.S. Department of Agriculture, Forest Service, Forest Products Laboratory; 1999.

13. Davis MJ, Mcgregor A. Assessing adhesive bond failures: mixed-mode bond failures explained. Canberra: ISASI Australian Safety Seminar; 2010:4–6 pp.

14. Davis MJ, Bond DA. The importance of failure mode identification in adhesive bonded aircraft structure and repairs. In: Aircraft structural integrity section; 2012.

Ridhwan Jumaidin*, Amirul Hazim Abdul Rahman,
Salit Mohd Sapuan and Ahmad Ilyas Rushdan

13 Effect of sugarcane bagasse on thermal and mechanical properties of thermoplastic cassava starch/beeswax composites

Abstract: The demand for biodegradable material has been an important issue, especially in food packaging applications. Among many biodegradable materials, starch biopolymer has been recognised as a completely biodegradable material that can be produced from various plants. It is one of the richest resources that are renewable, biodegradable, and available at low cost. However, starch biopolymers are often associated with poor mechanical properties. Hence, the main objective of this study is to evaluate the mechanical and thermal characteristics of sugarcane bagasse fibre (SBF) reinforced thermoplastic cassava starch (TPCS), which was prior modified with beeswax (BW). It was found that the mechanical properties such as tensile, flexural, and impact strength have improved significantly with the incorporation of SBF loading into the TPCS/BW matrix. The highest tensile strength (12.2 MPa) and modulus (2222.6 MPa) were exhibit by sample with 20 wt% SBF loading and further increment of fibre led to decrease in the strength of the materials. The thermal properties showed that higher SBF loading resulted in improved thermal stability of the material, i.e., higher glass transition and melting temperature than the polymer matrix. Overall, SBF has shown good potential as a reinforcing material which is able to improve the functional characteristics of TPCS/BW as a new potential biodegradable material.

Keywords: beeswax; sugarcane bagasse; thermoplastic starch.

*Corresponding author: Ridhwan Jumaidin, Fakulti Teknologi Kejuruteraan Mekanikal dan Pembuatan, Universiti Teknikal Malaysia Melaka, Hang Tuah Jaya, 76100 Durian Tunggal, Melaka, Malaysia; and Institute of Tropical Forestry and Forest Products, Universiti Putra Malaysia, Serdang 43400, Malaysia, E-mail: ridhwan@utem.edu.my
Amirul Hazim Abdul Rahman, Fakulti Teknologi Kejuruteraan Mekanikal dan Pembuatan, Universiti Teknikal Malaysia Melaka, Hang Tuah Jaya, 76100 Durian Tunggal, Melaka, Malaysia
Salit Mohd Sapuan, Advanced Engineering Materials and Composite Research Centre (AEMC), Department of Mechanical and Manufacturing Engineering, Universiti Putra Malaysia, Serdang 43400, Malaysia
Ahmad Ilyas Rushdan, Sustainable Waste Management Research Group (SWAM), School of Chemical and Energy Engineering, Faculty of Engineering, Universiti Teknologi Malaysia, 81310 UTM Johor Bahru, Johor, Malaysia; and Centre for Advanced Composite Materials (CACM), Universiti Teknologi Malaysia, 81310 UTM Johor Bahru, Johor, Malaysia

As per De Gruyter's policy this article has previously been published in the journal Physical Sciences Reviews. Please cite as: R. Jumaidin, A. H. Abdul Rahman, S. M. Sapuan and A. I. Rushdan Rushdan "Effect of sugarcane bagasse on thermal and mechanical properties of thermoplastic cassava starch/beeswax composites" *Physical Sciences Reviews* [Online] 2022. DOI: 10.1515/psr-2022-0047 | https://doi.org/10.1515/9783110769227-013

13.1 Introduction

Plastic has played an important role in human life, ranging from packaging, furniture, household tools, etc. Back in the 1950s, large scales of plastic usage have begun in the series of creating products from plastic. The quick growth of plastic usage and production is at mega-scale, outstanding any other man-made materials [1]. However, the presence of plastics has resulted in a large amount of waste accumulated on the Earth's atmosphere that can affect the society and economy of the global population, particularly when used as packaging materials.

Hence, the development of more environmentally friendly materials is highly important to overcome the catastrophic damage caused by conventional plastic waste. Biopolymers are one of the most promising alternatives for this issue since they are developed from natural resources and are biodegradable, thus, leaving no harm to the environment [2, 3].

Among biopolymers, starch is one of the most versatile materials due to its compelling characteristics, including low cost, biodegradable, renewable resource, and highly abundant in nature [4–8]. However, starch biopolymer, commonly known as thermoplastic starch, has several limitations, i.e., poor physical properties. Thus, numerous modification methods have been implemented in order to enhance the starch properties, such as plasticisation, blending, derivation, and graft co-polymerisation have been investigated [9–14]. Waxes is an ancient material and has been used for water resistance purposes such as coating for wood and paper. Modification of starch using waxes is a promising approach due to the excellent moisture resistance characteristics which can reduce the hygroscopic properties of starch.

Beeswax is a natural wax produced by honeybees in the beehives which also possess good hydrophobic characteristics. It is a part of a honeycomb can be separated via the centrifugation process. The quality and value of beeswax is highly depending on the colour of its appearance [15], hence, applying excessive heat to this material can lessen the aroma and changes its colours and quality [16]. Owing to its outstanding characteristics, this natural wax is widely used for textile industry, polish material, candle manufacturing, food processing, casting of metal, etc. As a coating material, beeswax shows good properties to hinder moisture, hence, able to enhance the shelf life of the coated product [17–20].

In recent, a study was reported on the effect of beeswax on thermoplastic cassava starch [21]. The authors reported significant decrement in the water affinity behaviour of the samples i.e. moisture absorption and water solubility, However, this improvement were accompanied with the decrement in the tensile and flexural strength of the materials. Apart from beeswax, recent study reported that palm wax were able to improve the mechanical properties of thermoplastic cassava starch while improving the moisture resistance characteristics of the biopolymer [22]. The authors associated this finding with good intermolecular bonding between palm wax and starch which enhance the properties of this material.

The effects of wax on natural polymers such as chitosan biofilm [18], pullulan [23] and gelatin [24] were reported. FTIR analysis on the samples show hydrogen bonding formation between the wax and the polymer matrix which then enhanced the mechanical properties and water resistance of the biocomposites film [23, 25, 26].

Sugarcane bagasse is a well-known waste produced from sugar production. This bagasse has the potential to be used as reinforcement due to the high amount of cel-lulose in its fibre. Several studies have been carried out on using sugarcane bagasse as reinforcement in polymer composites [27, 28]. However, no work has been reported on the characterisation of sugarcane bagasse as reinforcement in thermoplastic cassava starch/beeswax composites.

In this study, the polymer matrix was developed using a mixture of cassava starch/beeswax based on our previous study, which has proved that adding beeswax improves the characteristic of moisture absorption and mechanical properties of thermoplastic cassava starch [21]. The composites were characterised for their thermal and mechanical properties.

13.2 Materials and methods

13.2.1 Materials

Raw sugarcane bagasse fibre (SBF) was obtained as discarded waste from the sugarcane juice extraction process. After the juice extraction, the remaining waste was collected and water-retted for 3 weeks. The fibre was extracted using a manual separation process and sun-dried for 5 h. The dried fibre was ground to obtain 2–3 mm of fibre length and placed in a drying oven at 100 °C for 5 h. Analytical grade beeswax was acquired from Sigma-Aldrich (CAS = 8012-89-3, mp = 61–65 °C). The 99.5% AR grade glycerol ($C_3H_8O_3$) was procured from QRëC (Asia) with a molar mass of 92.1 g/mol. Cassava starch (CS) was purchased from Antik Sempurna company with pure white colour and in powder form.

13.2.2 Fabrication of composite

The preparation of the thermoplastic cassava starch (TPCS) mixture was conducted using the weight ratio of starch and glycerol at 100:30. Then, the mixture was further modified with beeswax (BW) to reduce the moisture sensitivity of the material by incorporating 10 wt% of beeswax into the TPCS mixture. The resulting TPCS/BW was used as the matrix for this study. The preparation TPCS/BW/SBF composites were carried out by incorporating 10–40 wt% of SBF into the mixture. The mixture was subjected to stirring at 3000 rpm for 5 min using BL1515 Dry Mixer from Khind (Shah Alam, Selangor, Malaysia). The resultant mixture is then compressed using a GT7014-P30 C plastic hydraulic moulding press from GOTECH Testing Inc (Taichung City, Taiwan) at 155 °C for 1 h. The pressed samples were subjected to conditioning parameter of 53% RH and 25 °C for 48 h before further characterisation.

13.2.3 Mechanical testing

13.2.3.1 Tensile test: The samples were cut according to the ASTM D638 standard, and the test was conducted using Instron 5969 (Norwood, MA, USA) with a 5 kN load cell and 5 mm/min cross-head speed. The environment parameters for testing were set at $23 \pm 1\,°C$ and relative humidity of $50 \pm 5\%$.

13.2.3.2 Impact test: The test was conducted according to ASTM D256 using digital INSTRON CEAST 9050 pendulum impact tests. The dimensions of the unnotched samples were 60 mm (L) × 13 mm (W) × 3 mm (T).

13.2.3.3 Fourier transform infrared spectroscopy (FTIR): The analysis of the Fourier transform infrared (FTIR) spectroscopy was performed using JASCO FT/IR-6100 (Germany). FTIR spectra of the specimens were collected in the range of $4000–400\ cm^{-1}$.

13.2.3.4 Scanning electron microscopy (SEM): The SEM model Zeiss Evo 18 Research, (Jena, Germany) was used to investigate the fracture morphology of the samples. The acceleration voltage used was 10 kV.

13.2.4 Thermal testing

13.2.4.1 Thermogravimetric analysis (TGA): TGA was conducted using Mettler–Toledo AG, Analytical (Switzerland). The specimen should weigh around 10–18 mg to be put in the ceramic crucible for testing. The analysis was carried out under a dynamic nitrogen atmosphere in the temperature range 25–900 °C and the heating rate of $10\ °C\ min^{-1}$.

13.2.4.2 Differential scanning calorimetry (DSC): The machine used for the DSC test was PerkinElmer Jade, made in the USA. The sample underwent heating process from 35 to 265 °C at a rate of 10 °C/min. Nitrogen was used to maintain an inert environment and to flush the DSC cell at a flow rate of 20 mL/min.

13.3 Results and discussion

13.3.1 Mechanical testing

13.3.1.1 Tensile testing

The mechanical properties of biopolymers were determined by measuring the quantitative values such as tensile strength, modulus, and percentage of elongation at break, which were used to measure the strength and flexibility of the material.

Figure 13.1 demonstrates the effect of SBF loading on tensile strength, modulus, and elongation at break. The tensile strength of the TPCS/BW/SBF composites showed a significant improvement with the increasing fibre loading in the composites. Figure 13.1(a) shows that the addition of 20 wt% of SBF into TPCS/BW has led to a significant improvement of tensile strength from 6.2 to 12.2 MPa. Similar trends were

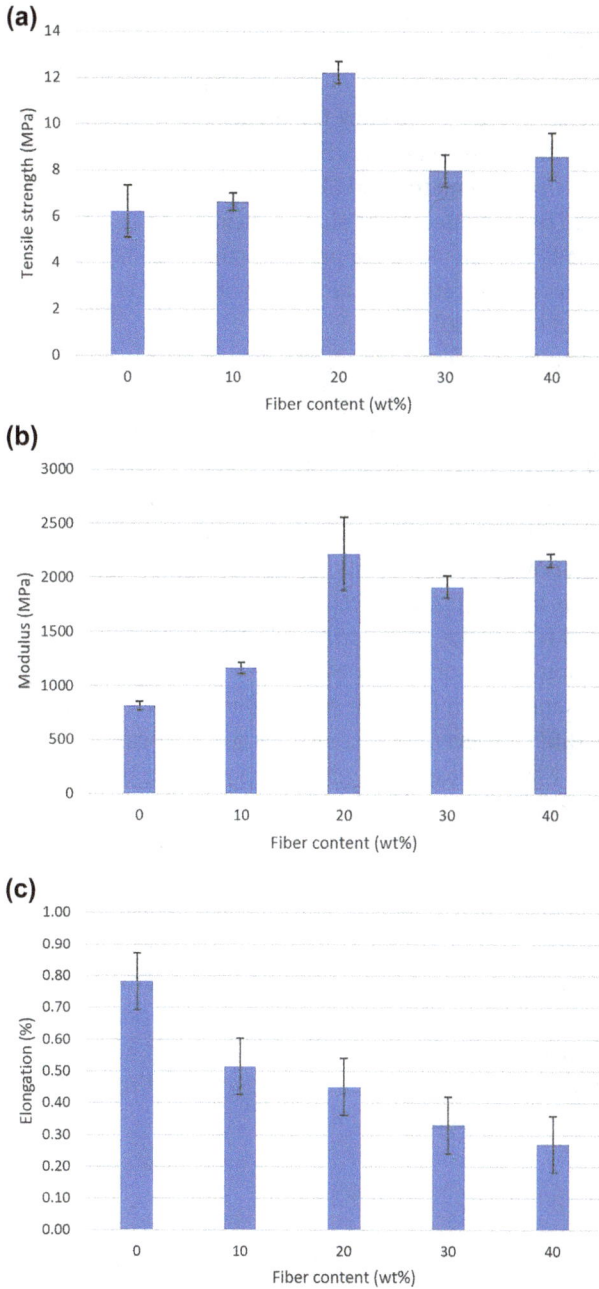

Figure 13.1: Tensile properties of TPCS/BW/SBF composites. (a) Tensile strength of TPCS/BW/SBF composites. (b) Tensile modulus of TPCS/BW/SBF composites. (c) The elongation at break of TPCS/BW/SBF composites.

observed for the tensile modulus in Figure 13.1(b). The improvement in the tensile strength and modulus for this material could be associated with several reasons. Firstly, the efficient matrix-fibre stress transfer were observed in the SEM image of fractured samples. This phenomenon can be seen by the occurrence of fibre fracture after the tensile test. Secondly, the improvement of the occurrence of fibre fractures revealed good compatibility between the matrix and SBF. The excellent compatibility between SBF and the matrix can be attributed to the similar chemical characteristic of TPCS and SBF. Similar arguments were made in the previous study [12, 29]. This was due to the similarity of the main cellulose structure composed of hydroxyl functional groups for both natural fibre and starch [30].

However, the tensile strength and tensile modulus were decreased with the incorporation of 30 wt% SBF. It can be seen from the SEM image that the morphology of the samples with 30 wt% fibre shows the appearance of void. This was due to the excessive SBF loading which led to matrix discontinuity. As the fibre loading was increased, the amount of matrix was reduced, leading to less adhesion of the matrix to the fibre [31]. According to Sarifuddin et al. [31], tensile strength tends to deplete because a higher amount of fibre will cause discontinuity of the matrix. Fraction of weight was regarded as one of the most important factors because the higher the binder loading, the lower dispersion of fibre inside the matrix could take place [32]. Overall, the composites have reach optimum composition at 20 wt% SBF loading, further addition at 30 wt% shows decrement in the tensile strength/modulus. Even though there are increments at 40 wt% SBF, however, the increment is not significant as shown by the error bar in Figure 13.1(a). Nevertheless, the slight increment might be attributed to the tensile behaviour of the compressed fibre which involved in the tensile stress of the materials.

Figure 13.1(c) shows the elongation at break for the TPCS/BW/SBF composites. It can be seen that the elongation at break was decrease from 0.78 to 0.27% as the fibre loading increased from 0 to 40 wt%. This finding contradicts with the improvement in the tensile strength and modulus of the composites. In general, the trend of decreasing elongation at break or also known as ductility of materials was often observed for composites experiencing increment in the tensile strength. This was attributed to increment in the rigidity of the material which reduces the ductility of the material, hence, led to lower elongation of the materials [33].

13.3.1.2 Impact testing

In this study, the izod impact test was used to determine the material's high-strain rate test with standards that exhibited the quantity of absorbed energy by a fracture to the material's surface. The material's toughness was measured by calculating the energy absorbed. Figure 13.2 shows the impact strength of TPCS/BW with various SBF wt% content. It can be observed that the incorporation of SBF up to 40 wt% has significantly increased the impact strength of TPCS/SBF by 175% compared to the matrix. However,

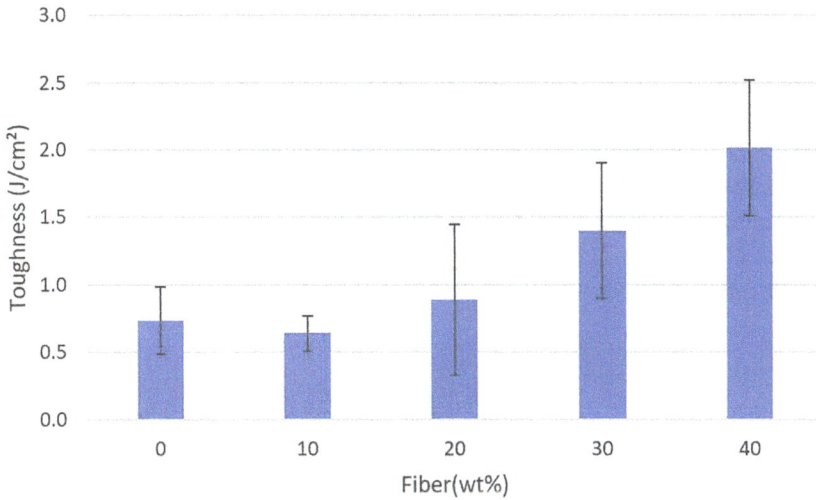

Figure 13.2: Impact strength of TPCS/BW/SBF composites.

this trend is quite different as shown by the tensile properties of the materials where the tensile strength were drop after the SBF loading achieve 30 wt%. The difference between these two results was due to the different nature of force exerted on the materials which led to different reaction of the materials. The increment in the impact strength might be attributed to the good fibre-matrix adhesion which enables efficient stress transfer between matrix and fibre, hence, improved the rigidity and ability of the TPCS to absorb the impact energy. In the previous study, it was proven that increasing fibre loading makes the composite stronger from the impact test results [34].

13.3.2 Fourier transform infrared spectroscopy (FTIR)

FTIR spectroscopy testing was used in this study to determine the structure of the TPS in the molecular range level and to investigate the latent interaction of plasticiser and starch that strained when more than two substances were mixed. The chemical interactions versus physical blends were reflected by characteristic changes of spectral bands [35]. It was found in the study carried out by Edhirej et al. [35] that most of the peaks in FTIR spectra were recognised as functional starch and fibre. The value of the O–H band was found in the range of 3100–3700 cm^{-1}. For the samples tested, it was found that the O–H band decreased from 3301 to 3291 cm^{-1}. Based on a study conducted by [35], a similar trend of the O–H band was observed that when the fibre loading was increased, the wavenumber of the O–H band decreased. According to Jumaidin et al. [5], when the O–H band decreased, the intermolecular hydrogen bonding increased. The O–H band shift in the FTIR spectra indicated its stability due to polymer compatibility between the two polymers, creating a good hydrogen bonding [36].

Sarifuddin et al. [31] stated that amylose and amylopectin existed in the starch exhibited an O–H group peak in the FTIR spectra. In Figure 13.3, the O–H group peak could relate to a similar finding, as amylose and amylopectin also existed in cassava starch. It was found in all samples that the wavenumber located at 1648 cm^{-1} was water. A parallel finding was also found in the past studies carried out by Edhirej et al. [35] and Sarifuddin et al. [31]. Sarifuddin et al. [31] reported the wavenumber location at 1646 cm^{-1}, as for study conducted by Mendes et al. [37], the wavenumber location was 1648 cm^{-1}.

13.3.3 SEM micrograph

Scanning electron microscopy (SEM) is a tool to study the microstructure of granules and surface characteristics of a material or composite. This tool provides observation of material's characteristics at nanometric maximum scale as compared to 0.2 μm scale of optical microscope [38]. Therefore, the granules, surface, dispersion and distribution of TPS matrix and fibre were further investigated using this tool [39].

Figure 13.4 shows the SEM micrograph of the TPCS/SBF surface and granules. The curves represent the samples with different fibre loadings. Figure 13.4(a) shows the pure TPCS, the only sample that was unmixed with fibre as the control experiment for comparison with other samples with fibre loadings of 10, 20 and 30%. Based on the observation made on pure TPCS, the surface was granular-like. However, it could be observed that the surface was bumpy and not smooth. Aside from that, there were small voids observed, as there was no reinforcement in the sample.

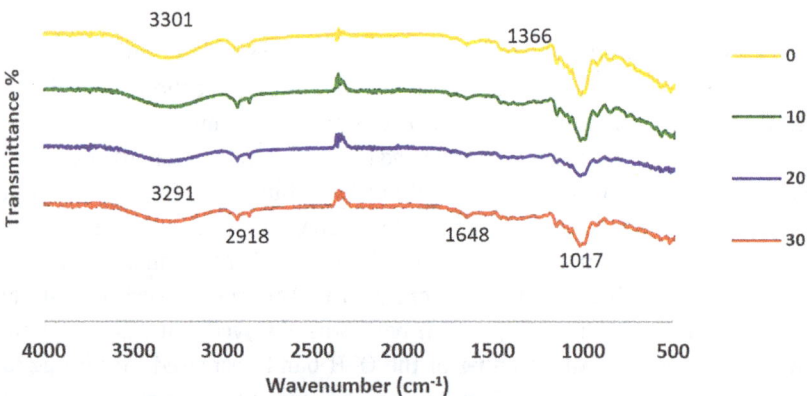

Figure 13.3: Results of FT-IR of TPCS/BW/SBF composites.

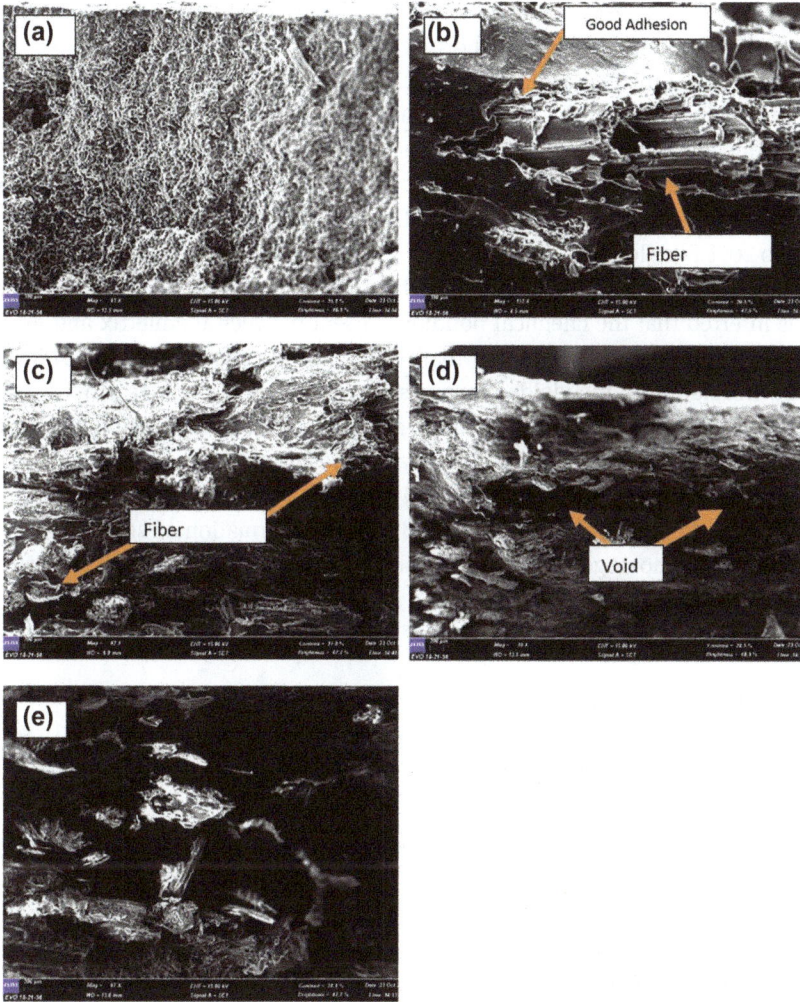

Figure 13.4: SEM image of TPCS/BW/SBF composites (a) pure TPCS/BW (b) 10% SBF (c) 20% SBF (d) 30% SBF (e) 40% SBF.

Figure 13.4(b) shows TPCS loaded with 10% SBF. It was found that the matrix and the fibre had minimal gaps in between them, proving that the matrix and fibre had good adhesion. This led to a good ratio of fibre and matrix from the observation of the specimens. In relation to the tensile strength result, the specimen with 10% fibre increased by 6.74% compared to pure TPCS. Figure 13.4(c) with fibre loading of 20% showed fibre breakage on its surface. It was found that the adhesion of the matrix and fibre was improved compared to the specimen with 10% fibre. Based on the tensile strength result, it was proven that the adhesion of 20% of fibre into the matrix significantly increased the strength of the specimen.

However, for Figure 13.4(d) and (e) with 30 and 40 wt% fibre loadings, it was observed that more voids were visible on the fracture surface of the specimen. This was due to less adhesion of fibre and matrix because the matrix was not able to interact to form bonding as a result of a high percentage of fibre loading. It was also found in the finding of tensile strength that the strength was significantly decreased when compared to 20% fibre loading.

A similar pattern of results was found in the study performed by Sarifuddin et al. [31]. It was inferred that the chemical bonding present between the matrix and fibre used was due to the hydrogen bonding that resulted in the good dispersion of fibre and matrix. In relation to the yielded tensile strength, the SEM findings proved that the fibre affinity towards the matrix was better with the increase of the fibre up until 20% loading. A higher percentage of fibre loading resulted in a decrease in tensile strength. The poor wettability of matrix and fibre as fibre loading increased was caused by non-uniform fibre distribution leading to the formation of stress concentration points. Thus, homogeneity and interfacial adhesion of matrix and fibre were decreasing as the fibre loading increased.

13.3.4 Thermal testing

13.3.4.1 Thermogravimetric analysis (TGA)

Thermogravimetric analysis is a method to study thermal degradation of TPS based material. It is essential to identify the limitation of the material in terms of process, treatment, and operation of a material [34].

Figure 13.5(a) shows the TGA curve for pure TPCS/SBF, 10–40 wt% of SBF loading where the percentage lost occurred to the sample weight and the derivative mass loss caused by the volatilisation of product that was examined as a function of temperature [34]. It was observed that based on Figure 13.5(a), the pattern shows weight decrement due to thermal degradation.

According to Sanyang et al. [40], the degradation process took place below 100 °C might be recognised as dehydration or evaporation of loosely bound H_2O and low molecular mass compound. Based on the study carried out by Ibrahim et al. [41], the degradation that occurred within the 130–250 °C temperature range could be related to the evaporation of the glycerol. Thermal degradation of plasticiser for SBF occurred at the temperature of 190–250 °C based on a study performed by Reddy & Rhim [42]. This was due to the boiling point of the glycerol at 198–200 °C, based on a previous study [32]. In the meantime, a huge degradation at above 310 °C can be related to hydrogen group elimination, depolymerisation and decomposition of the carbohydrate chains of starch [43].

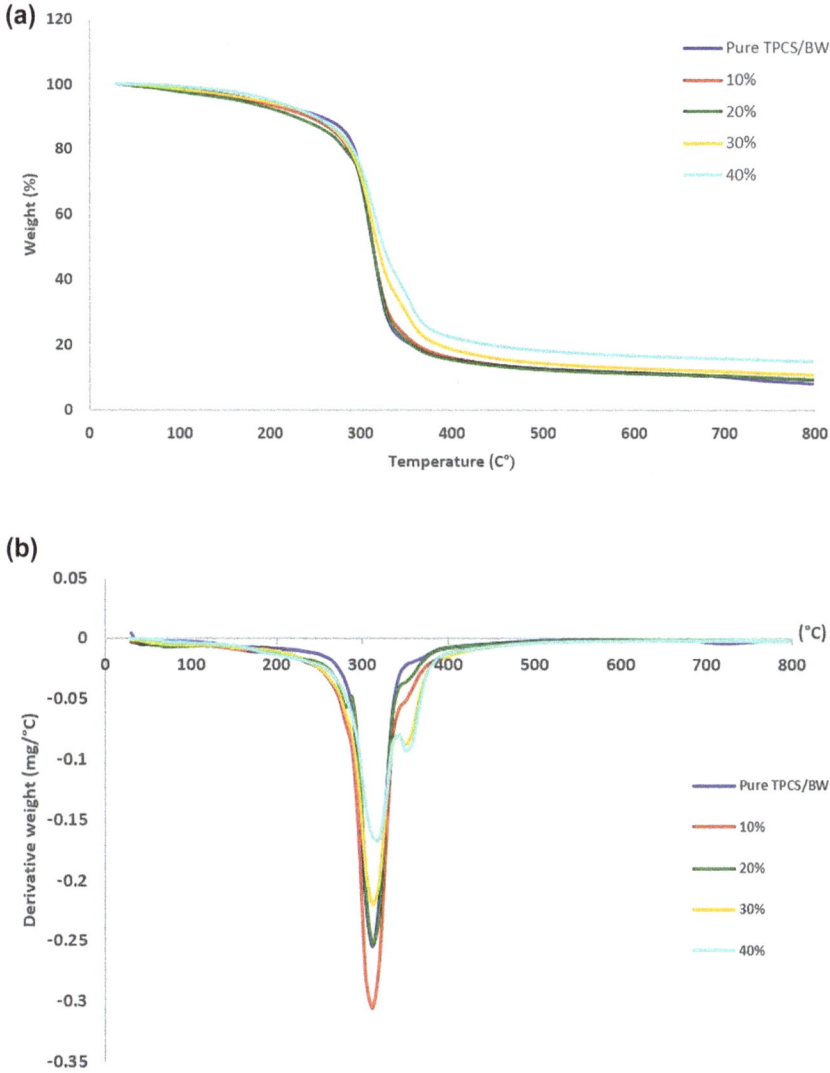

Figure 13.5: (a) TGA curve for pure TPCS/BW with 10–40 wt% of SBF. (b) DTG curve for pure TPCS/BW with 10–40 wt% of SBF.

Meanwhile, Figure 13.5(b) shows the DTG curve obtained from pure TPCS and TPCS composites comprising various SBF loadings. The purpose of DTG curve is to notice the thermal degradation transformation of the matrix and fibre characteristics [44]. Based on Figure 13.5(b), it was observed that the sample with the least weight loss was the sample with the highest fibre loading, 40 wt% and the sample with the most weight loss was the sample with 10 wt%. These peaks corresponded to the decomposition pattern of starch. As the starch content in the sample was higher, the peak formed was lower. The samples with high fibre loading had a lower starch ratio in the sample. At maximum decomposition, the weight loss decreased as the fibre loading was increased.

A similar finding could be found in a study by Lomelí-ramírez & Kestur [32] that increasing the loading of fibre into the matrix improved the composite thermal stability, decreasing the composite weight loss. Similar findings were also recorded by Teixeira et al. [44], where the thermoplastic starch foamed composites reinforced with cellulose nanofiber revealed TGA curve of a similar trend.

13.3.4.2 Differential scanning calorimetry (DSC)

DSC is an effective tool to analyse and investigate the thermo-analytical of material by scanning the difference in the amount of heat required to increase sample temperature, and reference is measured as temperature function [45]. When material reaches the glassy and changes to a rubbery state at a certain heating rate, it is called glass transition state, T_g. This study is important to determine the mechanical, thermal and dimensional stability of material when subjected to high temperatures [46].

Table 13.1 shows the T_g and T_m values for TPCS SBF with various SBF loadings. For a sample of pure TPCS, the T_g reached 49.87 °C, which was lower compared to pure TPS native corn starch (150 °C) from a study done by [45]. For 10 wt% TPCS SBF, the T_g was 43.37 °C, slightly lower than pure TPCS. For 20 wt%, the T_g was higher than pure TPCS. As for melting temperature, it was observed that the T_m demonstrated an increasing trend. It was known that T_g has a complex phenomenon where it depends on many influences such as intermolecular interaction, molecular weight, branching, chain flexibility, and other research regarding the TPS mixture and the cross-link density of fibre [45]. On T_m, the temperature increased as the fibre loading increased due to increased crystallinity form of TPS, changing the melting point towards greater temperatures [45].

Table 13.1: Glass transition and melting temperatures of TPCS/BW/SBF.

Sample (SBF wt%)	T_g (°C)	T_m (°C)
0	49.87	160.79
10	43.37	227.66
20	66.69	229.65
30	–	229.26
40	–	230.36

13.4 Conclusions

Incorporation of SBF into modified TPCS/BW was carried out, and the thermal and mechanical properties were evaluated. It was found that the incorporation of SBF into TPCS/BW improved the thermal stability of the material in terms of higher glass transition and melting temperature. Improvement in mechanical properties was also evident with the improvement in tensile, flexural, and impact properties. The highest tensile strength and modulus was exhibit by sample with 20 wt% SBF loading and further increment of fibre led to decrease in the strength of the materials. Overall, SBF has the potential to improve the functional characteristic of TPCS/BW matrix. This novel material has the possibility as an alternative to plastics for being more environmentally friendly and sustainable, hence, suitable for the single-use application.

References

1. Geyer R, Jambeck JR, Law KL. Production, use, and fate of all plastics ever made. Sci Adv 2017;3: e1700782.
2. Ilyas RA, Sapuan SM. The preparation methods and processing of natural fibre bio-polymer composites. Curr Org Synth 2020;16:1068–70.
3. Ilyas RA, Sapuan SM. Biopolymers and biocomposites: chemistry and technology. Curr Anal Chem 2020;16:500–3.
4. Abotbina W, Sapuan SM, Sultan MTH, Alkbir MFM, Ilyas RA. Development and characterization of cornstarch-based bioplastics packaging film using a combination of different plasticizers. Polymers 2021;13:3487.
5. Jumaidin R, Sapuan SM, Jawaid M, Ishak MR, Sahari J. Characteristics of thermoplastic sugar palm starch/agar blend: thermal, tensile, and physical properties. Int J Biol Macromol 2016;89:575–81.
6. Jumaidin R, Sapuan SM, Jawaid M, Ishak MR. Effect of agar on physical properties of thermoplastic starch derived from sugar palm tree. Pertanika J Sci Technol 2017;25:1235–48.
7. Jumaidin R, Adam NW, Ilyas RA, Hussin MSF, Taha MM, Mansor MR, et al. Water transport and physical properties of sugarcane bagasse fibre reinforced thermoplastic potato starch biocomposite. J Adv Res Fluid Mech Therm Sci 2019;61:273–81.
8. Tarique J, Sapuan SM, Khalina A, Sherwani SFK, Yusuf J, Ilyas RA. Recent developments in sustainable arrowroot (Maranta arundinacea Linn) starch biopolymers, fibres, biopolymer composites and their potential industrial applications: a review. J Mater Res Technol 2021;13: 1191–219.
9. de Farias JGG, Cavalcante RC, Canabarro BR, Viana HM, Scholz S, Simão RA. Surface lignin removal on coir fibers by plasma treatment for improved adhesion in thermoplastic starch composites. Carbohydr Polym 2017;165:429–36.
10. Domene-López D, Delgado-Marín JJ, Martin-Gullon I, García-Quesada JC, Montalbán MG. Comparative study on properties of starch films obtained from potato, corn and wheat using 1-ethyl-3-methylimidazolium acetate as plasticizer. Int J Biol Macromol 2019;135:845–54.
11. García MA, López OV, Barbosa SE, García MA, Castillo LA, Olivia VL, et al. Crystalline morphology of thermoplastic starch/talc nanocomposites induced by thermal processing. Heliyon 2019;5:e01877.

12. Jumaidin R, Khiruddin MAA, Asyul Sutan Saidi Z, Salit MS, Ilyas RA. Effect of cogon grass fibre on the thermal, mechanical and biodegradation properties of thermoplastic cassava starch biocomposite. Int J Biol Macromol 2020;146:746–55.

13. Nazrin A, Sapuan SM, Zuhri MYM, Tawakkal ISMA, Ilyas RA. Flammability and physical stability of sugar palm crystalline nanocellulose reinforced thermoplastic sugar palm starch /poly (lactic acid) blend bionanocomposites. Nanotechnol Rev 2022;11:86–95.

14. Ren J, Zhang W, Lou F, Wang Y, Guo W. Characteristics of starch-based films produced using glycerol and 1-butyl-3-methylimidazolium chloride as combined plasticizers. Starch/Staerke 2017;69:1–8.

15. Tinto WF, Elufioye TO, Roach J. Waxes. In: Pharmacognosy. Elsevier; 2017:443–55 pp.

16. Serra Bonvehi J, Orantes Bermejo FJ. Detection of adulterated commercial Spanish beeswax. Food Chem 2012;132:642–8.

17. Bergel BF, da Luz LM, Santana RMC. Comparative study of the influence of chitosan as coating of thermoplastic starch foam from potato, cassava and corn starch. Prog Org Coating 2017;106:27–32.

18. Eshetu A, Ibrahim AM, Forsido SF, Kuyu CG. Effect of beeswax and chitosan treatments on quality and shelf life of selected mango (Mangifera indica L.) cultivars. Heliyon 2019;5:e01116.

19. Goslinska M, Heinrich S. Characterization of waxes as possible coating material for organic aerogels. Powder Technol 2019;357:223–31.

20. Trevisani M, Cecchini M, Siconolfi D, Mancusi R, Rosmini R. Effects of beeswax coating on the oxidative stability of long-ripened Italian salami. J Food Qual 2017;2017:1–5.

21. Diyana ZN, Jumaidin R, Selamat MZ, Suan MSM. Thermoplastic starch /beeswax blend: characterization on thermal mechanical and moisture absorption properties. Int J Biol Macromol 2021;190:224–32.

22. Hafila KZ, Jumaidin R, Ilyas RA, Selamat MZ, Asyadi F. Effect of palm wax on the mechanical, thermal, and moisture absorption properties of thermoplastic cassava starch composites. Int J Biol Macromol 2022;194:851–60.

23. Khanzadi M, Jafari SM, Mirzaei H, Chegini FK, Maghsoudlou Y, Dehnad D. Physical and mechanical properties in biodegradable films of whey protein concentrate–pullulan by application of beeswax. Carbohydr Polym 2015;118:24–9.

24. Cheng Y, Wang W, Zhang R, Zhai X, Hou H. Effect of gelatin bloom values on the physicochemical properties of starch/gelatin–beeswax composite films fabricated by extrusion blowing. Food Hydrocolloids 2021;113:106466.

25. Chiumarelli M, Hubinger MD. Stability, solubility, mechanical and barrier properties of cassava starch – carnauba wax edible coatings to preserve fresh-cut apples. Food Hydrocolloids 2012;28:59–67.

26. Kamaruddin ZH, Sapuan SM, Mohamed Yusoff MZ, Jumaidin R. Rapid detection and identification of dioscorine compounds in dioscorea hispida tuber plants by LC-ESI-MS. Bioresources 2020;15:5999–6011.

27. Guimarães JL, Wypych F, Saul CK, Ramos LP, Satyanarayana KG. Studies of the processing and characterization of corn starch and its composites with banana and sugarcane fibers from Brazil. Carbohydr Polym 2010;80:130–8.

28. Mulinari DR, Voorwald HJC, Cioffi MOH, da Silva MLCP, da Cruz TG, Saron C. Sugarcane bagasse cellulose/HDPE composites obtained by extrusion. Compos Sci Technol 2009;69:214–9.

29. Prachayawarakorn J, Ruttanabus P, Boonsom P. Effect of cotton fiber contents and lengths on properties of thermoplastic starch composites prepared from rice and waxy rice starches. J Polym Environ 2011;19:274–82.

30. Prachayawarakorn J, Sangnitidej P, Boonpasith P. Properties of thermoplastic rice starch composites reinforced by cotton fiber or low-density polyethylene. Carbohydr Polym 2010;81:425–33.

31. Sarifuddin N, Ismail H, Ahmad Z. The effect of kenaf core fibre loading on properties of low density polyethylene/thermoplastic sago starch/kenaf core fiber composites. J Phys Sci 2013;24:97–115.

32. Lomelí-ramírez MG, Kestur SG. Bio-composites of cassava starch-green coconut fiber: Part II-structure and properties. Carbohydr Polym 2014;102:576–83.
33. Cheng W. Preparation and properties of lignocellulosic fiber/CaCO₃/thermoplastic starch composites. Carbohydr Polym 2019;211:204–8.
34. Jumaidin R, Sapuan SM, Jawaid M, Ishak MR, Sahari J. Thermal, mechanical, and physical properties of seaweed/sugar palm fibre reinforced thermoplastic sugar palm starch/agar hybrid composites. Int J Biol Macromol 2017;97:606–15.
35. Edhirej A, Sapuan SM, Jawaid M, Zahari NI. Cassava/sugar palm fiber reinforced cassava starch hybrid composites: physical, thermal and structural properties. Int J Biol Macromol 2017;101:75–83.
36. Prachayawarakorn J, Pomdage W. Effect of carrageenan on properties of biodegradable thermoplastic cassava starch/low-density polyethylene composites reinforced by cotton fibers. Mater Des 2014;61:264–9.
37. Mendes JF, Paschoalin RT, Carmona VB, Sena Neto AR, Marques ACP, Marconcini JM, et al. Biodegradable polymer blends based on corn starch and thermoplastic chitosan processed by extrusion. Carbohydr Polym 2016;137:452–8.
38. Horovitz O, Cioica N, Jumate N, Pojar-Fanesan M, Balea A, Liteanu V, et al. SEM characterization of starch granules. Studia Ubb Chemia 2015;1:211–19.
39. Ajiya DA, Jikan SS, Talip BHA, Badarulzaman NA, Derawi D, Yahaya S. The influence of glycerol on mechanical, thermal and morphological properties of thermoplastic tapioca starch film. J Sci Technol 2017;9:24–9.
40. Sanyang ML, Sapuan SM, Jawaid M, Ishak MR, Sahari J. Recent developments in sugar palm (Arenga pinnata) based biocomposites and their potential industrial applications: a review. Renew Sustain Energy Rev 2016;54:533–49.
41. Ibrahim H, Farag M, Megahed H, Mehanny S. Characteristics of starch-based biodegradable composites reinforced with date palm and flax fibers. Carbohydr Polym 2014;101:11–9.
42. Reddy JP, Rhim JW. Characterization of bionanocomposite films prepared with agar and paper-mulberry pulp nanocellulose. Carbohydr Polym 2014;110:480–8.
43. Nascimento TA, Calado V, Carvalho CWP. Development and characterization of flexible film based on starch and passion fruit mesocarp flour with nanoparticles. Food Res Int 2012;49:588–95.
44. Teixeira EDM, Pasquini D, Curvelo AAS, Corradini E, Belgacem MN, Dufresne A, et al. PBAT/thermoplastic starch blends: effect of compatibilizers on the rheological, mechanical and morphological properties. Carbohydr Polym 2018;101:24–9.
45. Ghanbari A, Tabarsa T, Ashori A, Shakeri A, Mashkour M. Thermoplastic starch foamed composites reinforced with cellulose nano fibers: thermal and mechanical properties. Carbohydr Polym 2018;197:305–11.
46. Jumaidin R, Sapuan SM, Jawaid M, Ishak MR, Sahari J. Effect of seaweed on mechanical, thermal, and biodegradation properties of thermoplastic sugar palm starch/agar composites. Int J Biol Macromol 2017;99:265–73.

Nor Amira Izzati Ayob, Nurul Fazita Mohammad Rawi*,
Azniwati Abd Aziz, Baharin Azahari and
Mohamad Haafiz Mohamad Kassim

14 The properties of 3D printed poly (lactic acid) (PLA)/poly (butylene-adipate-terephthalate) (PBAT) blend and oil palm empty fruit bunch (EFB) reinforced PLA/PBAT composites used in fused deposition modelling (FDM) 3D printing

Abstract: Poly (lactic acid) (PLA) is amongst the preferable materials used in 3D printing (3DP), especially in fused deposition modelling (FDM) technique because of its unique properties such as good appearance, higher transparency, less toxicity, and low thermal expansion that help reduce the internal stresses caused during cooling. However, PLA is brittle and has low toughness and thermal resistance that affect its printability and restricts its industrial applications. Therefore, PLA was blended with various content of polybutylene adipate terephthalate (PBAT) at 20, 50 and 80 wt% via twin-screw extruder to improve the ductility and impact properties of PLA. The addition of PBAT increased the elongation at break of PLA with a linear increasing amount of PBAT. However, 20 wt% PBAT was selected as the most promising and balance properties of PLA/PBAT because although it has a slight increment in its elongation at break but it exhibits higher impact strength than that of PLA. The tensile strength and tensile modulus of sample with 20 wt% PBAT is greater than 50 and 80 wt% PBAT. Then, PLA/PBAT (80/20, 50/50 and 20/80) and PLA/PBAT/EFB (80/20/10) were printed using FDM machine and were characterized in tensile, impact and morphological properties. The tensile result indicated that the addition of PBAT decreased the tensile strength and tensile modulus of PLA/PBAT-3DP. The terephthalate group in the PBAT affects the mechanical properties of PLA/PBAT-3DP, resulting in high elongation at break but relatively low tensile strength. Besides, the tensile strength and tensile modulus of PLA/PBAT/EFB-3DP decreased and lower than PLA-3DP and PLA/PBAT-3DP. The impact test resulted in high impact strength in PLA/PBAT-3DP, where 50/50-3DP and 20/80-3DP are unbreakable. The impact strength of PLA/PBAT/EFB-3DP is also increased from PLA-3DP but lower than PLA/PBAT-3DP. The scanning electron microscopy (SEM) results revealed that the

*Corresponding author: Nurul Fazita Mohammad Rawi, School of Industrial Technology, Universiti Sains Malaysia, Gelugor, 11800 Penang, Malaysia, E-mail: fazita@usm.my
Nor Amira Izzati Ayob, Azniwati Abd Aziz, Baharin Azahari and Mohamad Haafiz Mohamad Kassim, School of Industrial Technology, Universiti Sains Malaysia, Gelugor, 11800 Penang, Malaysia

As per De Gruyter's policy this article has previously been published in the journal Physical Sciences Reviews. Please cite as:
N. A. I. Ayob, N. F. Mohammad Rawi, A. Abd Aziz, B. Azahari and M. H. Mohamad Kassim "The properties of 3D printed poly (lactic acid) (PLA)/poly (butylene-adipate-terephthalate) (PBAT) blend and oil palm empty fruit bunch (EFB) reinforced PLA/PBAT composites used in fused deposition modelling (FDM) 3D printing" *Physical Sciences Reviews* [Online] 2022. DOI: 10.1515/psr-2022-0048 | https://doi.org/10.1515/9783110769227-014

filament layering on 80/20-3DP was oriented than 50/50-3DP and 20/80-3DP. Besides, the SEM images of PLA/PBAT/EFB-3DP revealed the inhomogeneous and large agglomeration of EFB particle in PLA/PBAT matrix. Therefore, in the future, the polymer blend and polymer blend composite from PLA, PBAT and EFB can be developed where the properties will be based on the study and this study also shed light on the importance of extrusion settings during the manufacture of filament for 3D printing.

Keywords: oil palm empty fruit bunch; poly (butylene-adipate-terephthalate); poly (lactic acid); polymer blend composite and 3D printing.

14.1 Introduction

Recently, the global market is moving towards the fourth industrial revolution (IR 4.0) and 3D printing is one of them. 3D printing is good at reducing product development times, costs and most importantly it can fabricate designs and features unmatched by other methods of manufacturing. Generally, a material extrusion-based 3D printing technology is widely used, and the costs are very low [1]. FDM is an example of a material extrusion technique that uses thermoplastic in the filament form. It is a process of joining materials directly from 3D computer model data by extruding the thermoplastic filament with temperature control units and depositing the semi-molten filament onto the bed platform in a layer-by-layer manner [2]. The types of filaments used are limited to thermoplastic polymers with a suitable melt viscosity, have a low melting point and are limited to amorphous or low crystallinity polymers such as poly (lactic acid) (PLA) [2, 3].

PLA is a commonly used material in 3D printing applications as it has a low degree of polymer shrinkage [3] and warping effect [4] that is crucial to the accuracy of components produced during 3D printing [5]. Besides, 3D printed PLA with a low glass transition and melting temperature are widely used in 3D printing technology due to its low cost and weight, and have flexible process-ability [6]. In addition, PLA is one of the most promising thermoplastic materials with unique properties such as good appearance, higher transparency, glossy feel, good mechanical strength and less toxicity and virtuous barrier properties (barrier or permeability performance against gases transfer, water vapour and aroma molecules) [7, 8]. Unfortunately, PLA is relatively brittle, has a low impact strength and thermal stability thus limit its application in certain areas that require high-stress levels in plastic deformation [9]. Thus, extensive research has been done on the development of PLA.

Recently numerous studies have been focused on polymer blending PLA with any other biodegradable polymer such as poly (butylene adipate-co-terephthalate) (PBAT) to increase PLA's ductility and thermal stability without compromising their degradability and thus expand its application. This is possible as PBAT is a copolymer made up of adipic acid, butanediol and terephthalic acid. The terephthalic acid gives rise to high thermal stability and mechanical properties while adipic acid and butanediol impart flexibility and biodegradability [10]. PBAT is a commercially available aliphatic-co-aromatic-co-polyester [11] and it is a fully biodegradable copolymer [12, 13] which offers a sustainable disposal alternative [14]. PBAT has high stability and mechanical properties which results in high elongation properties (700%) but relatively low tensile strength (32 MPa), [13]. Thus, those high elongation properties make it a good choice to help toughen the brittle PLA.

Several researchers have reported the success-ability of polymer blending PLA and PBAT without compromising their degradability and the polymer blending can be used with natural fibres such as Empty Fruit Bunch (EFB) particle to help reduce the cost. EFB is one of the most valuable plants in Malaysia [15] and one of the abundant residues that is widely utilised in polymer composites [16]. It is an interesting cellulosic biomass to be utilized as a reinforcing material in the polymer composite into a beneficial and higher commercial value-added product because it has low density, real strength, relatively high toughness and most importantly it is biodegradable [17]. Bakri et al. [18] reported that the tensile strength of EFB fibres is comparable to other natural fibres like kenaf, sugar palm, coir and sisal. They also found that the tensile modulus of EFB fibres is higher than kenaf, sugar palm, coir and sisal. Recently, in the research conducted by Sekar et al. [19], they have successfully produced FDM filament from EFB reinforced PLA composites. Their motivation to produce this type of filament is because they are cost-effective and exhibit lower environmental impacts. However, limited research can be found on the properties of FDM filament from PLA blend and PLA/PBAT reinforced with EFB fibres in the literature.

Therefore, the aim of this research is to explore the feasibility of producing filament materials from PLA/PBAT blend and EFB reinforced PLA/PBAT composites for FDM 3D printing. The properties of 3D printed PLA/PBAT blend and EFB reinforced PLA/PBAT blend composites were also investigated.

14.2 Materials and methodology

14.2.1 Materials

The polymers chosen were PLA and PBAT. PLA, Ingeo™ Biopolymer 2003D manufactured by Nature Works, USA was obtained from Innovative Pultrusion Sdn. Bhd, Negeri Sembilan, Malaysia while PBAT (Ecoflex F Blend C1200) was obtained from BASF, Malaysia. Both PLA and PBAT were in a pellet form. The mechanical and physical properties of PLA and PBAT were shown in Table 14.2.1.

The natural fibre used was oil palm empty fruit bunch (EFB) that was supplied by United Oil Palm Industries Sdn. Bhd, Nibong Tebal, Penang. The EFB was washed with distilled water to remove any impurities and was dried for a few days until it was fully dried. Then, the EFB was grounded for a few times using grinder. After that, the EFB was sieved using a sieve to get a particles size in the range of 32–75 µm.

Table 14.2.1: Physical and mechanical properties of PLA and PBAT (according to the material datasheet).

Materials	PLA	PBAT
Trade name	PLA 2003D	Ecoflex C1200
Density (g/cm^3)	1.24	1.25–1.27
Tensile strength (MPa)	53	35/44
Tensile modulus (MPa)	3500	95
Elongation at break (%)	6	560/710
Notched izod impact (J/m)	16	–
Clarity	Transparent	Transparent to translucent

14.2.2 Preparation of PLA/PBAT blend filament

Prior to extrusion process, the auto feed hopper of the extruder was cleaned with compressed air to remove any traces of contaminants. Then, polyethylene (PE) was used as a purging material to clean the barrel of the extruder, removing and cleaning any remaining purge from the die. Prior to compounding, PLA and PBAT pellets were weighed and dried for 24 h at 70 °C in a vacuum oven to avoid possible moisture-degradation reaction. Both materials were then manually premixed in a plastic zip-lock bag and before being fed into co-rotating twin screw extruder for melt blending. The extrusion temperature was controlled in 10 zones along the extruder barrel, which a temperature profile ranging from 150 to 180 °C (150/155/160/165/170/175/180/180/180/180). The screw speed was kept constant at 60 rpm. After exiting the die, the compounding PLA/PBAT blend filament was cooled in water and dried in a room temperature environment before being use in 3D printing. The composition of PLA/PBAT blend and PLA/PBAT composite were tabulated in Table 14.2.2.

During production, the diameter of the filament must be controlled to a consistent diameter of ~1.75 mm (±0.10 mm). This is because filament with a diameter smaller or greater than this diameter will cause printability issues later during the FDM 3D printing process. Therefore, two important steps have been implemented to achieve a consistent extrudate filament diameter in the range of ~1.75 mm (±0.10 mm) for the smooth and easy 3D printing process. First, the die nozzle of the twin-screw extruder was modified (Figure 14.2.1) to avoid entanglement of the double filament produced from the extrusion, which would make it difficult to separate it later for use in 3D printing. Second, an electric spooling machine (Figure 14.2.2) was custom-built to avoid producing an irregular

Table 14.2.2: Composition of PLA/PBAT blend and PLA/PBAT/EFB composite.

Sample	PLA (wt%)	PBAT (wt%)	EFB (wt%)
PLA	100	–	–
80/20	80	20	–
50/50	50	50	–
20/80	20	80	–
80/20/10	80	20	10

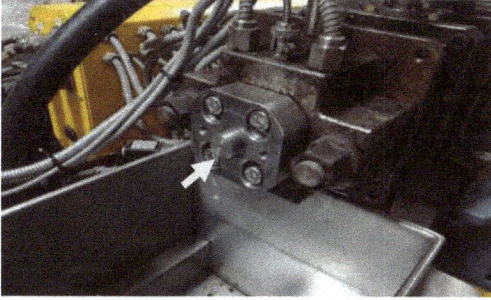

Figure 14.2.1: Single die nozzle of extruder.

Figure 14.2.2: Electric spooling machine.

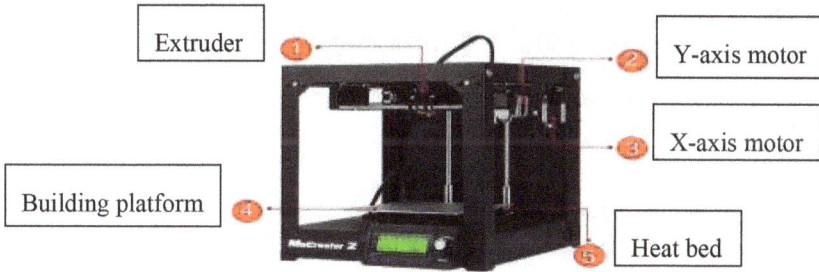

Figure 14.2.3: FDM 3D printing machine (MeCreator2).

filament diameter (Figure 14.2.3) that was unsuitable for 3D printing, and the filament size can be controlled by controlling the speed of the electric spooling machine. For example, when the first layer was extruded onto the printing bed using the irregular filament diameter, the cohesion between the strands was visually inspected. If the cohesion between the strands is insufficient, the samples become porous during printing as the filament strands are laid perpendicular to one another layer by layer. However, if the diameter was larger than the average recommended, the filament was more likely to become stuck inside the throat tube of the 3D printing machine or even before entering it. Thus, a micrometre was used to measure the diameter of the sample filament before it was extruded into the 3D printing machine.

14.2.3 Preparation of 3D printed PLA/PBAT blend and PLA/PBAT/EFB composite

The extrudate PLA/PBAT blend filament produced using the method described in Section 14.2.2 was extruded through the 1.0 mm die of the FDM 3D printing machine using Model MeCreator2 (Figure 14.2.3). The Repetier-Host is a host software that is used to add the STL files onto the stimulated print bed and then slice them all together with built-in Slic3r slicer. The printing process began once the nozzle and the printing bed were pre-heated, and the file or associated nozzle paths were delivered to the machine by the host software. The tensile sample was printed in a dog-bone shape (according to ASTM D638 Type 1), as were the flexural (according to ASTM D790) and impact samples (accordance to ASTM D256). Following that, each specimen was measured to ensure that it was within specification for each experimental group.

14.2.4 FDM 3D printing software preparation

The slicing parameters (Table 14.2.3) were manually set for a better printing effect. It should be noted that while the effects of printing parameters are not the primary goal of this study, it was not examined in depth.

To achieve good printing quality, levelling was performed in the printing process Figure 14.2.4(a) for each experimental group using a piece of paper to evaluate whether the gap between the extruder's nozzle and the four corners of the bed plates was the same. Levelling was an important part of the printed specimen's outcome because poor levelling causes the specimen to be weak. During the printing process, there are two aspects of levelling that should be avoided: when the nozzle is too far away from the printing bed and when it is too close to the printing bed. The first issue was that when the nozzle was too far away from the printing bed Figure 14.2.4(b), the filament was inconsistently sticking to the printing bed, causing certain parts to be visible to the printing bed and eventually falling off in the middle of printing. Second, placing the nozzle too close to the printing bed causes the filament to bunch up next to each other, forming mountains and, ultimately, resulting in irregular thickness of the specimen Figure 14.2.4(c).

Table 14.2.3: Parameter printing setting.

Parameter Printing Setting	Used
Layer height (mm)	0.2
Infill density (%)	100
Solid infill pattern	Rectilinear
Infill pattern	Rectilinear
Raft, brims, supports	None
Travel feed rate (mm/min)	80
Print cooling fan	Yes
Build plate adhesion type	Blue Tape
Bed temperature (°C)	65
Extruder temperature (°C)	200

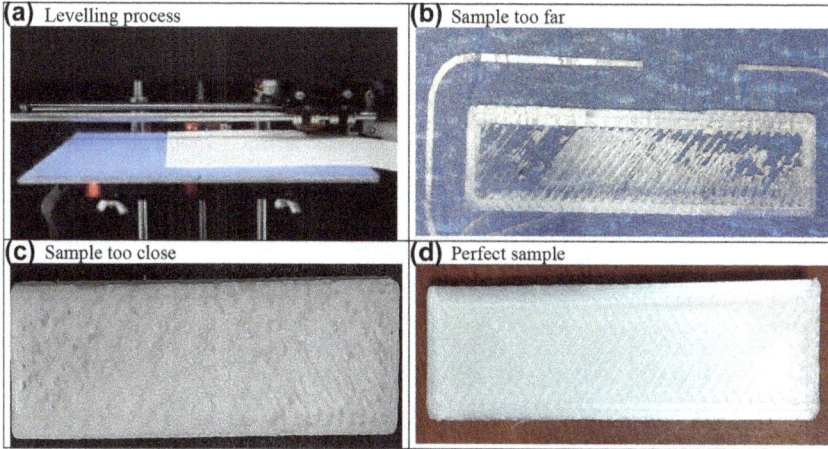

Figure 14.2.4: Levelling process and observation of different levelling parameters on FDM 3D Printed samples: (a) Levelling process, (b) sample too far, (c) sample too close and (d) perfect sample of PLA.

14.2.5 Characterizations of 3D printed PLA/PBAT blend and PLA/PBAT/EFB composite

The 3D printed PLA, PLA/PBAT blend (80/20, 50/50 and 20/80) and PLA/PBAT/EFB composite (80/20/10) were characterized with mechanical (tensile and impact) and morphological (SEM) analysis.

14.2.5.1 Mechanical properties

14.2.5.1.1 Tensile properties: Tensile test was carried out according to ASTM D638 Type I using Instron model 5528. The samples were printed according to the specification for Type I specimen in Figure 14.2.5. Universal mechanical testing machine fitted with a 50 kN load cell and operated at crosshead speed of 5 mm/min with deflections measured using an extensometer.

14.2.5.1.2 Impact properties: Impact test was carried out according to ASTM D256 using an Izod GOTECH impact tester model GT-7045-MDL. The samples were printed in accordance with the standard specifications. The impact resistance of the composite was determined in vertical notched test samples

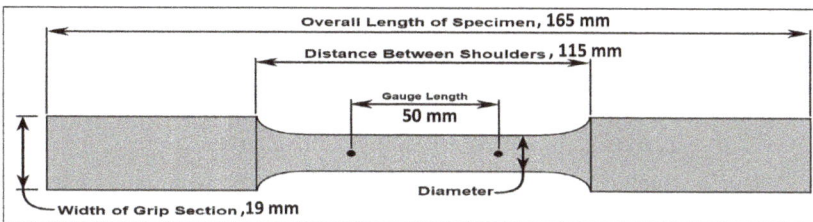

Figure 14.2.5: Measurement of specimen for ASTM D638 type 1.

according to the standard using a pendulum impact tester. The impact velocity used was 3.46 m/s with hammer weight 1 J.

14.2.5.2 Morphological analysis
14.2.5.2.1 *Scanning electron microscopy (SEM):* SEM analysis was performed to examine the fractured surfaces of impact test samples using Quanta FEG 650 model. Prior to SEM observation, the samples were mounted on aluminum and then evaporatively sputter coated with gold to make them conductive. Prior to SEM observation, the samples were mounted on aluminum and then sputter coated with gold by evaporation to make them conductive.

14.3 Result and discussion

14.3.1 Mechanical properties

14.3.1.1 Tensile properties

Figure 14.3.1 displays the tensile strength and tensile modulus, as well as the elongation at break (Figure 14.3.2), respectively of 3D printed PLA, PLA/PBAT blend (80/20, 50/50 and 20/80) and PLA/PBAT/EFB composite (80/20/10).

PLA is known as a brittle and stiff material [20] due to its molecular structure with a relatively rigid backbone and small methyl side groups [21]. Thus, PLA has high tensile strength and tensile modulus but low elongation at break when compared to PBAT. Meanwhile, as previously mentioned in Section 14.2.1 (Table 14.2.1), PBAT has high elongation at break (560–710%) but a relatively low tensile strength (35 MPa) and tensile

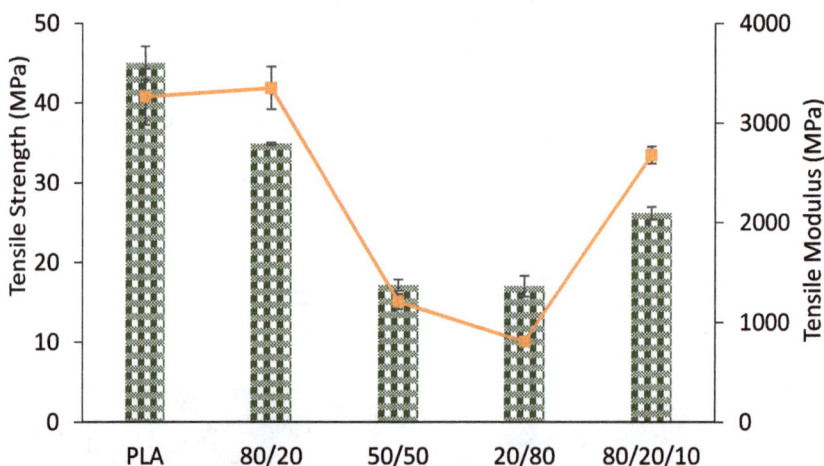

Figure 14.3.1: Tensile strength and tensile modulus of 3D printed PLA, PLA/PBAT blend (80/20, 50/50 and 20/80) and PLA/PBAT/EFB composite (80/20/10).

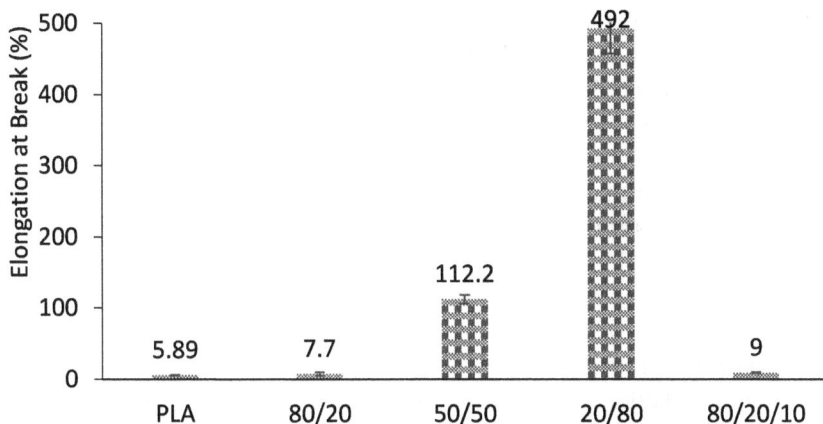

Figure 14.3.2: Elongation at break of 3D printed PLA, PLA/PBAT blend (80/20, 50/50 and 20/80) and PLA/PBAT/EFB composite (80/20/10).

modulus (95 MPa). The PLA-3DP, on the other hand, has a lower tensile strength (45 MPa) and tensile modulus (3262.5 MPa) but a slightly higher elongation at break (5.89%) than PLA as mentioned in Section 14.2.1 (Table 14.2.1) due to the microstructural difference. Besides, the decrement in tensile strength of PLA-3DP could be caused by the sudden cooling using a fan in 3D printing, which causes internal stress in the 3D printed sample. According to Shanmugam [22], they have reported that the annealing process in 3D printing is an effective technique for improving layer bonding, which could lead to increase in mechanical strength of the 3D printed sample. The addition of 20 wt% PBAT content into PLA decreased the tensile strength of 80/20-3DP decreases from 45 to 34.90 MPa, but its tensile modulus increases slightly from 3262.50 to 3350 MPa when compared to PLA-3DP. PLA and PBAT can acts either as a dispersed component or polymer matrix (or also known as continuous phase) in the PLA/PBAT blend [17, 19]. Thus, the reduction in tensile strength and tensile modulus of 80/20 could be attributed to the presence of soft elastics phase of PBAT, which results in increasing the levels of soft and flexible material to a hard and rigid of PLA. Furthermore, PLA and PBAT are an immiscible polymer blend with lower mechanical properties than pure polymers [24]. In addition, PBAT has low tensile strength and tensile modulus than that PLA [25]. This indicated that the addition of PBAT content as a dispersed component causes the dispersed PLA component to change and acts as a matrix in the dispersed PLA. Deng et al. [23] and Teamsinsungvon et al. (2013) reported similar findings. However, increasing the PBAT content up to 50 wt% and 80 wt% into PLA dramatically reduced the tensile strength and tensile modulus of 50/50-3DP and 20/80-3DP, respectively. The dramatic reduction when PBAT content exceeds 50 wt% suggests that the upper limit of the co-continuous phase is surpassed by 50 wt% PBAT content, as reported by Deng et al. [23]. The tensile strength of 80/20-3DP was slightly better than the 50/50-3DP and 20/80-3DP. This might be due to the uniformity and continuity of consistency diameter of the filament used during 3D printing which is critical to the strength and durability of the 3D printing sample [26]. Similarly, the tensile modulus of 80/20-3DP was higher than 50/50-3DP and 20/80-3DP must be due to the stiffening effect of the slightly textured structure formed due to the angle orientation during 3D printing process as reported by Kaynak and Varsavas [27]. Furthermore, Gomez-Gras et al. [28] stated that the infill density influences the stiffness of the 3D printed sample which increases as the air gaps between the rasters decrease. While Rahim et al. [29] and Melenka et al. [30] revealed that the increasing in infill percentage of the printed sample will significantly increase both the elastic modulus and tensile strength of printed parts.

Meanwhile, the addition of 10 wt% EFB particles into 80 wt% PBAT and 20 wt% PLA reduced the tensile strength and tensile modulus of 80/20/10-3DP compared to 80/20-3DP. This could be due to the inadequate amount of EFB particles which may result in ineffective stress transfer between filler and matrix. The other reason could be due to the lower level of interaction between the PLA and PBAT polymers in the presence of EFB particles which has effect on the morphology of the blend, as found by Shahlari and Lee [31]. Aside from that, inadequate melt-mixing during processing leads to EFB particle agglomeration at the interface, which can be seen in SEM pictures in Section 14.3.2.1. Similar findings have been made by others [32]. In addition, the length of the EFB particle used is too short to be able to produce a significant increment in tensile strength of 80/20/10-3DP due to the lower area of the interfacial interaction. Besides, the length of the fibre might get shorter during the pelletizing process after the extrusion for the compression moulding. However, due to the complexity associated with the use of long and continuous fibres, short and discontinuous fibres are preferred in 3D printing [31].

Figure 14.3.2 shows the elongation at break of 3D printed PLA, PLA/PBAT blend and PLA/PBAT/EFB composite. PLA-3DP has lower elongation at break than PBAT which is around 560–710% (Section 14.2.1, Table 14.2.1). This is because PLA is brittle, whereas PBAT is ductile, thus PBAT has high elongation at break contrary to PLA. As a result, the increment in the elongation at break of PLA/PBAT blend is expected indicating a sign of brittle-to-ductile transition due to the addition of soft elastic phase of PBAT that effectively reduces the brittleness characteristics of PLA [33]. In addition, the terephthalate group in the PBAT affects the mechanical properties of PLA/PBAT blend, resulting in high elongation at break but relatively low tensile strength [31]. Figure 14.3.2 also shows an increasing trend in the elongation at break of PLA with the addition of PBAT into PLA with their elongation at break are increasing as the amount of PBAT in PLA increases from 20 wt% to 50 wt% and 80 wt%. indicating a sign of brittle-to-ductile transition. This could be owing to the addition of soft elastic phase of PBAT into PLA that contributes to the reduction of brittleness characteristics of PLA [33].

The addition of 10 wt% EFB particle into PLA and PBAT slightly increased the elongation at break of 80/20/10-3DP than 80/20-3DP suggesting that the EFB particles might provide higher surface area, interfacial interaction, wettability and stress transfer amongst the fibre and polymers [33].

14.3.1.2 Impact properties

PLA, as a brittle and rigid plastic exhibits a lower impact strength when compared to PBAT. Hence, the presence of PBAT in PLA/PBAT blend successfully reduced the brittleness characteristics of PLA which was the evident from the increasing impact strength. Figure 14.3.3 displays the impact strength of 3D printed PLA, PLA/PBAT blend and PLA/PBAT/EFB composite. The impact strength of 80/20-3DP rose marginally over the PLA-3DP. The impact strength of 80/20-3DP increases when 20 wt% of PBAT is added into PLA matrix as a dispersed component, indicating a sign of brittle-to-ductile transition similarly as reported by [12], Su et al. [21] and Farsetti et al. [20]. In addition, the increasing amount of 50 and 80 wt% of PBAT into PLA led to the ductility of 50/50-3DP and 20/80-3DP based on its unbreakable sample. Chaiwutthinan et al. [33] reported a similar result, stating that the sample is unbreakable when the PBAT content reaches 40 wt% and above. Even after changing the hammer with higher energy (5.5 J), the unbreakable sample suggests that their impact strength values are too high to be detected from Izod impact test.

The impact strength of the 80/20/10-3DP is reduced when compared with 80/20-3DP. The rigidity of the EFB particles limits the polymer chain mobility, and the inhomogeneous distribution of EFB particles resulted in agglomeration of EFB particles on the polymer blend composite, weakening the properties of 80/20/10-3DP.

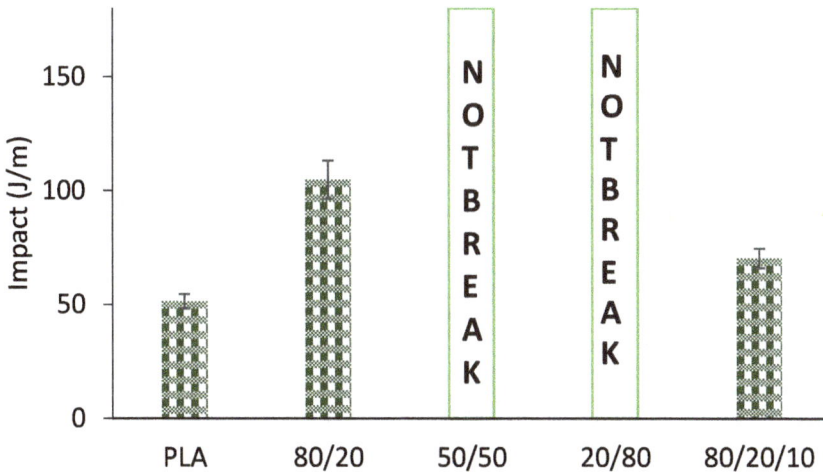

Figure 14.3.3: Impact strength of 3D printed PLA, PLA/PBAT blend (80/20, 50/50 and 20/80) and PLA/PBAT/EFB composite (80/20/10).

14.3.2 Morphological analysis

14.3.2.1 Scanning electron microscopy (SEM)

SEM analysis was performed on the impact fractured samples of 3D printed PLA, PLA/ PBAT blend (80/20, 50/50 and 20/80) and PLA/PBAT/EFB composite (80/20/10) [20]. The images were taken at the magnifications of 100× and 1000×. The 100× magnification in Figure 14.3.4 simply means for the layering of 3D printing. Few identifiers with various coloured arrows are used to make a subtle discussion.

Figure 14.3.4 show the surface morphology of PLA-3DP that showed the impact fractured images with a flat, featureless, and faceted texture surface after the break. The white arrow in Figure 14.3.4 (a) shows the joining line for PLA-3DP, where the crack line does not always run precisely along the bonding lines connecting the individual layers of the filament, indicating the formation of a very stable diffusion joint, similar as reported by Andrzejewski et al. [34]. Besides, the faceted texture fracture surface for PLA-3DP might be due to the fracture cracks splits (cleavage) that pass through the grains in most brittle crystalline materials.

The green arrow in PLA-3DP Figure 14.3.4(a), 80/20-3DP Figure 14.3.4(b), 20/80-3DP Figure 14.3.4(d) and 80/20/10-3DP Figure 14.3.4(e) shows the appearance of air voids between layers, similar images are reported by Prasong et al. [25] and Andrzejewski et al. [34]. The smaller air voids inside the printing are attributed by the layer height of 0.2 mm used during 3D printing [35]. However, the air voids are getting bigger in Figure 14.3.4(e) might be due to the insufficient cohesion between the strands suggesting that the filament diameter used on that part is irregular or smaller than the supposedly used diameter, similar as reported by Cale [36]. This is because the increasing of PBAT content in PLA/PBAT blend is believed to increase the coefficient of volume expansion of PLA/PBAT blend, allowing the melted layers to adhere more easily during 3D printing and continue to immerse layer by layer [25]. Besides, the air voids, which are responsible for the reduction of their material strength owing to unequal stress distribution as reported by Deng et al. [23]. In addition, Cale [36] also highlighted that air voids or air gaps can occur during the printing process when using 100% infill. Furthermore, layer-by layer filament placement did not prevent the formation of gaps between the adjacent rasters on the same layer, resulting in large voids or porosity in the 3D printed product, and lowering density and mechanical performance [29]. As a result, Diederichs et al. [37] suggested that the cohesion can be improved by changing the melt temperature, where by increasing the melt or nozzle temperature, resulted in smaller void sizes between layers. The extruded filament must be heated enough to partially melt the preceding layer in order to achieve maximal diffusion and cohesiveness of a material between layers. Rapid cooling, especially in the outer shell region, caused larger voids between layers.

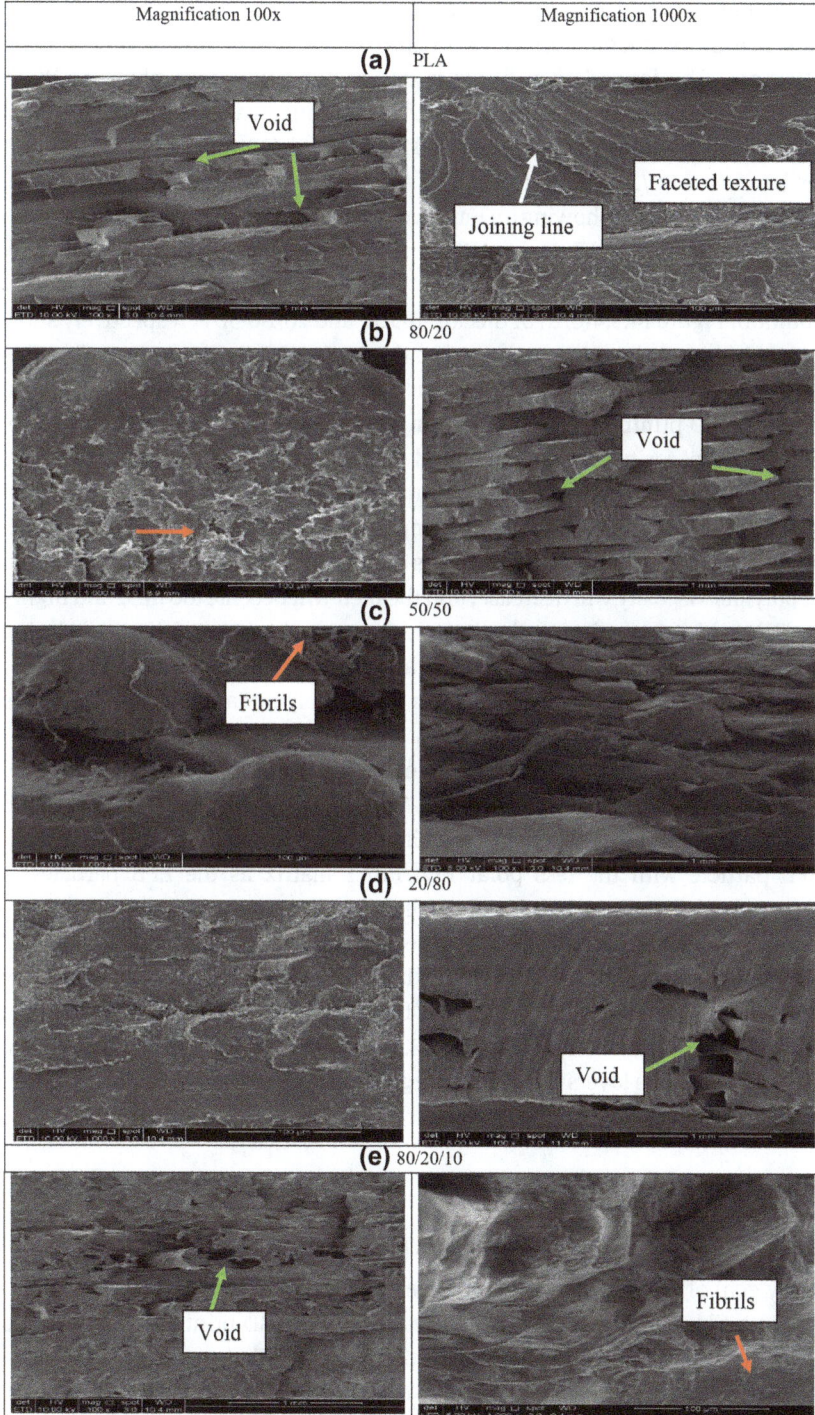

Figure 14.3.4: SEM images of 3D printed (a) PLA, PLA/PBAT blend (b) 80/20, (c) 50/50 and (d) 20/80, and PLA/PBAT/EFB composite (e) 80/20/10 with different magnification.

Next, the red arrow in Figure 14.3.4(b, c), represent the formation of fibrils which is a common feature of ductile surface and contributed to the improvement in their impact strength. The formation of fibrils is due to the continuous phase of PBAT because it has a slower yield stress than PLA thus resulting in a plastic deformation at lower stress. Similar result was revealed by Nakayama et al. [38] and Deng et al. [23]. On the other hand, the 80/20-3DP Figure 14.3.4 (b) are showing a rough wavy surface means that their ductility is showing an improvement too and it is in an agreement with the elongation at break and impact strength. Similar SEM images are reported by Chaiwutthinan et al. [33].

In addition, Figure 14.3.4(c, d) of the 50/50-3DP and 20/80-3DP, respectively show the disordered filament layering compared to the 80/20-3DP in Figure 14.3.4(b). This might be due to the flowability of the polymer blending caused by the temperature selection during 3D printing [29]. This is because the temperature selection is greatly influenced by the viscosity of the polymer and should be adjusted in conjunction with the right printing speed. The temperature set might be too high for the 50/50-3DP and 20/80-3DP into PLA because the melting temperature of the PBAT is around 110 °C while the temperature set for the 3D printing is around 200 °C. If the temperature is too high, the polymer viscosity will decrease, and the melt will become too fluid and highly flowable. This will cause a lot of plastic to leak out of the hot end (nozzle) during printing, affecting both dimensional accuracy and surface finishing. On the contrary, if the temperature is too low, the resulting layer will simply not adhere to the previous layer, and the surface of the thread could be a little rough.

Lastly, the yellow arrow in Figure 14.3.4(e) shows the inhomogeneous and large agglomeration of EFB particle dispersed in PLA/PBAT matrix for PLA/PBAT/EFB composite (80/20/10-3DP). This could be owing to the incompatibility of the hydrophilic EFB particle with the less polar PLA/PBAT matrix as the EFB particles is dispersed poorly visible. This slightly affects their tensile properties and causing huge reduction in their impact strength. Besides, the aggregation of EFB particles in the 80/20/10-3DP clearly shows that there is a rough layering orientation may be induced by 3D printing process, which leads to numerous formations of voids (green arrow) at the matrix-filler interaction as shown in Figure 14.3.4(e). With the addition of EFB particles, the air voids between the layers are getting bigger which might be due to the reduction of coefficient of volume expansion [25].

14.4 Conclusions

The primary goal of this research is to produce an EFB particles reinforced PLA/PBAT composite in a filament form that can be used in an FDM 3D printing machine. The decreased in tensile strength and tensile modulus when PBAT was added to PLA could be attributed to the lower tensile properties of PBAT (35 MPa) than PLA (53 MPa). Besides, PLA and PBAT are an immiscible polymer blend that is said to be having a weaker mechanical property than pure polymers. However, the elongation at break and impact strength of PLA/PBAT were decreased, indicating a sign of brittle to ductile transition due to the soft elastic phase of PBAT that reduced the brittleness characteristics of PLA and increased the toughness of PLA/PBAT blend. The tensile strength of 80/20-3DP, 50/50-3DP and 20/80-3DP are 34.5 MPa, 17.18 and 17.06 MPa, respectively. While, the tensile modulus of 80/20-3DP, 50/50-3DP and 20/80-3DP are 3.35 GPa, 1.21 and 0.81 GPa, correspondingly. The tensile strength and tensile modulus of 80/20-3DP was higher than 50/50-3DP and 20/80-3DP could be attributed to the PLA has high tensile strength and tensile modulus when compared to PBAT. The other reason might be due to the uniformity and continuity of consistency diameter of the filament used for 80/20-3DP, which is critical to the strength and durability of the 3D printed samples. The decreased impact strength of PLA/PBAT/EFB composite (70.30 J/m) compared to PLA/PBAT blend (104.73 J/m) might be due to the rigidity of the EFB particles, that restricted the mobility of the polymer chains, and the inhomogeneous distribution of EFB particles, which resulted in agglomeration of EFB particles on the polymer blend composite, weakening the properties of 80/20/10-3DP because no linkage existed in the interface, as proven in the SEM images. Besides, this might be due to consequence of depositing thin semi-melted fibres side by side and layer by layer in forming the 3D printed samples, resulting in the formation of porous structures with air gaps between deposited fibres that weaken their properties.

Acknowledgment: The authors are gratefully acknowledged the financial support from the Malaysian Ministry of Higher Education under Fundamental Research Grant Scheme (FRGS) (FRGS/1/2021/STG05/USM/02/5) and the Universiti Sains Malaysia (USM) under Research University Grant [1001.PTEKIND.8011098].

References

1. Melnikova R, Ehrmann A, Finsterbusch K. 3D printing of textile-based structures by fused deposition modelling (FDM) with different polymer materials. IOP Conf Ser Mater Sci Eng 2014;62. https://doi.org/10.1088/1757-899x/62/1/012018.
2. Mohan N, Senthil P, Vinodh S, Jayanth N. A review on composite materials and process parameters optimisation for the fused deposition modelling process. Virtual Phys Prototyp 2017;12:47–59.
3. Stoof D, Pickering K. Sustainable composite fused deposition modelling filament using recycled pre-consumer polypropylene. Compos B Eng 2018;135:110–8.
4. Cardoso PHM, Coutinho RRTP, Drummond FR, da Conceição MdoN, Thiré RMdaSM. Evaluation of printing parameters on porosity and mechanical properties of 3D printed PLA/PBAT blend parts. Macromol Symp 2020;394:1–12.
5. Liu Z, Wang Y, Wu B, Cui C, Guo Y, Yan C. A critical review of fused deposition modeling 3D printing technology in manufacturing polylactic acid parts. Int J Adv Manuf Technol 2019;102:2877–89.
6. Iwata T. Biodegradable and bio-based polymers: future prospects of eco-friendly plastics. Angew Chem Int Ed 2015;54:3210–5.

7. Farah S, Anderson DG, Langer R. Physical and mechanical properties of PLA, and their functions in widespread applications — a comprehensive review. Adv Drug Deliv Rev 2016;107:367–92.
8. Mohan S, Oluwafemi OS, Kalarikkal N, Thomas S, Songca SP. Biopolymers: application in nanoscience and nanotechnology. In: Recent advanced in biopolymers. London, UK: IntechOpen; 2016:47–72 pp.
9. Bates-Green K, Howei T. Materials for 3D printing by fused deposition. Edmonds Community College: Advanced Technology; 2017.
10. Singamneni S, Smith D, LeGuen MJ, Truong D. Extrusion 3D printing of polybutyrate-adipate-terephthalate-polymer composites in the pellet form. Polymers 2018;10. https://doi.org/10.3390/polym10080922.
11. Mohapatra AK, Mohanty S, Nayak SK. Study of thermo-mechanical and morphological behaviour of biodegradable PLA/PBAT/layered silicate blend nanocomposites. J Polym Environ 2014;22: 398–408.
12. Jiang L, Wolcott PM, Zhang J. Study of biodegradable polylactide/poly(butylene adipate-co-terephthalate) blends. Biomacromolecules 2005;7:199–207.
13. Jian J, Xiangbin Z, Xianbo H. An overview on synthesis, properties and applications of poly(butylene-adipate-co-terephthalate)–PBAT. Adv. Ind. Eng. Polym. Res. 2020;3:19–26.
14. Correa-pacheco ZN, Daniel J, Ortega-gudiño P, Antonio M, Barajas-cervantes A. PLA/PBAT and cinnamon essential oil polymer fibers and life-cycle assessment from hydrolytic degradation. Polymers 2020;12:3–32.
15. Uppal N, Pappu A, Gowri VKS, Thakur VK. Cellulosic fibres-based epoxy composites: from bioresources to a circular economy. Ind Crop Prod 2022;182:114895.
16. Das O, Babu K, Shanmugam V, Sykam K, Tebyetekerwa M, Neisiany RE, et al. Natural and industrial wastes for sustainable and renewable polymer composites. Renew Sustain Energy Rev 2022;158: 1–22.
17. Faizi MK, Shahriman AB, Majid MSA, Shamsul BMT, Ng YG, Basah SN, et al. An overview of the oil palm empty fruit bunch (OPEFB) potential as reinforcing fibre in polymer composite for energy absorption applications. MATEC Web Conf 2016;90. https://doi.org/10.1051/matecconf/20179001064.
18. Bakri B, Naharuddin M, Medi A, Padang L. A review of oil palm fruit fiber reinforced composites. IOP Conf Ser Mater Sci Eng 2022;1212:012050.
19. Sekar V, Zarrouq M. Development and characterization of oil palm empty fruit bunch fibre reinforced polylactic acid filaments for fused deposition modeling. J Mech Eng 2021;18:89–107.
20. Farsetti S, Cioni B, Lazzeri A. Physico-mechanical properties of biodegradable rubber toughened polymers. Macromol Symp 2011;301:82–9.
21. Su S, Duhme M, Kopitzky R. Uncompatibilized pbat/pla blends: manufacturability, miscibility and properties. Materials 2020;13:1–17.
22. Shanmugam V, Rajendran DJJ, Babu K, Rajendran S, Veerasimman A, Marimuthu U, et al. The mechanical testing and performance analysis of polymer-fibre composites prepared through the additive manufacturing. Polym Test 2021;93:106925.
23. Deng Y, Yu C, Wongwiwattana P, Thomas NL. Optimising ductility of poly(lactic acid)/poly(butylene adipate-co-terephthalate) blends through co-continuous phase morphology. J Polym Environ 2018;26:3802–16.
24. Ibrahim BA, Kadum KM. Influence of polymer blending on mechanical and thermal properties. Mod Appl Sci 2010;4:157–61.
25. Prasong W, Muanchan P, Ishigami A, Thumsorn S, Kurose T, Ito H. Properties of 3D printable poly(lactic acid)/poly(butylene adipate-co-terephthalate) blends and nano talc composites. J Nanomater 2020;16. https://doi.org/10.1155/2020/8040517.

26. Matos BDM, Rocha V, da Silva EJ, Moro FH, Bottene AC, Ribeiro CA, et al. Evaluation of commercially available polylactic acid (PLA) filaments for 3D printing applications. J Therm Anal Calorim 2019; 137:555–62.
27. Kaynak C, Varsavas SD. Performance comparison of the 3D-printed and injection-molded PLA and its elastomer blend and fiber composites. J Thermoplast Compos Mater 2019;32:501–20.
28. Gomez-Gras G, Jerez-Mesa R, Travieso-Rodriguez JA, Lluma-Fuentes J. Fatigue performance of fused filament fabrication PLA specimens. Mater Des 2018;140:278–85.
29. Rahim TNAT, Abdullah AM, Akil H. Recent developments in fused deposition modeling-based 3D printing of polymers and their composites. Polym Rev 2019;59:589–624.
30. Melenka GW, Schofield JS, Dawson MR, Carey JP. Evaluation of dimensional accuracy and material properties of the makerbot 3D desktop printer. Rapid Prototyp J 2015;21:618–27.
31. Shahlari M, Lee S. Mechanical and morphological properties of Poly(butylene adipate-co-terephthalate) and poly(lacticacid) blended with organically modified silicate layers. Polym Eng Sci 2012;52:420–1428.
32. Akindoyo JO, Beg MDH, Ghazali SB, Islam MR, Mamun AA. Preparation and characterization of poly(lactic acid)-based composites reinforced with poly dimethyl siloxane/ultrasound-treated oil palm empty fruit bunch. Polym Plast Technol Eng 2015;54:1321–33.
33. Chaiwutthinan P, Chuayjuljit S, Srasomsub S, Boonmahitthisud A. Composites of poly(lactic acid)/ poly(butylene adipate-co-terephthalate) blend with wood fiber and wollastonite: physical properties, morphology, and biodegradability. J Appl Polym Sci 2019;136:8–10.
34. Andrzejewski J, Cheng J, Anstey A, Mohanty KA, Misra M. Development of toughened blends of poly(lactic acid) and poly(butylene adipate-co-terephthalate) for 3D printing applications: compatibilization methods and material performance evaluation. ACS Sustainable Chem Eng 2020;8:6576–89.
35. Wang L, Gramlich WM, Gardner DJ. Improving the impact strength of Poly(lactic acid) (PLA) in fused layer modeling (FLM). Polymer (Guildf) 2017;114:242–8.
36. Rauch CB. Effect of Infill Angle on Tensile Strength of Solid 3D-printed Carbon Fiber Reinforced Acrylonitrile Butadiene Styrene (ABS). Doctoral dissertation, Ball State University 2018.
37. Diederichs EV, Picard MC, Chang BP, Misra M, Mielewski DF, Mohanty AK. Strategy to improve printability of renewable resource-based engineering plastic tailored for FDM applications. ACS Omega 2019;4:20297–307.
38. Nakayama D, Wu F, Mohanty AK, Hirai S, Misra M. Biodegradable composites developed from PBAT/PLA binary blends and silk powder: compatibilization and performance evaluation. ACS Omega 2018;3:12412–21.

Asniza M, Alia Syahirah Y, Mohd Fahmi A, Lee SH and Anwar UMK*

15 Properties of plybamboo manufactured from two Malaysian bamboo species—

Abstract: The aim of the presented investigation was to explore the physical and mechanical properties of plybamboo manufactured from two Malaysian commercial bamboo species – *Gigantochloa levis* (beting) and *Gigantochloa scortechinii* (semantan). Three-layer-plybamboo with dimensions of $12 \times 350 \times 350$ mm was fabricated using melamine urea formaldehyde (MUF) as a binder. The moisture content, thickness swelling, water absorption, linear expansion parallel and perpendicular to the grain were done for determination of the physical properties. Whereas, mechanical properties were evaluated via bonding shear strength, modulus of rupture (MOR) and modulus of elasticity (MOE) and compression strength parallel to the grain. Results indicated that semantan plybamboo exhibited excellent mechanical properties along with better bond shear strength than beting plybamboo. MOR and MOE value for semantan and beting plybamboo were 130.86 N/mm^2, 19661.30 N/mm^2, 115.92 N/mm^2 and 25331.5 N/mm^2, respectively. Moreover, the findings proved that the properties of both plybamboo satisfy the minimum standard requirement stipulated in the Malaysian Standard.

Keywords: bonding; *Gigantochloa levis*; *Gigantochloa scortechinii*; laminated bamboo; mechanical properties; physical.

15.1 Introduction

Bamboo is a fast-growing plant with high yield renewable resource. In contrast to traditional timbers which requires decades between planting and harvesting, bamboo can be harvested in 3–4 years from the time of planting [1, 2]. Researches proved that bamboo is stronger than timber, and the ratio of strength-to-weight is greater than common timber, cast iron, structural steel and aluminium alloy [3, 4]. However, the use of round bamboo is limited due to its handling process is influenced by the structure of the bamboo such as diameter, size and thickness. One way to solve the limitation as

*Corresponding author: Anwar UMK, Forest Products Division, Forest Research Institute Malaysia (FRIM), 52109 Kepong, Malaysia, E-mail: mkanwar@frim.gov.my. https://orcid.org/0000-0002-3787-9907
Asniza M, Alia Syahirah Y and Mohd Fahmi A, Forest Products Division, Forest Research Institute Malaysia (FRIM), Kepong, Selangor, Malaysia.
Lee SH, Laboratory of Biocomposite Technology, Institute of Tropical Forestry and Forest Products (INTROPS), Universiti Putra Malaysia, Serdang, Selangor, Malaysia.

As per De Gruyter's policy this article has previously been published in the journal Physical Sciences Reviews. Please cite as: "Properties of plybamboo manufactured from two Malaysian bamboo species—" *Physical Sciences Reviews* [Online] 2022. DOI: 10.1515/psr-2022-0049 | https://doi.org/10.1515/9783110769227-015

well as to improve strength and uniformity is through disassembling bamboo culm into strips for laminae production. The laminae are then laminated together using selected adhesive to form certifiable structural material known as plybamboo [5].

Plybamboo is an excellent engineered material with high strength, low deformation and stable form. It is one of the promising green-construction materials that can be used in both residential and industrial structures [6]. The mechanical properties of plybamboo are comparable to the common wood, making it as competitive as commonly used building materials, in addition to having renewable characteristics [7, 8]. For instance, studies on plybamboo and laminated panels using bamboo have been reported by Anwar et al. [9], Sulastiningsih et al. [10] and Rahman et al. [11]. It was found that plybamboo have strength and stiffness which comparable to plywood. Most plybamboos are made using assembled bamboo strips either in a lengthwise or crosswise direction alternately with a hot pressing technique using adhesive [12]. Plybamboo is primarily used indoors for surface applications or furniture. However, its use is not limited to internal applications, depending on the type of adhesive used.

Adhesive is one of the important factors affecting physical and mechanical properties of wood-based composites. In the manufacture of conventional bio-composites, thermoset resins such as phenol formaldehyde (PF), urea formaldehyde (UF) and melamine urea formaldehyde (MUF) are generally used as a binder. MUF based plywood requires a low temperature and short pressing time to achieve better bonding strength. Some studies also reported that MUF resin can enhance water resistance and reduce formaldehyde emissions due to its high functionality and stability of the molecular structure [13–16].

While the potential of bamboo is promising, more widespread development and use of bamboo is needed. Based on the study by Nordahlia et al. [17], beting and semantan bamboo have potential for structural application since these species have higher density, MOR, and thicker culm walls. In addition, semantan bamboo was also reported as suitable for making plybamboo, particleboard, laminated bamboo boards and bamwood [18–21]. In this study, beting and semantan bamboo were used to fabricate plybamboo using MUF as a binder. The moisture content, density, thickness swelling, water absorption, linear expansion parallel and perpendicular to the grain were done for determination of the physical properties. Meanwhile, mechanical properties were assessed via bonding shear strength, modulus of rupture (MOR), modulus of elasticity (MOE) and compression strength parallel to the grain. Determination of the properties is important to understand its behavior and performance in structural design and applications.

15.2 Materials and methods

Bamboo culms were harvested from Forest Research Institute Malaysia (FRIM), Kepong, Selangor. The bamboo culms at approximately 4 years old were cut at 15 cm above ground level. Boric acid which was used for bamboo preservation was supplied by—Celcure Chemicals, Selangor, Malaysia. Polyvinyl acetate (PVAc) used for edges joint and melamine urea formaldehyde (MUF) used for surface bonding of plybamboo were supplied by Aica Adtek, Selangor, Malaysia.

15.2.1 Preparation of plybamboo

The bamboo culms (beting and semantan) were cut into bamboo splits of 20 mm wide (based on outer layer) using sizing and splitting machine. The bamboo splits were first soaked in 2% borax solution for two weeks and dried in a kiln until the overall moisture content reaches 12%. Treated bamboo splits were planed to 4 mm thick with final dimension of $4 \times 20 \times 350$ mm. Subsequently, the edge of bamboo strips was glued and joined using polyvinyl acetate (PVAc), clamped and left to dry for 6 h to produce laminae. Next, three layers of bamboo laminae were stacked and glued using MUF resin to produce three-layer-plybamboo. Each layer of bamboo laminae was glued in opposite manner between the layers (perpendicular to the grain). The glue was spread at a rate of 140 g/m^2 and cold pressed at 5 kg/cm^2 for 15 min to set. Following this, the assembled laminae was hot pressed at 140 °C with specific pressure of 14 kg/cm^2 for another 15 min to produce a plybamboo with the final thickness of 12 mm.

15.2.2 Evaluation of bonding properties

A total of 10 test specimens were cut to $12 \times 25 \times 50$ mm, following the Malaysian Standard MS 2693:2020 [22]. The standard requires the samples to be tested in dry and wet conditions. Samples were first immersed in water at 20 ± 3 °C for 24 h. Samples were then immersed in boiling water for 6 h and dried at room temperature for 1 h. The samples were again boiled for 4 h followed by cooling in water at <30 °C for at least 1 h. Initial dimension of each test samples were measured and the test was conducted by using Shimadzu Universal Testing machine. The loading rate was set at 7 mm/min in which the maximum load of the sample was reached within 30 ± 10 s as specified in the standard. Once the test is completed, the samples were inspected to find out the estimated percentage of bamboo failures along the glue line. Bamboo failures of individual specimens was recorded and the average shear strength and wood failure were compared with the minimum standard requirement.

15.2.3 Evaluation of physical properties

Dimensional stability of the plybamboo was rated by determining the thickness swelling and linear expansion after immersion in water, in accordance with ASTM standard test methods [23]. Test specimens with dimension of $12 \times 50 \times 50$ mm were prepared. The initial thickness and width of all samples were measured. The final dimension of samples after horizontally immersed in water for 24 h was then recorded. Water absorption determination was the evaluated by measuring the initial and final weight of the sample after immersion in water for 24 h. The weight of the sample was measured as soon as the excess water was wiped off with a dry cloth.

15.2.4 Evaluation of mechanical properties

Plybamboo was conditioned in a conditioning room maintained at 20 ± 2 °C and 60% relative humidity for a week before the test was carried out. Compression strength parallel to grain was conducted according to ASTM standards for wood-based structural panels [23, 24]. Test specimens of $60 \times 50 \times 12$ mm were placed in a vertical position between two parallel metal plates. A constant load was applied to the top of the sample at a rate of 6.5 mm/min [23]. Moreover, static bending tests were performed according to Chinese Standards, following the similar procedure as described in Anwar et al. [25]. Test specimens with dimension of $300 \times 30 \times 12$ mm were tested with support span at 240 mm. The load is applied at the mid-span with a constant load rate of 3500 N/min.

15.2.5 Statistical analysis

Statistical analysis was carried out using a statistical analysis system (SAS 9.4). The means were separated by the least significant difference (LSD). Means which are followed by the same letter(s) indicate that the studied factor has no significant effect on the observed values at $p \leq 0.05$.

15.3 Results and discussion

15.3.1 Shear strength and bamboo failure percentage

The bond shear strength and bamboo failure percentage of beting and semantan plybamboo were assessed in different conditions according to Malaysian Standard MS 2693:2020. The average values of bond shear strength and bamboo failure percentage of beting and semantan plybamboo are tabulated in Table 15.1. Meanwhile, image profiles of bamboo failure observations are illustrated in Table 15.2. The shear strength of semantan plybamboo was significantly higher than that of beting plybamboo. Moreover, the values of shear strength and bamboo failure percentage of both plybamboo were slightly decreased when samples were soaked in water for 24 h. However, there was a sharp decrease in the values after the samples were exposed to boiling water (BDB) where both beting and semantan plywood experienced a higher reduction in shear strength. This phenomenon occurred as it is suggested that the boiling and drying treatment created extensive stresses in bamboo and glue lines which, upon shearing, were easily broken due to the release of internal stresses.

For bamboo failure observation, semantan plybamboo shows high percentage of bamboo failure for dry (76%) and wet (60%) shear samples, except for boiling condition where it gave low percentage of bamboo failure (27.32%). Samples that exposed to

Table 15.1: Average values of shear strength and bamboo failure of beting and semantan plybamboo

Pre-treatment	Beting		Semantan	
	Shear strength, N/mm²	Bamboo failure (%)	Shear strength, N/mm²	Bamboo failure (%)
Dry/Control	2.67[a] (0.46)	35.00 (11.78)	3.36[b] (0.41)	76.00 (11.37)
Soak 24 h	2.28[a] (0.21)	44.00 (11.30)	2.93[b] (0.51)	60.00 (11.87)
Boil-dry-boil (BDB)	0.29[a] (0.17)	45.00 (16.43)	0.46[b] (0.41)	27.32 (12.45)

Values in parentheses are standard deviations. Means followed with the same letters [a,b] in the same row are not significantly different at $p \leq 0.05$ according to LSD.

Table 15.2: Images of bamboo failure observation after bond shear strength test.

Pre-treatment	Beting	Semantan
Dry/Control	Bamboo failure: 35%	Bamboo failure: 76%
Soak 24 h	Bamboo failure: 44%	Bamboo failure: 60%
Boil–dry–boil (BDB)	Bamboo failure: 45%	Bamboo failure: 27.3%

BDB seemed to have tendency to fail due to glue failure and resulted in decrement in the strength of glue joints. However, the result is contradicted for beting plywood, where the bamboo failure was found to increase slightly from dry to BDB samples ranging from 35% to 45%. This occurrence can be associated by fibre characteristics for each bamboo species including fibre length, fibre diameter, cell wall dimension, microfibril angle and lumen diameter. For example, study by Razak et al. [26] reported that semantan bamboo exhibited larger lumen diameter (8.66 µm) compared to beting bamboo (4.01 µm). Higher lumen diameter will give more room for adhesive to penetrate into the bamboo strips forming a good anchorage between the bamboo layers. Theoretically, good bonding has occurred when both the shear strength and the bamboo failure percentage value are high. In contrast, if one of the values is high and the other is low, it designates either the adhesive or the bamboo is poor [25]. Therefore, from the findings, both beting and semantan bamboo are appropriate to be used for plybamboo application as the results indicated that both plybamboo met the minimum shear strength requirement of the Malaysian standard which was 0.35 to 2.5 N/mm^2 for dry condition [22, 27].

15.3.2 Physical properties

Average values of thickness swelling, water absorption, linear expansion parallel to the grain and linear expansion perpendicular to the grain of the beting and semantan plybamboo are given in Table 15.3. It seemed that beting plybamboo resulted in the lower thickness swelling value compared to semantan plybamboo. After 24 h of soaking, the thickness of beting plybamboo swelled 5.76%, while semantan plybamboo swelled 13.21%. When soaked in water, semantan plybamboo absorbed more water (27.7%) than beting plybamboo (21.86%). On the other hand, semantan plybamboo demonstrated a low value in linear expansion parallel to the grain compared to beting plybamboo after immersion in water. However, the value for expansion perpendicular to the grain was higher in beting plybamboo (0.73%) than in semantan plybamboo (0.47%).

The results obtained suggest that the beting plybamboo is less susceptible to water absorption as well as to thickness swelling. The differences in thickness swelling attained by both plybamboo might attribute to the proportion of cellulose and hemicellulose in the bamboo species. Due to the presence of free –OH group in the molecular structure, cellulose and hemicelluloses are responsible for water absorption [28, 26]. In general, cellulose content of semantan is 46.9%, whereas beting contains 33.8%, as reported by Razak et al. [26]. Thus, lower cellulose and/or hemicellulose content of matured beting bamboo might restrict the absorption of water as well as thickness swelling in beting plybamboo.

15.4 Mechanical properties

The moisture content (MC) of the beting plybamboo and semantan plybamboo were 13.94% and 13.57% with density of 783.80 kg/m^3 and 783.14 kg/m^3, respectively. The average values of the mechanical properties for beting and semantan plybamboo are given in Table 15.4. It was found that the strength properties of semantan plybamboo

Table 15.3: Physical properties of beting and semantan plybamboo

Species	Thickness swelling (%)	Linear expansion parallel to grain (%)	Linear expansion perpendicular to grain (%)	Water absorption (%)
Beting	5.76[a] (1.54)	1.46[a] (0.80)	0.73[a] (1.49)	21.86[a] (2.97)
Semantan	13.21b (1.94)	1.09[a] (0.77)	0.47[a] (0.17)	27.77[b] (4.46)

Values in parentheses are standard deviations. Means followed with the same letters [a,b] in the same row are not significantly different at p ≤ 0.05 according to LSD.

were significantly higher than beting plybamboo. The modulus of rupture (MOR), modulus of elasticity (MOE) and compression strength parallel to the grain for semantan plybamboo were 130.86 N/mm^2, 19661.3 N/mm^2 and 53.21 N/mm^2, respectively. This was an unexpected result, since Nordahlia et al. [17] reported that beting bamboo in split form exhibited higher MOR and MOE compared to semantan bamboo split. In this case, it somehow proved that different species, parts and form of bamboo influence the values of MOR and MOE.

Table 15.4: Mechanical properties of beting and semantan plybamboo

Species	MC (%)	Density (kg/m^3)	Mechanical properties (N/mm^2)		
			MOR	MOE	Compression strength parallel to grain
Beting	13.94[a] (1.79)	783.80[a] (48.52)	115.92[a] (19.1)	25331.5[a] (1755.2)	46.15[a] (7.11)
Semantan	13.57[a] (1.50)	783.14[a] (47.50)	130.86[b] (14.28)	19661.3[b] (2323.7)	53.21[b] (4.21)

Values in parentheses are standard deviations. Means followed with the same letters [a,b] in the same column are not significantly different at p ≤ 0.05 according to LSD.

MOR and MOE are the most important mechanical properties of load-bearing plywood applications (i.e., construction and industrial applications). Apparently, the MOR, MOE and compression strength parallel to the grain of the plybamboo obtained in this study were significantly higher compared to commercial plywood. In other study, it was found that MOR, MOE and compression strength parallel to the grain for semantan plybamboo were 68 N/mm^2, 9642 N/mm^2 and 35 N/mm^2, respectively. Authors stated that the MOR and MOE values of the plybamboo were 36% and 57% higher than that of merawan plywood [25]. Basically, fibre length and fibre density play an important role in plywood manufacturing [29]. It has been reported that bamboo with high density has MOR, MOE and shear values that are almost identical to commercial tropical hardwood plywood [9, 30]. An earlier study by Bhat et al. [31] has also discovered that bamboo is suitable for the plywood industry due to its high elasticity and strength.

15.5 Conclusions

This study investigated the potential of using two Malaysian bamboo species in the manufacture of 3-layer-plybamboo. On the basis of the physico-mechanical properties, it appears that both beting and semantan bamboo is technically feasible for plybamboo fabrications. Slight differences in the physico-mechanical properties among the plybamboo can be associated to the characteristics of the bamboo species. Therefore, it can be noted that the properties of plybamboo are species dependent. Based on the findings, it also shows that the bonding properties of the plybamboo satisfies the minimum standard requirement stipulated in Malaysian Standard MS 2693:2020.

References

1. Lakkad SC, Patel JM. Mechanical properties of bamboo, a natural composite. Fibre Sci Technol 1981;14:319–22.
2. Amada S, Ichikawa Y, Munekata T, Nagase Y, Shimizu K. Fibre texture and mechanical graded structure of bamboo. Composites Part B 1997;28:13–20.
3. Zhang QS, Jiang SX, Tang YY. Industrial utilization on bamboo: technical report no. 26. People's Republic of China: The International Network for Bamboo and Rattan (INBAR); 2002:24 p.
4. Chaowana P. Bamboo: an alternative raw material for wood and wood-based composites. J Mater Sci Res 2013;2:90–102.
5. Mahdavi M, Clouston PL, Arwade SR. Development of laminated bamboo lumber: review of processing, performance, and economical considerations. J Mater Civ Eng 2011;23:1036–42.
6. Xiao Y, Shan B, Yang RZ, Li Z, Chen J. Glue laminated bamboo (GluBam) for structural applications. In: Vol. 9 of Materials and joints in timber structures: Recent developments of technology. Dordrecht: Springer; 2014. p. 589–95.
7. Ahmad M, Kamke FA. Properties of parallel strand lumber from Calcutta bamboo (*Dendrocalamus strictus*). Wood Sci Technol 2011;45:63–72.
8. Verma CS, Chariar VM. Development of layered laminate bamboo composite and their mechanical properties. Compos B Eng 2012;43:1063–9.
9. Anwar UMK, Paridah MT, Hamdan H, Abd Latif M, Zaidon A. Adhesion and bonding properties of plybamboo manufactured from *Gigantochloa scortechinii*. Am J Appl Sci 2005;53–8.
10. Sulastiningsih IM, Nurwati SA. Pengaruh lapisan kayu terhadap sifat bambu lamina. JPHH 2003; 23:15123–22.
11. Rahman KS, Nazmul Alam DM, Islam MN. Some physical and mechanical properties of bamboo mat-wood veneer plywood. ISCA J Biol Sci 2012;1:61–4.
12. Qisheng Z, Shenxue J, Yongyu T. Industrial utilization on bamboo. Beijing, China: International Network for Bamboo and Rattan; 2002.
13. No BY, Kim MG. Syntheses and properties of low-level melamine-modified urea melamine-formaldehyde resins. J Appl Polym Sci 2004;93:2559–69.
14. No BY, Kim MG. Evaluation of melamine-modified urea-formaldehyde resins as particleboard binders. J Appl Polym Sci 2007;106:4148–56.
15. Zhang J, Wang X, Zhang S, Gao Q, Li J. Effects of melamine addition stage on the performance and curing behavior of melamine-urea-formaldehyde (MUF) resin. Bioresources 2013;8:5500–14.
16. Murata K, Watanabe Y, Nakano T. Effect of thermal treatment of veneer on formaldehyde emission of popular plywood. Materials 2013;6:410–20.
17. Nordahlia AS, Anwar UMK, Hamdan H, Abd Latif M, Mohd Fahmi A. Anatomical, physical and mechanical properties of thirteen malaysian bamboo species. Bioresources 2019;14:3925–43.
18. Razak W, Hashim WS, Wan Tarmeze WA, Mohd Tamizi M. Industri pembuatan pepapan laminasi buluh [Manufacture of bamboo laminate). Kepong, Selangor, Malaysia: Forest Research Institute Malaysia; 1997.
19. Zaidon A, Paridah MT, Sari CKM, Razak W, Yuziah MYN. Bonding characteristics of *Gigantochloa scortechinii*. J Bamboo Rattan 2004;3:57–68.

20. Hanim AR, Zaidon A, Abood F, Anwar UMK. Adhesion and bonding characteristics of preservative-treatment bamboo (*Gigantochloa scortechinii*) laminates. J Appl Sci 2010;10:1435–41.
21. Anwar UMK, Paridah MT, Hamdan H, Zaidon A, Roziela Hanim A, Nordahlia AS. Adhesion and bonding properties of low molecular weight phenol formaldehyde-treated plybamboo. J Trop For Sci 2012;24:379–86.
22. Malaysian Standard MS 2693:2020. Laminated bamboo - Specifications for general use. Cyberjaya, Selangor, Malaysia: Department of standards Malaysia; 2020.
23. Standard test methods of for evaluating properties of wood based fibre and particle panel materials Annual Book of ASTM Standard ASTM Des D 1037-96a. Philadelphia, PA: ASTM; 1996, vol 4: 4–9 pp.
24. Standard test methods for wood based structural panels in compression Annual Book of ASTM Standard ASTM Des D 3501-94. Philadelphia, PA: ASTM; 1995, vol 4-10.
25. Anwar UMK, Zaidon A, Paridah MT, Razak W. The potential of utilising bamboo culm (*Gigantochloa scortechinii*) in the production of structural plywood. J Bamboo Rattan 2004;3:393–400.
26. Razak W, Mustafa MT, Abdus Salam M, Sudin M, Samsi HW, Sukhairi M. Chemical composition of four cultivated tropical bamboo in Genus Gigantochloa. J Agric Sci 2013;5:66–75.
27. Malaysian Standard MS 228:1991. Specification for plywood (1st revision). Malaysia: Standard and Industrial Research Institute of Malaysia; 1991.
28. Wardrop AB. The phase of lignification in the differentiation of wood fibres. Tappi 1957;40:225–43.
29. Freeman H. Properties of wood and adhesion. For Prod J 1959;9:451–8.
30. Loh YF, Paridah MT, Hoong YB. Density distribution of oil palm stem veneer and its influence on plywood mechanical properties. J Appl Sci 2011;11:824–31.
31. Bhat IUH, Mustafa MT, Mohmod AL, Abdul Khalil HPS. Spectroscopic, thermal and anatomical characterization of cultivated bamboo (*Gigantochloa spp.*). Bioresources 2011;6:1752–63.

Moustafa Alaa*, Khalina Abdan, Lee Ching Hao, Ammar Al-Talib, Muhammad Huzaifah and Norkhairunnisa Mazlan

16 Fundamental study of commercial polylactic acid and coconut fiber/polylactic acid filaments for 3D printing

Abstract: This study aims to provide an alternative fully green biodegradable 3D printing filament other than polylactic acid (PLA) with better properties and lower prices using a fully environmentally friendly process. Two filaments [polylactic acid (PLA) and polylactic acid/coconut fiber (PLA-CF)] to be purchased and used to prepare a similar samples under the same conditions which to undergo the same testing to obtain and compare their properties as well as for further comparison with other filaments. The samples are to be designed using SOLIDWORKS software according to the American Society for Testing and Materials (ASTM)standards. The prepared designs are then to be converted to gcode using CURA software. FDM Creality 3D printer (Model: CR10S-PRO) to be used printing a set of specimens for each required test. The prepared samples then undergo several mechanical tests to specify their exact properties. PLA 3D filament roll had been purchased from Fabbxible Technology; Crystallized nature based NatureWorks made from corn starch. While Magma PLA-CF roll had been purchased from 3D Gadgets Malaysia. Both rolls had an average diameter of 1.75 mm and average length of 300 m.

Keywords: 3D printing; American society for testing and materials (ASTM); filament; Polylactic Acid (PLA).

*Corresponding author: Moustafa Alaa, Department of Biological and Agricultural Engineering, Faculty of Engineering, Universiti Putra Malaysia, Serdang, 43400, Malaysia,
E-mail: Moustafaalaa1@outlook.com
Khalina Abdan, Department of Biological and Agricultural Engineering, Faculty of Engineering, Universiti Putra Malaysia, Serdang, 43400, Malaysia; and Institute of Tropical Forestry and Forest Products (INTROP), Universiti Putra Malaysia, Serdang, 43400, Malaysia
Lee Ching Hao, Department of Biological and Agricultural Engineering, Faculty of Engineering, Universiti Putra Malaysia, Serdang, 43400, Malaysia
Ammar Al-Talib, Department of Mechanical Engineering, Faculty of Engineering, Technology & Built Environment, UCSI University, Cheras, 56000, Malaysia
Muhammad Huzaifah, Department of Crop Science, Universiti Putra Malaysia, Serdang, 43400, Malaysia
Norkhairunnisa Mazlan, Department of Aerospace Engineering, Universiti Putra Malaysia, Serdang, 43400, Malaysia

As per De Gruyter's policy this article has previously been published in the journal Physical Sciences Reviews. Please cite as: M. Alaa, K. Abdan, L. C. Hao, A. Al-Talib, M. Huzaifah and N. Mazlan "Fundamental study of commercial polylactic acid and coconut fiber/polylactic acid filaments for 3D printing" *Physical Sciences Reviews* [Online] 2022. DOI: 10.1515/psr-2022-0050 | https://doi.org/10.1515/9783110769227-016

16.1 Introduction

The use of renewable, recyclable, and biodegradable materials to reduce waste while lowering carbon emissions is considered one of the most important aspects of implementing a green economy. As a result of its properties, natural fibers will play a significant role in the emergence of a green economy. Natural fibers are long-lasting materials that have been replenished by nature and humans throughout time. The use of biodegradable, renewable, and recyclable materials to reduce waste while lowering carbon emissions is considered one of the most important aspects of implementing a green economy. As a result of its properties, natural fibers will play a significant role in the emergence of a green economy. Natural fibers are long-lasting materials that have been replenished by nature and humans throughout time. It offers a number of benefits, including ease of recycling, low cost, biodegradability, minimal environmental effect, low density, and strong mechanical qualities. Natural fibers have lately been an appealing choice for academics and scientists as an alternative filler for fiber-reinforced composites to apply the meaning of a green economy, owing to their properties. As a result, bio-composites are projected to play a role in the development of ecologically friendly goods. As a result, bio-composites are intended to contribute to the production of ecologically sound goods that help to execute and accomplish various United Nations-set Sustainable Development Goals – SDG's (see Figure 16.1), which are backed by both UPM management and the government of Malaysia.

If at least one of the elements is obtained from natural/biological material, the material is to be characterized as a bio-composite. Plant fibers, which are made up

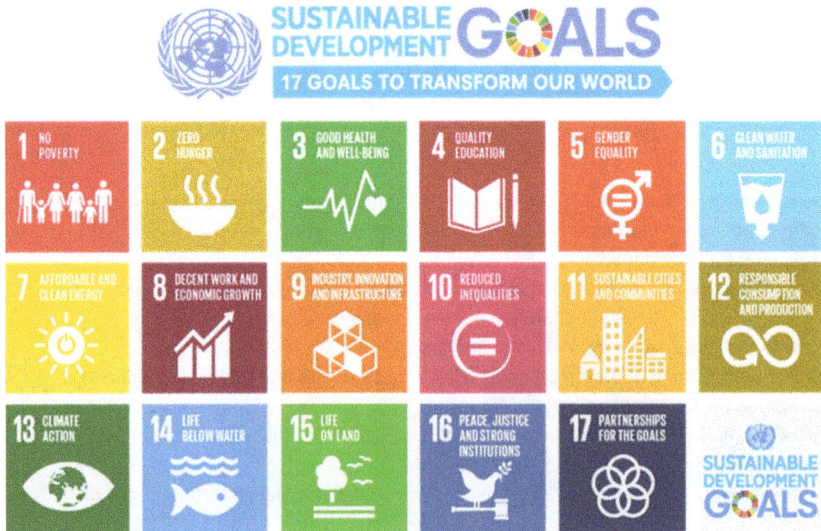

Figure 16.1: Sustainable development goals [1].

of carbohydrate polymers (cellulose and hemicellulose) and an aromatic polymer (lignin), are the most plentiful raw material for the generation of biofuels on the planet. Due to that unique structure of plant fibers are also known as lignocellulose fibers. Lignocellulose fibers such as kenaf fibers (KF) became a promising materials due to its abundance, flexibility, lightness, environmentally friendly, as well as its low price and respectable mechanical properties. Owing to the good mechanical properties and thermal properties as well as its abundant local supply with relatively low cost, it will be a perfect choice for bio-composite products used in high-tech applications [2]. One of the most common applications for bio-composites is 3D printing which keeps on getting involved in several industries such as automotive and biomedical. The research area to create complex geometries using 3D printing had recently emerged with the help of developed computer-aided design (CAD) software.

Additive manufacturing (AM) has moved beyond its specific uses in the roughly 30 years since its beginnings, transforming a wide variety of production techniques [3]. Various governmental and private institutes are encouraging scientists and researchers to carry on developing new high-performance green materials which will help to provide additional benefits such reducing polymer waste, minimal chemical use, high precision, cost effectiveness, and the ability to produce complex geometries due to the current limitation of green 3D printing materials choices as well as the relatively low mechanical properties of virgin polymers. However, due to the limitations in green 3D printing materials and mechanical properties of virgin polymers, researchers are encouraged to continue developing green high-performance 3D printing materials that provide benefits such as cost effectiveness, high precision, reduced polymer waste, minimal chemical use, and usable to produce complex geometries. As a result of the aforementioned factors, as well as the availability of fiber resources, technical innovation, and competition, the demand for bio-composites has grown urgent [4].

Fused deposition modeling (FDM) is one of the most common rapid prototyping techniques in 3D printing, with the main advantage of multiple material usage. FDM is a 3D printing technique that involves melting a thermoplastic filament and extruding it through a circular nozzle. A 3-axis system controls the nozzle movement, allowing the molten plastic to be deposited onto a print bed. However, the only 3D printing material which could be considered as an environmentally friendly is the polylactic acid (PLA); which makes it costly to use such material [5]. Plastic strands or pellets made from polylactic acid (PLA), polycarbonate (PC), and acrylonitrile butadiene styrene (ABS), are the most common materials to be used in FDM based 3D printing [4]. This paper to describe the mechanical performance of several 3D printed samples throughout various mechanical tests such as tensile, flexural, impact, compression, and dynamic mechanical analysis (DMA).

16.2 Methodology and experimental setup

16.2.1 Research methodology and bio-composite fabrication

There are general procedures to follow in every research work to provide comparable and acceptable results that are valuable to both educational and scientific communities. First and foremost, acceptable specimens must be created as necessary under the same conditions, with the expectation that any significant changing of any parameter would influence the mechanical characteristics of the prepared specimens, which will be incomparable in the end. Two filaments [polylactic acid (PLA) and polylactic acid/coconut fiber (PLA-CF)] to be purchased and used to prepare a similar samples under the same conditions which to undergo the same testing to obtain and compare their properties as well as for further comparison with other filaments. The mechanical properties of the prepared composites will be studied and analyzed throughout different testing stages. This study aims to provide an alternative fully green biodegradable 3D printing filament other than polylactic acid (PLA) with better properties and lower prices using a fully environmentally friendly process. The overall experimental steps to be as described in Figure 16.2 below.

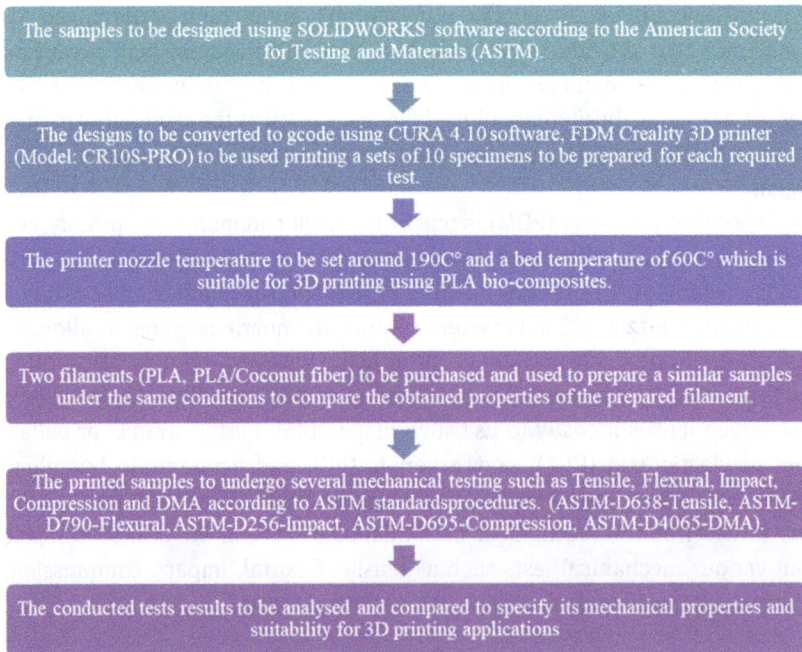

The samples to be designed using SOLIDWORKS software according to the American Society for Testing and Materials (ASTM).

⬇

The designs to be converted to gcode using CURA 4.10 software, FDM Creality 3D printer (Model: CR10S-PRO) to be used printing a sets of 10 specimens to be prepared for each required test.

⬇

The printer nozzle temperature to be set around 190C° and a bed temperature of 60C° which is suitable for 3D printing using PLA bio-composites.

⬇

Two filaments (PLA, PLA/Coconut fiber) to be purchased and used to prepare a similar samples under the same conditions to compare the obtained properties of the prepared filament.

⬇

The printed samples to undergo several mechanical testing such as Tensile, Flexural, Impact, Compression and DMA according to ASTM standardsprocedures. (ASTM-D638-Tensile, ASTM-D790-Flexural, ASTM-D256-Impact, ASTM-D695-Compression, ASTM-D4065-DMA).

⬇

The conducted tests results to be analysed and compared to specify its mechanical properties and suitability for 3D printing applications

Figure 16.2: Flow chart of research methodology.

16.2.2 Materials

The average density of PLA is practically around 1.25 g/. PLA 3D filament roll had been purchased from Fabbxible Technology; Crystallized nature based NatureWorks made from corn starch. While Magma PLA-CF roll had been purchased from 3D Gadgets Malaysia. Both rolls had an average diameter of 1.75 mm and average length of 300 m.

16.2.3 Characterization

The printed samples to undergo several mechanical testing such as tensile, flexural, impact, compression and DMA. Various samples to be prepared using the prepared filament to specify its exact properties. The samples to be designed using SOLIDWORKS (Figure 16.3) software according to the American Society for Testing and Materials (ASTM) standards (ASTM-D638-Tensile, ASTM-D790-Flexural, ASTM-D256-Impact, ASTM-D695-Compression, ASTM-D4065-DMA) with testing parameters as mentioned in Table 16.1 below. The designs to be converted to gcode using CURA 4.0 software (Figure 16.4), FDM Creality 3D printer (Model: CR10S-PRO) to be used printing a sets of 10 specimens to be prepared for each required test, the printer nozzle temperature to be set around 190 °C and a bed temperature of 60 °C.

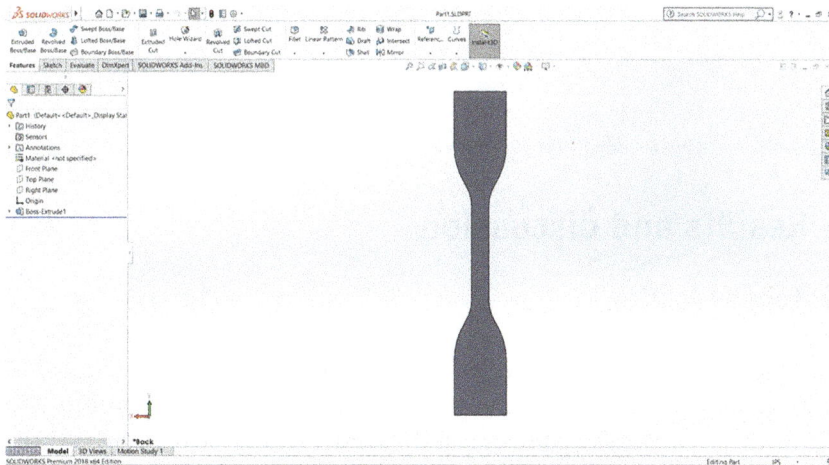

Figure 16.3: ASTM tensile sample design using SOLIDWORKS software.

Table 16.1: Parameters of the prepared samples.

Test	Standard	Dimensions (l × w × t) mm	Controlled parameters
Tensile	ASTM-D638	120 × 19 (6) × 3	Speed: 5 mm/min
Flexural	ASTM-D790	125 × 12.7 × 3.2	Speed: 1.37 mm/min
Impact	ASTM-D256	63.5 × 12.7 × 3.2 Notch: 22.5°	Force: 15 Nm
Compression	ASTM-D695	12.7 × 12.7 × 25.4	Speed: 1 mm/min
DMA	ASTM-D4065	56 × 13 × 3	Heat: 5 °C/min From: Room-150 °C

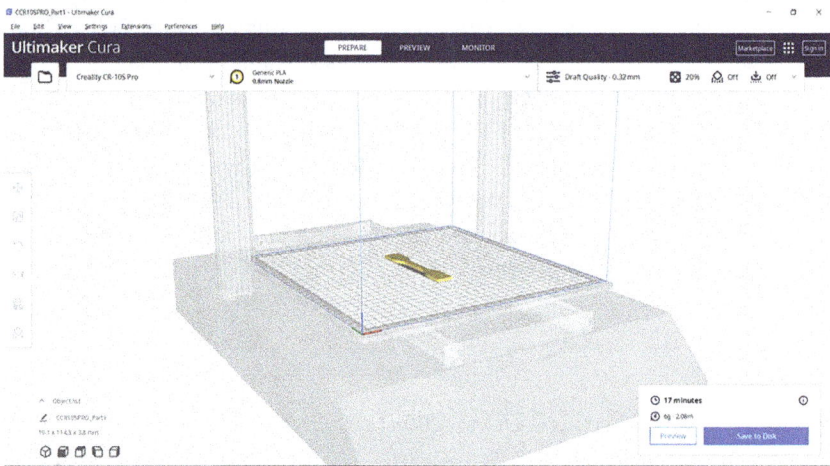

Figure 16.4: ASTM Tensile sample gcode conversion using CURA 4.0 software.

16.3 Results and discussion

16.3.1 Tensile testing

Tensile testing was the first test carried out as part of this study. Tensile test could be utilized to determine various mechanical characteristics such as ultimate tensile strength, yield tensile strength, maximum elongation, and breaking loads [4]. The ASTM-D638 (Type IV) standard was used to print the 3D samples [6]. One of the mechanical qualities discovered during tensile testing is stress at the peak, often known as ultimate strength. The ultimate tensile strength of a material is the utmost stress that the material can sustain before breaking when stretched or pulled [7]. Two different materials (PLA and PLA-CF) had been used so far to print 3D tensile samples which had been analyzed and compared in terms of the Stress at Peak as shown in Figure 16.5 below.

Tensile Strength

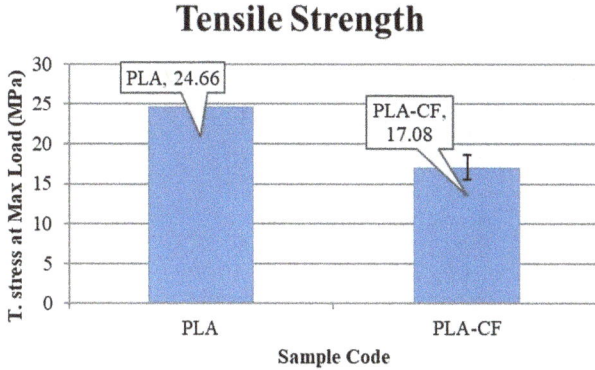

Figure 16.5: Tensile stress at peak.

It could be noticed that the PLA which is the control sample had the highest ultimate strength among the all other composites with 24.66 MPa Tensile stress at maximum load compared to 17.08 MPa for the PLA-CF.

One of the mechanical characteristics identified during tensile testing is strain at break, also known as tensile elongation at break, or fracture strain [8] (Figure 16.6). The strain at break is a ratio between the final elongated at break length to the initial specimen length within a constant controlled surrounding temperature. As a result, strain at break tensile testing could be considered to represent the capability of the tested specimens to resist dimensional changes without cracking. Figure 16.7 shows the strain at break for all examined specimens which describes that the PLA samples had the greatest fracture strain value of all the composites tested, at roughly 3.6% compared to 3.3% for the PLA-CF. Clearly, the coconut fibers reinforced into the PLA-CF reduced the ability of PLA to stretch before breaking which explains the decreasing strain behavior.

Tensile stress at Break

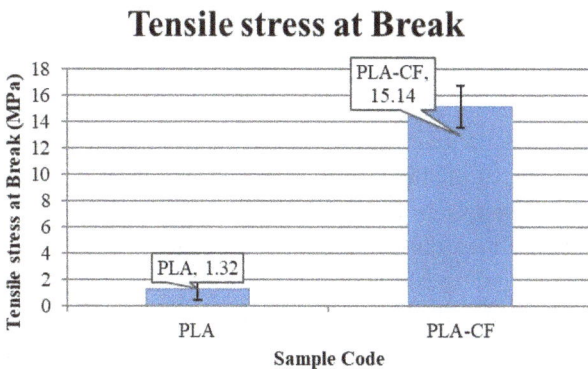

Figure 16.6: Tensile stress at break.

Tensile strain at Break (%)

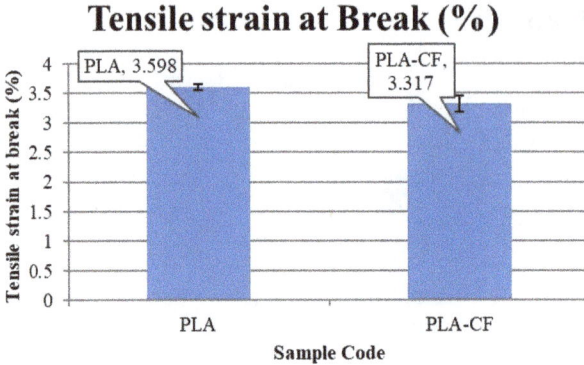

Figure 16.7: Tensile strain at break.

16.3.2 Flexural testing

Flexural testing was the second test carried out as part of this study. The flexural test, also known as a transverse beam test could be used to determine a variety of mechanical characteristics for the composites under test, specifically material bending capabilities. The flexural test is typically performed by simply supporting the testing specimens from both edges before applying load in the midpoint of the tested specimen using one or two load points. Thus, test is also commonly known as three-points and four-points flexural testing, respectively, based on the number of load points to be used [9].

However, because the three-points bend testing approach is easier to prepare, apply and more common than the four-points bend testing, it will be used throughout this study. Overall, the three-points flexural testing could be used to determined various bending properties including flexural stress, flexural strain, and the flexural modulus of elasticity, among other mechanical parameters. The flexural samples had been also printed using the same 3D printer under the same conditions to prepare the all required specimens with dimensions according to ASTM-D790 standards. Ten samples had been prepared for each category to be tested and compared to the other prepared specimens. Meanwhile, the midpoint had been marked to be placed directly below the load point [9].

One of the mechanical characteristics determined during flexural testing is bending strength, also known as flexural strength, transverse rupture strength, or modulus of rupture [10]. Bending strength is a material characteristic that describe the amount of bending stress that a material can endure before yielding. The PLA composite exhibited the greatest bending strength value of 45.09 MPa among all tested bio-composites thus far, followed by the PLA-CF composite at around 34.74 MPa, according to the flexural strength values shown in Figure 16.8. This unusual

Flexural Strength

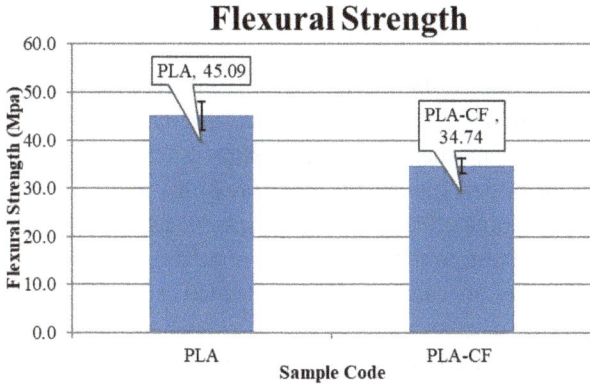

Figure 16.8: Flexural strength at break.

behavior might have been caused by the random coconut distribution used in the PLA-CF filament preparation.

The bending modulus, also known as flexural modulus, is another characteristic that could be identified using the bending test. Generally, the bending modulus is an important characteristic that is computed as the ratio of strain to stress during flexural deformation. It can also be represented as the material's inclination to resist bending. The normal PLA had the greatest bending modulus value of 2610.67 MPa among the all tested samples, according to the flexural modulus values shown in Figure 16.9 below. Meanwhile, the bending modulus of the PLA-CF was found to be lower, at around 2425.78 MPa.

Flexural Modulus

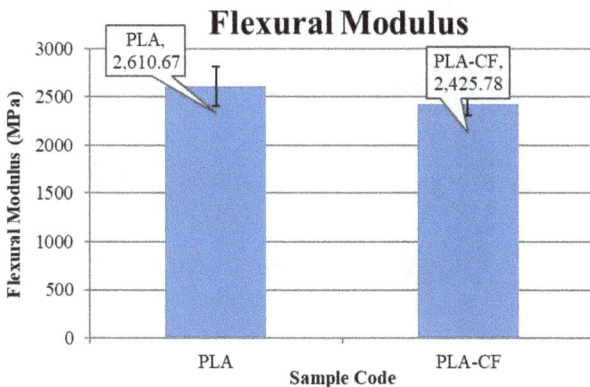

Figure 16.9: Bending modulus.

16.3.3 Compression testing

The third mechanical test which had been conducted so far for this investigation was the compression testing. Along with tensile and flexion tests, compression testing is one of the most fundamental types of mechanical testing. Compression tests are used to examine a material's behavior under applied crushing pressures and are often performed on a universal testing machine utilizing platens or specific fixtures to apply compressive pressure to a test specimen (normally of either a cuboid or cylindrical geometry). Various material properties are calculated and presented as a stress-strain diagram throughout the test, which is used to determine attributes such as elastic limit, proportional limit, yield point, yield strength, and compressive strength for some materials [11].

Another 10 cubic samples were 3D printed under the same conditions with dimensions according to the ASTM-D695 standards [6]. In mechanics, compressive strength or compression strength is the capacity of a material or structure to withstand loads tending to reduce size. In other words, compressive strength resists compression, whereas tensile strength resists tension. Compressive strength is defined as the maximum compressive load, which a sample can bear before its fracture, divided by its cross-sectional area. Two different materials (PLA and PLA-CF) had been used so far to print 3D compression samples so far which had been analyzed and compared in terms of the compression strength as shown in Figure 16.10 below. Compression strength showed a behavior similar to it in tensile strength. The PLA-CF composite showed a lower strength value at 13.04 MPa compared to 17.46 MPa for the regular PLA.

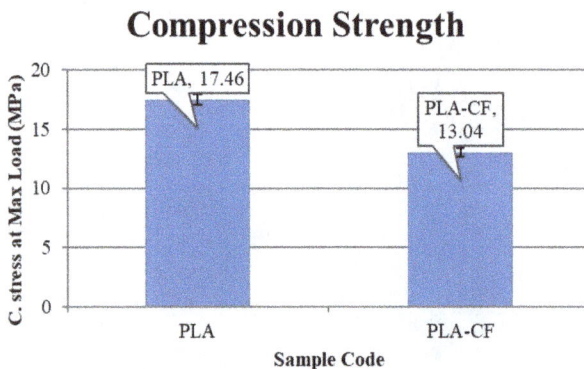

Figure 16.10: Compression strength.

Compression strain at Break

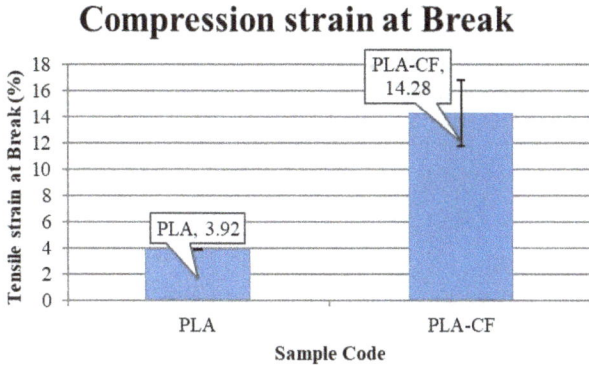

Figure 16.11: Compression strain at break.

One of the mechanical characteristics that may be identified by compression testing is strain at break, which is also known as compression strain at break, or compressive strain at break. The strain at break is generally a ratio between the initial length and the final length following compression of the specimen which to be conducted at a controlled temperature. It is mostly used to represent the capability of the tested specimens to resist the dimensional changes without cracking. Figure 16.11 shows the compression strain at break for the all tested samples. It is obvious that the PLA had the lowest compression strain value, at around 3.92% compared to 14.28% for the PLA-CF composite. It could be referred to the addition of coconut fibers into the bio-composite which enhanced the material capability to tolerate compression before breaking, which explains the higher strain behavior.

16.3.4 Impact testing

The toughness of a material could be tested by impact testing. Basically, the capability of a material to absorb energy during plastic deformation is referred to as toughness. However, because brittle materials can only endure a little amount of plastic deformation, they have a relatively poor toughness. Izod impact testing is an ASTM standard for measuring the impact resistance of materials [12]. A pivoting arm is lifted to a defined height before being released with a constant potential energy. The arm swoops down, crashing against and breaking the notched Izod sample. The height at which the arm swings after hitting the sample is used to assess the sample's energy absorption. Additionally, the notched sample is widely used to evaluate impact energy and notch sensitivity. Overall, the impact test determines the amount of energy that a material absorbs during fracture. This absorbed energy is a metric for a material's toughness and can be used to investigate its brittle–ductile transition. Its purpose is to identify whether a material is brittle or ductile.

Energy Lost at the Notch

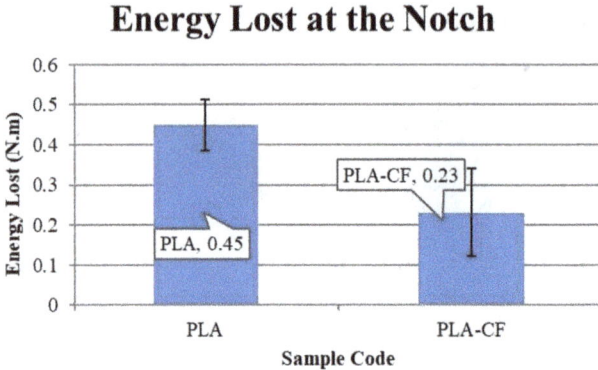

Figure 16.12: Energy loss at the notch.

The dimensions of the impact testing samples are printed according to ASTM-D256 standards. For every material to be tested, 10 identical samples are printed using 3D printer filament of the required materials. The obtained results are then to be analyzed and compared to the other prepared combinations. By measuring the height loss of the pendulum's swing during the Izod test, the material's impact strength could be evaluated as a function of the pendulum energy loss. In the Izod impact test, the samples will be held vertically while the specified weight at the end of the pendulum arm is released down to strike the testing sample. From Figure 16.12 above, it was clear that PLA had the greatest value for pendulum energy loss at the notch. Therefore, based on the data that has been analyzed it is convincing to say that the addition of coconut fiber as a filler in PLA filament reduced its impact modulus.

16.3.5 Dynamic mechanical analysis testing (DMA)

Dynamic mechanical analysis (DMA) which is also called DMTA for dynamic mechanical thermal analysis is a technique for studying and characterizing materials. It's best used to investigate polymer viscoelastic behavior. The complex modulus is determined by applying a sinusoidal load and measuring the strain in the material. The temperature of the sample or the frequency of the stress are frequently changed, resulting in changes in the complex modulus; this method may be used to find the material's glass transition temperature (T_g) as well as transitions corresponding to other molecular motions [13].

DMA applies an oscillatory force at a set frequency to the sample and reports changes in stiffness and damping. DMA data is used to obtain modulus information. Additionally, DMA is much more sensitive in detecting transitions compared to other available testing techniques. DMA works by applying a sinusoidal deformation to a sample of known geometry. The sample can be subjected to controlled stress or a controlled strain. For known stress, the sample will then deform a certain amount. In

DMA this is done sinusoidally. How much it deforms is related to its stiffness. A force motor is used to generate the sinusoidal wave, and this is transmitted to the sample via a driven shaft. One concern has always been the compliance of this drive shaft and the effect of any stabilizing bearing to hold it in position [14].

DMA evaluates stiffness and damping, which are expressed as modulus and tan delta. Overall, DMA properties will be describing the modulus as an in-phase component, the storage modulus, and an out-of-phase component, the loss modulus, by applying a sinusoidal force. The elastic behavior of a sample is measured by the storage modulus, which usually being referred to by either E' or G'. The tan delta, or loss-to-storage ratio, is sometimes referred to as damping. It is a measurement of a material's energy dissipation. Transitions in materials could be seen as variations in the E' or tan delta curves as the values vary with temperature. Those curves do not only describe the glass transition and the melting temperature range but shows any other transitions that occur in the glassy or rubbery plateau including any transitions of subtler material's changes [15].

Three identical samples from each material were 3D printed under the same conditions according to ASTM D-4065 standard dimensions [11]. By looking at Figures 16.13–16.15 it is easy to notice the improvement of all storage modulus, loss modulus as well as the tan delta (damping) at relatively similar temperatures. The storage modulus for PLA-CF was around 1536.01 MPa compared to 1330.1 MPa for the normal PLA. Loss Modulus was shown a value around 21.49 MPa for the normal PLA while the PLA-CF had a loss modulus value around 31.66 MPa. Meanwhile, the normal PLA damping value was around 1.03 compared to 1.103 for the PLA-CF composite. Meanwhile, it could be also noticed from Figures 16.13–16.15 the teeny reduction of the glass transition temperature of the PLA-CF compared to the normal PLA. By comparing the obtained results, the value obtained from storage modulus curve showed a reduction from 53.71 °C to 53.38 °C, and the loss modulus curve showed a reduction of the from 58.49 °C to 57.77 °C, while Tan delta curve showed a reduction from 63.71 °C to

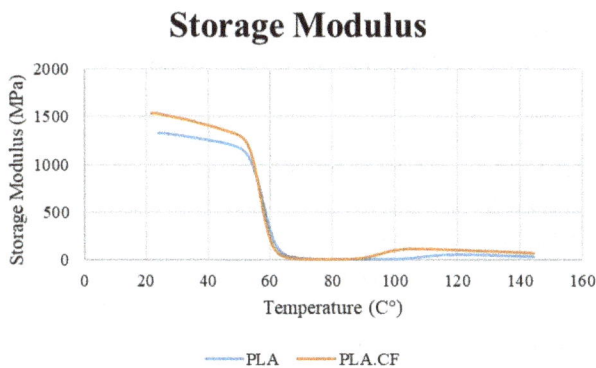

Figure 16.13: Storage modulus.

Loss Modulus

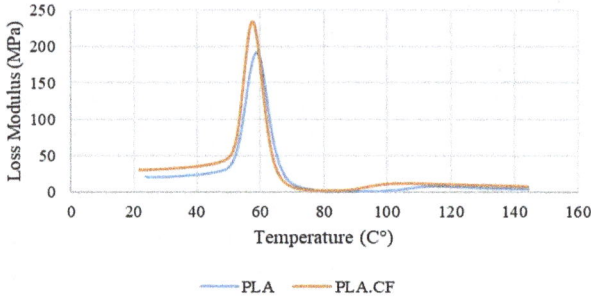

Figure 16.14: Loss modulus.

Tan Delta

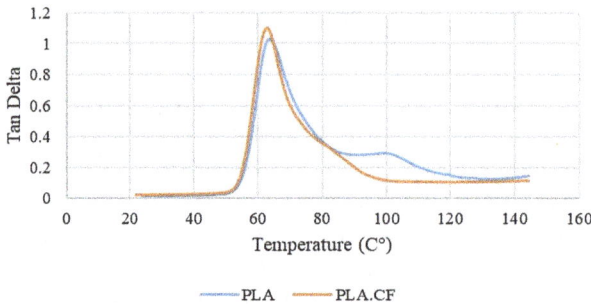

Figure 16.15: Tan delta (damping).

63.10 °C of the PLA to PLA-CF, respectively. This slight change of the DMA testing results will be referred to the addition of coconut fiber to the PLA which gave the composite more rigidity and shifted its glassy stage.

The DMA curves continued to drop until they approached the rubbery plateau, at which stage they leveled out. The addition of coconut fiber to the vitreous and rubbery phases generated a small shift in the curves. This is referred to happen due to the presence of static hydrogen bonds between hydrophilic filler and matrix amide groups, this increase is linked to the transference of stresses between fibers and matrix in the glassy state. According to the Figures below, both PLA and PLA-CF exhibited similar behavior, with the PLA-CF showing a modest improvement in terms of storage modulus and loss modulus when compared to the neat PLA, it is expected that happened due to that the inclusion of CF acted as extra topological nodes in the bio-composites. Furthermore, according to previous studies the improvement of the loss modulus could be due to improved stress transfer at the fiber interface, which resulted in an increased modulus when CF was added to the plain polymer. Overall, it could be notable that the addition of coconut fiber slightly affected the glass transition behavior, however, there was no major shifting in the in the peak of the tan delta. This could be explained also by the strong interactions between the coconut fiber and the PLA particles in the bio-composite, but the addition of fiber still caused the neat PLA to have a better behavior in the rubbery phase.

16.4 Conclusions

The conducted mechanical tests so far on the purchased PLA and PLA-CF bio-composite showed comparable mechanical properties. Overall, the regular PLA showed a better behavior in terms of tensile strength, flexural strength, compression strength, and energy loss of the impact test, while, PLA-CF showed an improvement in terms of tensile and compression strain at break as well as an improvement of the material crystallinity as showed by DMA test results. It is expected that the new PLA-KF bio-composite to be prepared will be having a comparable and improved properties compared to the tested polymers especially with implementing the SHS treatment which to improve the fiber water absorption and as well as the bio-composite suitability for 3D printing applications with relatively low cost.

References

1. United Nations Department of Economic and Social Affairs, Sustainable Development Goals. Sustainable Development Goals. In: The sustainable development goals report 2017. United Nations Department of Economic and Social Affairs; 2017. https://doi.org/10.18356/4d038e1e-en.
2. Raj SSR, Dhas JER, Jesuthanam CP. Challenges on machining characteristics of natural fiber-reinforced composites – a review. J Reinforc Plast Compos 2021;40:41–69.
3. Stoof D, Pickering K. 3D printing of natural fibre reinforced recycled polypropylene. In: Processing and Fabrication of Advanced Materials; 2017:668–91 pp. https://researchcommons.waikato.ac.nz/handle/10289/11095.
4. Cisneros-López EO, Pal AK, Rodriguez AU, Wu F, Misra M, Mielewski DF, et al. Recycled poly(lactic acid)–based 3D printed sustainable biocomposites: a comparative study with injection molding. Mater Today Sustain 2020;7–8:100027.
5. Chiarakorn S, Permpoonwiwat CK, Nanthachatchavankul P. Cost Benefit Analysis of Bioplastic Production in Thailand. Jan; 2011:2554 p. https://so01.tci-thaijo.org/index.php/econswu/article/view/75041.
6. ASTM. Standard Test Method for Compressive Properties of Rigid Plastics. In: Standard test method for the properties of rigid plastics. American Society for Testing and Materials; 2015:19428–2959 pp. https://doi.org/10.1520/D0695-15.
7. Fernandes J, Deus AM, Reis L, Vaz MF, Leite M. Study of the influence of 3D printing parameters on the mechanical properties of PLA. In: Proceedings of the International conference on progress in additive manufacturing; 2018:547–52 pp. https://doi.org/10.25341/D4988C.
8. Tang C, Liu J, Yang Y, Liu Y, Jiang S, Hao W. Effect of process parameters on mechanical properties of 3D printed PLA lattice structures. Compos Part C: Open Access 2020;3:1–12.

9. Latif R, Wakeel S, Khan NZ, Noor Siddiquee A, Lal Verma S, Akhtar Khan Z. Surface treatments of plant fibers and their effects on mechanical properties of fiber-reinforced composites: a review. J Reinforc Plast Compos 2019;38:15–30.

10. Le Duigou A, Correa D, Ueda M, Matsuzaki R, Castro M. A review of 3D and 4D printing of natural fibre biocomposites. Mater Des 2020;194:108911.

11. Wei C, Gu H, Sun Z, Cheng D, Chueh YH, Zhang X, et al. Ultrasonic material dispensing-based selective laser melting for 3D printing of metallic components and the effect of powder compression. Addit Manuf 2019;29:100818.

12. Nordin NIAA, Ariffin H, Hassan MA, Shirai Y, Ando Y, Ibrahim NA, et al. Superheated steam treatment of oil palm mesocarp fiber improved the properties of fiber-polypropylene biocomposite. Bioresources 2017;12:68–81.

13. Bashir MA. Use of dynamic mechanical analysis (DMA) for characterizing interfacial interactions in filled polymers. Solids 2021;2:108–20.

14. Mustapa IR, Shanks RA, Kong I, Daud N. Morphological structure and thermomechanical properties of hemp fibre reinforced poly(lactic acid) Nanocomposites plasticized with tributyl citrate. Mater Today Proc 2018;5:3211–8.

15. Torrado AR, Shemelya CM, English JD, Lin Y, Wicker RB, Roberson DA. Characterizing the effect of additives to ABS on the mechanical property anisotropy of specimens fabricated by material extrusion 3D printing. Addit Manuf 2015;6:16–29.

Muhammad Aizuddin Mohamad, Aidah Jumahat* and
Napisah Sapiai

17 Flexural analysis of hemp, kenaf and glass fibre-reinforced polyester resin

Abstract: Natural fibres have a high potential to replace synthetic fibres such as glass in a variety of applications. However, natural fibre-reinforced composites still have some limitations with respect to the mechanical performance especially in high load bearing capabilities. The hybridization of natural fibres with synthetic fibres in the same matrix has proven to create a balancing effect and enhanced the composites performance. Besides that, fibre architectures that include fibres continuity, fibres orientation, fibres arrangement and fibres interlocking are also considered to enhance the overall performance of the composites. In this study, the hemp mat, kenaf mat and glass chopped strand mat were hybridised with woven glass fibres, respectively in polyester resin to form 12 systems of the composites. The hybridization effects of different fibre core material, fibre core thickness and fibre arrangement on flexural response were investigated according to ASTM D7264. The results indicated that hybrid CSM glass/woven glass composite showed the highest flexural strength and modulus compared to hemp/woven glass and kenaf/woven glass composites, with about 377.15 ± 48.41 MPa and 16.74 ± 7.15 GPa. Among natural fibres, kenaf fibre (2WG/K/2WG) composite showed better flexural properties compared to hemp fibre (2WG/H/2WG) composite. 2WG/2G/2WG composites with two plies of CSM glass showed maximum flexural properties. As for hemp/woven glass and kenaf/glass hybrid composites, the flexural properties reached a maximum value in system arrangement of (2:1:2) but it reduced in the system arrangement of (2:2:2) and (2:4:2). On the evaluation effect of fibre arrangement, hemp, kenaf and glass mat used as core (arrange in the middle; (2:2:2)) showed higher flexural properties as compared to the use as skin (arrange in outer; (1:4:1)). (2WG/2K/2WG) showed better flexural properties than (2WG/2H/2WG) as the core, while (H/4WG/H) showed better flexural properties than (K/4WG/K) as skin.

Keywords: flexural properties; hybridization; natural fibre; polyester resin; synthetic fibre.

---

*Corresponding author: Aidah Jumahat, School of Mechanical Engineering, College of Engineering, Universiti Teknologi MARA (UiTM), 40450, Shah Alam, Selangor, Malaysia,
E-mail: aidahjumahat@uitm.edu.my
Muhammad Aizuddin Mohamad and Napisah Sapiai, School of Mechanical Engineering, College of Engineering, Universiti Teknologi MARA (UiTM), 40450, Shah Alam, Selangor, Malaysia

As per De Gruyter's policy this article has previously been published in the journal Physical Sciences Reviews. Please cite as:
M. A. Mohamad, A. Jumahat and N. Sapiai "Flexural analysis of hemp, kenaf and glass fibre-reinforced polyester resin" *Physical Sciences Reviews* [Online] 2022. DOI: 10.1515/psr-2022-0051 | https://doi.org/10.1515/9783110769227-017

17.1 Introduction

The considerable use of natural fibres in composite materials for various applications is due to several benefits such as accessibility, inexpensive, renewability, low density, greater specific properties, biodegradability and nontoxicity [1–3]. Natural fibres are lighter than glass fibre, which can reduce the density of a composite by 10–15%; moreover, natural fibres are compatible with the most frequently used polymers, such as polyester, epoxy, and viny-ester [4]. Despite of the desirable properties, the main drawbacks have been reported, which include poor interaction between the fibre and matrix as well as low toughness, along with lower mechanical properties when compared to synthetic fibres [5–7]. Several conditions influence natural fibres, causing variation in fibre quality, and hence changing the performance of natural fibre-reinforced polymer composites. As a consequent, natural fibres that are hybridised with synthetic fibres seem to be more cost effective, environmentally friendly, and capable of achieving the excellent performance of fibre composites [8–10].

Among the natural fibres that exist, hemp and kenaf have gained interest due to their advantages over conventional fibres. Hemp fibres with tensile strength of 550–1110 MPa and 30–70 GPa modulus, and kenaf fibres with tensile strength of 223–930 MPa and 14.5–73 GPa modulus are comparable to E-glass fibres that have tensile strength of 1000–3000 MPa and 70–76 GPa modulus [11, 12]. Hemp is a plant species that is mostly cultivated in Europe and Asia [11, 13]. Since a decade ago, hemp fibres have been used as rope, textiles, garden mulch, various construction materials, and animal beddings [5, 14–16]. Kenaf is a plant with fibrous structure that is rigid, robust, and resilience. Kenaf plants have been cultivated for about 4000 years in Africa, Asia, America, and Europe. Kenaf fibres are one of the most significant bast fibres and are primarily utilized in manufacturing paper and rope [12, 16–19]. Currently, hemp and kenaf fibres can be utilized to make various composites in recent advancements as insulators, sound-damping and resonance materials, construction engineering materials, containers for food, coarse fabric, liquid and oil-absorbent materials, etc. [11–13, 15–21].

In the composite industry, glass fibre as the synthetic reinforcement is commonly utilized due to its excellent mechanical properties, resistance to corrosion and affordable cost [8]. Glass fibre is a remarkable material because of its excellent strength-to-weight ratio, extensive availability, and lower cost than other fibres. In general, there are different types of glass fibres being used based on the usage such as E-glass, E-CR glass, S-glass and AR glass [22]. Glass fibre has the properties of being lightweight as a quarter of steel and has a high compressive strength because it is widely used as a supporting structure.

From the literature, hybridization techniques are preferred to synergistically employ natural and synthetic fibre reinforcements to overcome the limitations of the natural fibre-reinforced composites. Sapiai et al. [12, 16, 18, 19] in their studies found that the hybridisation kenaf-reinforced epoxy composites with glass fibre increases the tensile, flexural, compression, impact properties. The development of flax/carbon fibres hybrid composites by Dhakal et al. [23] revealed that each of these fibres has a unique specialty which contributes to the final composite performances. It has been suggested that flax fibres improve toughness properties by increasing crack propagation, whereas carbon fibres improve water absorption behaviour, thermal stability, and overall strength and stiffness of the hybrid composites. The similar result was also found by Koradiya et al. [24] in their study of hybrid jute/glass fibres composites.

Even though the hybridisation of different fibre gives impact to mechanical performance of the composite, the strength of hybrid composites also depends on a lot of factors. Fibre architectures is one of the factors that influence the mechanical performance. Fibre architectures includes fibres continuity, fibres orientation, fibres arrangement and fibres interlocking which must be considered in enhancing the overall performance of the composites. Therefore, in this present study, the hybrid natural/synthetic composites will be developed. The hemp, kenaf and glass mat fibres will be hybridised with woven type of glass fibres and polyester as a matrix resin. The arrangement of hemp, kenaf and glass mat fibres will be varied, which is arranged in the middle (core) and outer (skin). In addition, the different plies of hemp, kenaf and glass mat fibres in the middle arrangement (core) are also investigated.

17.2 Methods

17.2.1 Materials

The hybrid composites were developed using natural fibres; hemp and kenaf mat, and synthetic fibre; glass fibre in woven and chopped strand mat type as reinforcement materials and polyester resin as polymer matrix. Woven and chopped strand mat (CSM) types of glass fibres were supplied by Vistec Technology, Puchong, Selangor and Carbon Tech Global Sdn. Bhd., Rawang, Selangor, Malaysia, respectively. Meanwhile, both hemp and kenaf fibres were supplied by Innovative Pultrusion Sdn. Bhd.,

Table 17.1: Properties of polyester resin from manufacturer.

Properties	Values
Appearance	Yellowish
Viscosity at 25 °C	3.5 P
Specific gravity at 25 °C	1.10
Water absorption at 23 °C	18 mg
Tensile strength	79 MPa
Elongation at break	4.5%

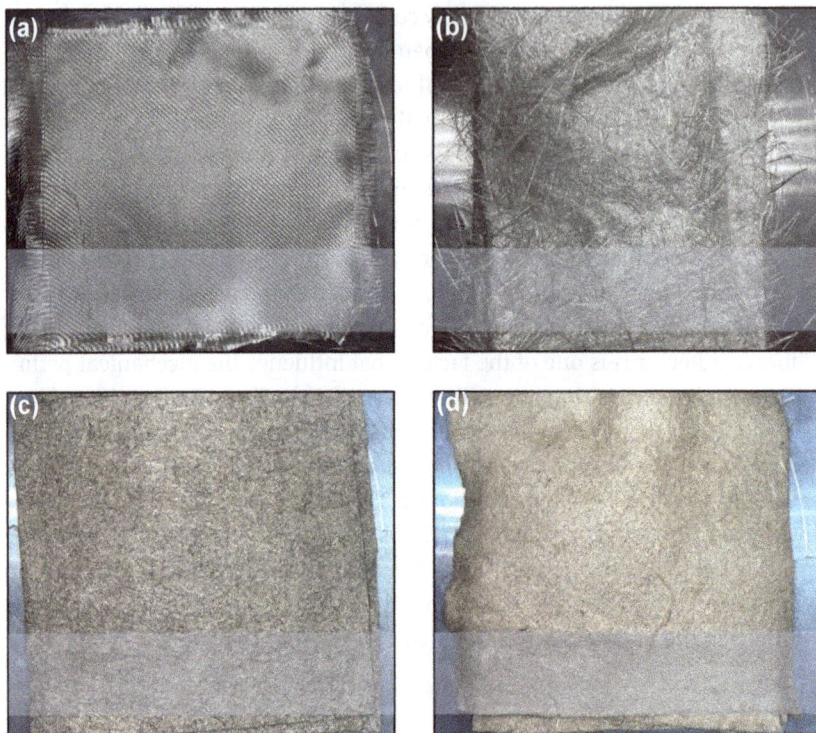

Figure 17.1: Fibres reinforcement (a) woven glass (b) chopped strand mat glass (c) hemp mat and (d) kenaf mat.

Seremban, Negeri Sembilan, Malaysia. Polyester resin, namely Crystic® 272, was obtained from Carbon Tech Global Sdn. Bhd, Rawang, Selangor, Malaysia. Table 17.1 displays the properties of the polyester resin according to the company data sheet. Figure 17.1 shows the reinforcement materials used in the fabrication of hybrid composites.

17.2.2 Fabrication of hybrid composites

The hybrid composites were prepared using resin infusion technique. The hemp, kenaf and glass fibres were layered according to different stacking sequence as shown in Table 17.2 to form 12 composite systems. The polyester resin was prepared in the ratio of 100(resin):2(hardener) before pouring it onto the layered fibres. The layered fibres and polyester resin then experienced vacuum bagging process to spread the polyester resin and eliminate entrapped air. The cured hybrid composites were cut into the desired dimension required for flexural test using vertical bandsaw machine. In this study, the systems were categorised into four different stacking sequences, which were (2:1:2), (2:2:2), (2:4:2) and (1:4:1) based on number of fibres plies. The (2:1:2), (2:2:2), (2:4:2) fibre stacking sequences had three different types of fibres mat i.e., hemp, kenaf and glass were placed in the middle (core) and woven glass fibre in outer layer. Meanwhile in (1:4:1), fibre stacking sequence, hemp, kenaf and glass mat were placed in outer layers and woven glass in the middle. The hybrid composite systems designation and schematic illustration for all types of fibre are displayed in Table 17.2.

Table 17.2: Designation of hybrid composites.

System of 2:1:2	System of 2:2:2	System of 2:4:2	System of 1:4:1
2WG/G/2WG	**2WG/2G/2WG**	**2WG/4G/2WG**	**G/4WG/G**
WG	WG	WG	G
WG	WG	WG	WG
G	G	G	WG
WG	G	G	WG
WG	WG	G	WG
	WG	G	G
		WG	
		WG	
2WG/H/2WG	**2WG/2H/2WG**	**2WG/4H/2WG**	**H/4WG/H**
WG	WG	WG	H
WG	WG	WG	WG
H	H	H	WG
WG	H	H	WG
WG	WG	H	WG
	WG	H	H
		WG	
		WG	
2WG/K/2WG	**WG/2K/2WG**	**2WG/4K/2WG**	**K/4WG/K**
WG	WG	WG	K
WG	WG	WG	WG
K	K	K	WG
WG	K	K	WG
WG	WG	K	WG
	WG	K	K
		WG	
		WG	

References	
WG	Woven glass
G	Chopped strand mat glass
H	Hemp mat
K	Kenaf mat

17.2.3 Flexural test

The flexural test was performed according to ASTM D7264. The hybrid composites with the dimension of 80 × 13 mm were tested using Instron 3382 Universal Testing Machine (UTM) with a crosshead speed of 2 mm/min. The three-point bending mode with support span of 60 mm was used to perform the test as shown in Figure 17.2. Approximately five specimens of hybrid composites for each system were tested to get an average of flexural properties: flexural strength and modulus.

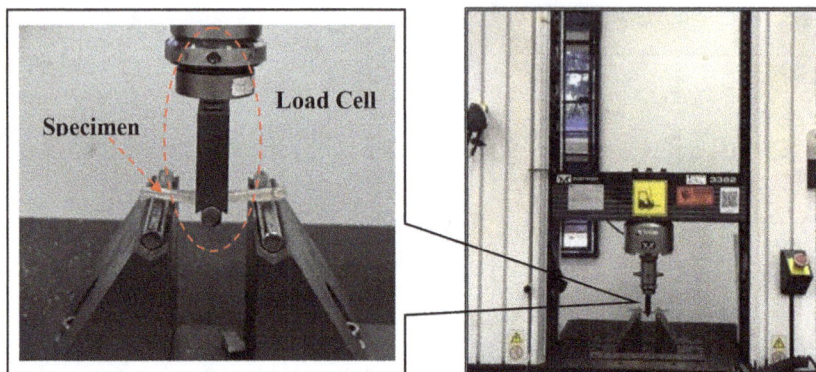

Figure 17.2: Instron 3382 universal testing machine with flexural test set up.

To obtain the average flexural properties, the values of flexural stress, flexural strain and flexural modulus were calculated by using of Eqs. (17.1)–(17.3), respectively.

$$\sigma = \frac{3PL}{2bd^2} \tag{17.1}$$

$$\epsilon = \frac{6Dd}{L^2} \tag{17.2}$$

$$E_f = \frac{mL^3}{4bd^3} \tag{17.3}$$

σ = Stress (MPa), ϵ = Strain (unitless), E_f = Flexural Modulus of Elasticity (MPa), P = Load (N), L = support span (mm), b = Specimen width, d = Specimen thickness (mm), D = Instantaneous deflection (mm), m = Slope of load-deflection curve (N/mm).

17.3 Results and discussion

The flexural properties of hybrid composites were evaluated based on the parameters of (1) different types of fibre core material; either hemp (H), kenaf (K) or glass (G) mat was arranged in the middle of composite, (2) fibre core thickness; either 1, 2 or 4 plies of hemp (H), kenaf (K) and glass (G), and (3) fibre arrangement; either hemp (H), kenaf (K) and glass (G) was arranged in the middle (core) or outer layer (skin) of the composites. The flexural modulus, flexural strength, and flexural strain of the composites, as derived from the stress-strain curves, were analysed and summarized.

17.3.1 Effect of different fibre core material on flexural properties of hybrid composites

Table 17.3 exhibits the flexural properties of 2:1:2 arrangement of hybrid composites by referring to Figure 17.3 as typical stress–strain curves. The hemp (H), kenaf (K) or glass

Table 17.3: Flexural properties of hybrid composites of 2WG/G/2WG, 2WG/H/2WG and 2WG/K/2WG.

Systems	Flexural modulus (GPa)	Flexural strength (MPa)	Flexural strain (%)
2WG/G/2WG	16.74 ± 7.15	377.15 ± 48.41	2.94 ± 0.75
2WG/H/2WG	6.00 ± 0.21	125.58 ± 3.76	2.42 ± 0.17
2WG/K/2WG	12.87 ± 2.03	160.17 ± 5.20	1.40 ± 0.19

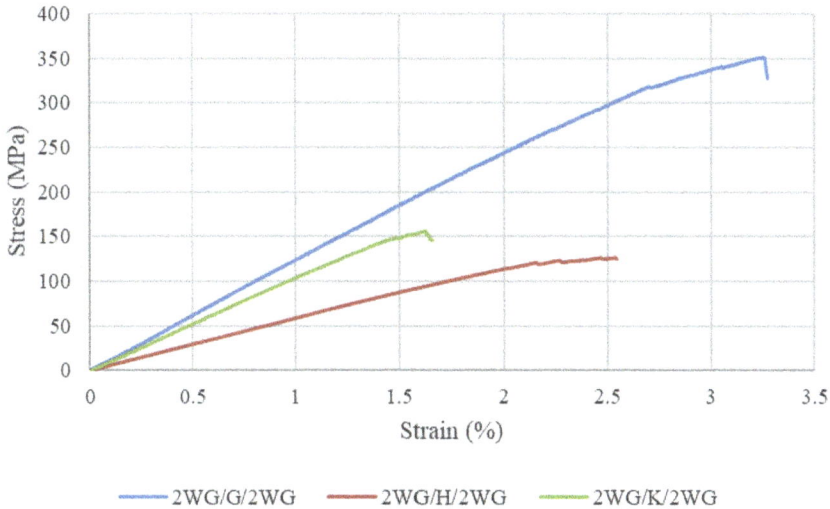

Figure 17.3: Typical stress–strain curves of hybrid composites.

(G) mat fibres were arranged in the middle of composite as core material, while woven glass fibre was arranged in outer layer as skin. The three systems were coded as 2WG/G/2WG, 2WG/H/2WG, and 2WG/K/2WG for CSM glass/woven glass, hemp/woven glass, and kenaf/woven glass hybrid composites, respectively.

Figure 17.4 displays that 2WG/G/2WG composite owned the highest flexural strength and modulus values compared to 2WG/H/2WG, and 2WG/K/2WG. The flexural strength and modulus value for 2WG/G/2WG composite were recorded at about 377.15 ± 48.41 MPa and 16.74 ± 7.15 GPa. Flexural strength of 125.58 ± 3.76 MPa and 160.17 ± 5.20 MPa were recorded for 2WG/H/2WG, and 2WG/K/2WG, respectively. For flexural modulus, 2WG/H/2WG composite recorded about 6.00 ± 0.21 GPa, 53.40% lower than 2WG/K/2WG which recorded about 12.87 ± 2.03 GPa. As expected, the CSM glass fibre as a core resulted in a higher flexural value when compared to hemp and kenaf as core. The evidence presented thus far supports the idea of Shireesha et al. [25] that the orientation of chopped strand mat glass composes excellent hybridization to have better bonding between fibre and matrix with higher strength and stiffness compared to natural fibres that are more specific to certain stiffness due to its

(a) Flexural Strength (MPa)

(b) Flexural Modulus (GPa)

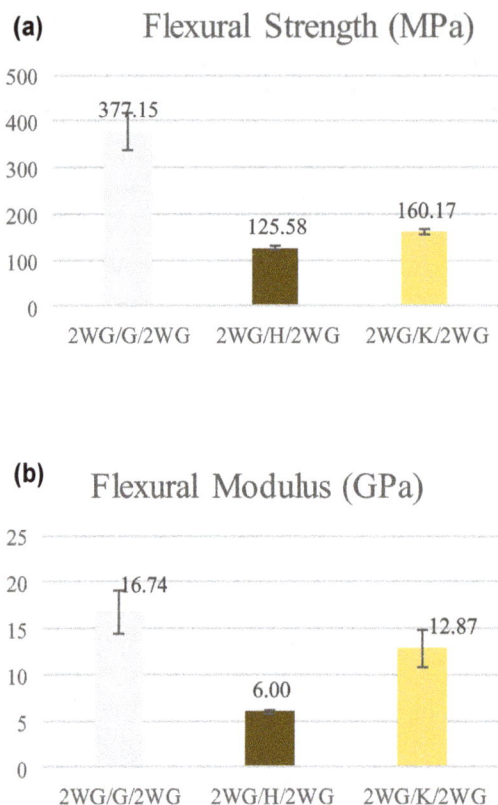

Figure 17.4: (a) Flexural strength and (b) flexural modulus of 2WG/G/2WG, 2WG/H/2WG, and 2WG/K/2WG hybrid composites.

biodegradability properties as reported by Rajak et al. [26]. In addition, the findings clearly demonstrated that the natural fibres' hydrophilic nature has resulted in poor interfacial interaction between fibre and matrix, resulting in weaker flexural properties as compared to glass fibre which was reported by Thyavihalli et al. [27]. Nevertheless, among natural fibre, 2WG/K/2WG showed better flexural properties compared to 2WG/H/2WG composite. These reveal that the incorporation of kenaf fibres into the matrix of hybrid composites is quite effective for the reinforcement as compared to hemp fibres.

17.3.2 Effect of different core thickness on flexural properties of hybrid composites

Table 17.4 exhibits the flexural properties of the hybrid composites according to the thickness of the core used. Core thickness refers to laminate ply between 1, 2, and 4 layers of CSM glass, hemp, and kenaf in hybrid composites. Figure 17.5 reveals that the stress-strain curves of CSM glass/woven glass composites exhibited the highest trend

Table 17.4: Flexural properties of hybrid composites for systems configuration of (2:1:2), (2:2:2) and (2:4:2).

Systems	Laminate plies	Flexural modulus (GPa)	Flexural strength (MPa)	Flexural strain (%)
2WG/G/2WG	2:1:2	16.74 ± 7.15	377.15 ± 48.41	2.94 ± 0.75
2WG/H/2WG		6.00 ± 0.21	125.58 ± 3.76	2.42 ± 0.17
2WG/K/2WG		12.87 ± 2.03	160.17 ± 5.20	1.40 ± 0.19
2WG/2G/2WG	2:2:2	20.96 ± 1.33	493.25 ± 50.10	2.74 ± 0.14
2WG/2H/2WG		4.55 ± 0.25	82.38 ± 28.11	2.15 ± 0.51
2WG/2K/2WG		7.41 ± 0.91	118.66 ± 9.67	1.98 ± 0.32
2WG/4G/2WG	2:4:2	11.46 ± 0.86	276.34 ± 53.75	2.58 ± 0.55
2WG/4H/2WG		3.67 ± 0.78	47.42 ± 6.76	2.18 ± 0.22
2WG/4K/2WG		5.29 ± 0.71	86.18 ± 21.02	1.73 ± 0.2

among other hybrid composites in either single ply, double plies or four plies core used. Figure 17.6 presents that the composite of hybrid CSM glass/woven glass had the highest flexural properties compared to hemp/woven glass and kenaf/glass hybrid composites for all systems arrangement i.e. (2:1:2), (2:2:2) and (2:4:2). The results indicated that 2WG/2G/2WG composites with two plies of CSM glass showed maximum flexural strength of 493.25 ± 50.10 MPa and modulus of 20.96 ± 1.33 GPa. Meanwhile, for system arrangement of (2:4:2), the use of four plies of CSM glass (2WG/4G/2WG composite) reduced the flexural strength by 43.98% and flexural modulus by 45.33% as compared to 2WG/2G/2WG. As for hemp/woven glass and kenaf/glass hybrid composites, the flexural properties reached the maximum value in system arrangement of (2:1:2), in which only one ply of hemp and kenaf mat was used. Obviously, the hemp/woven glass and kenaf/glass hybrids were found to have reduction in flexural properties of both strength and modulus when the core thickness increased as 2 and 4 plies of hemp and kenaf mat were used. A possible explanation for these results might be that the difficulty of the resin to penetrate between the layers as the core thickness or number of plies increases. Apparently, between natural fibres used for hybrid composites, the kenaf/glass hybrid composites showed greater flexural properties compared to hemp/glass hybrid composites. These findings demonstrated that kenaf fibre-reinforced the hybrid composites better in all conditions i.e. core thickness as compared to hemp fibres which has been reported by Akil et al. [28].

17.3.3 Effect of fibre arrangement on flexural properties of hybrid composites

Table 17.5 exhibits the flexural properties of hybrid composites in system arrangement of (2:2:2) and (1:4:1), in which hemp (H), kenaf (K) and glass (G) were arranged in the

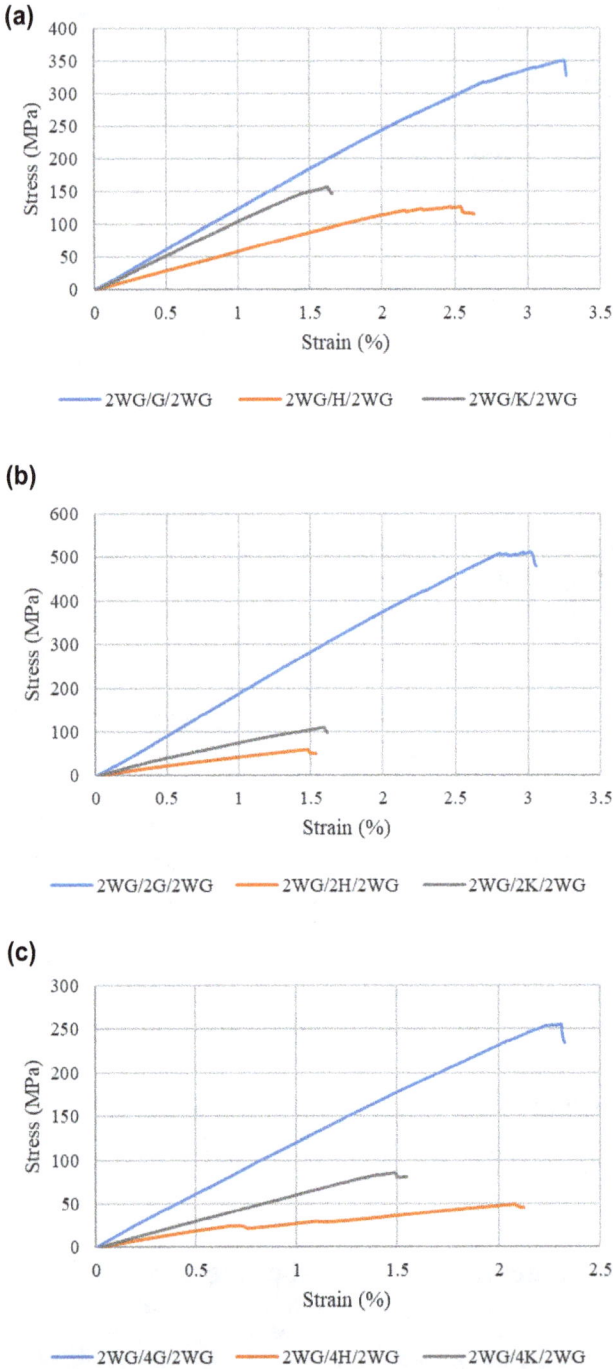

Figure 17.5: Typical stress-strain curves of (a) single ply (2:1:2) configuration (b) double plies (2:2:2) configuration and (c) four plies (2:4:2) configuration of hybrid composites.

(a)

Flexural Strength (MPa)

CSM Glass/Woven Glass Composite ■Hemp/Woven Glass Composite ■ Kenaf/Woven Glass Composite

(b)

Flexural Modulus (GPa)

CSM Glass/Woven Glass Composite ■Hemp/Woven Glass Composite ■ Kenaf/Woven Glass Composite

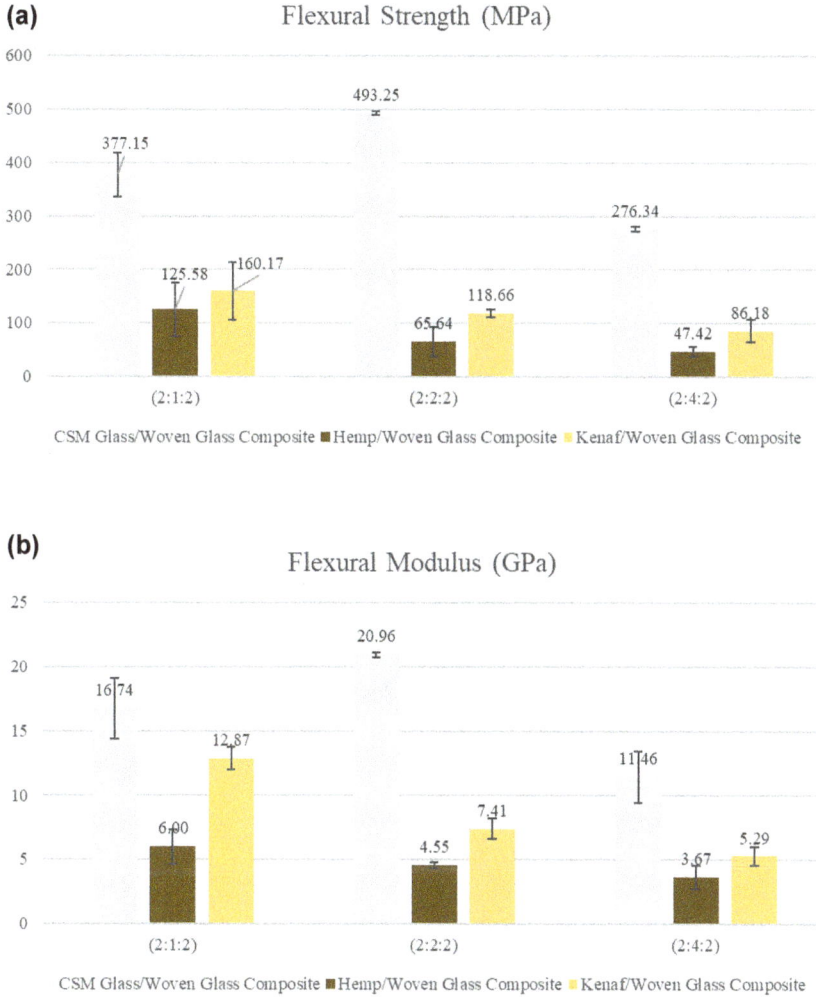

Figure 17.6: (a) Flexural strength and (b) flexural modulus of hybrid composites with systems configuration of (2:1:2), (2:2:2) and (2:4:2).

Table 17.5: Flexural properties of hybrid composites for (2:2:2) and (1:4:1) fibre arrangement.

Systems	Laminate plies	Flexural modulus (GPa)	Flexural strength (MPa)	Flexural strain (%)
2WG/2G/2WG	2:2:2	20.96 ± 1.33	493.25 ± 50.10	2.74 ± 0.14
2WG/2H/2WG		4.55 ± 0.25	82.38 ± 28.11	2.15 ± 0.51
2WG/2K/2WG		7.41 ± 0.91	118.66 ± 9.67	1.98 ± 0.32
G/4WG/G	1:4:1	4.07 ± 1.04	133.19 ± 16.88	3.81 ± 0.52
H/4WG/H		2.19 ± 0.22	58.77 ± 1.93	4.25 ± 0.40
K/4WG/K		1.77 ± 0.25	42.94 ± 2.55	3.39 ± 0.44

middle (core) or outer layer (skin), respectively. In (2:2:2) and (1:4:1), the hybrid composite systems were fabricated using four plies of woven glass fibres and two plies of hemp, kenaf and glass mat. Generally, as illustrated in Figure 17.7, the graph showed

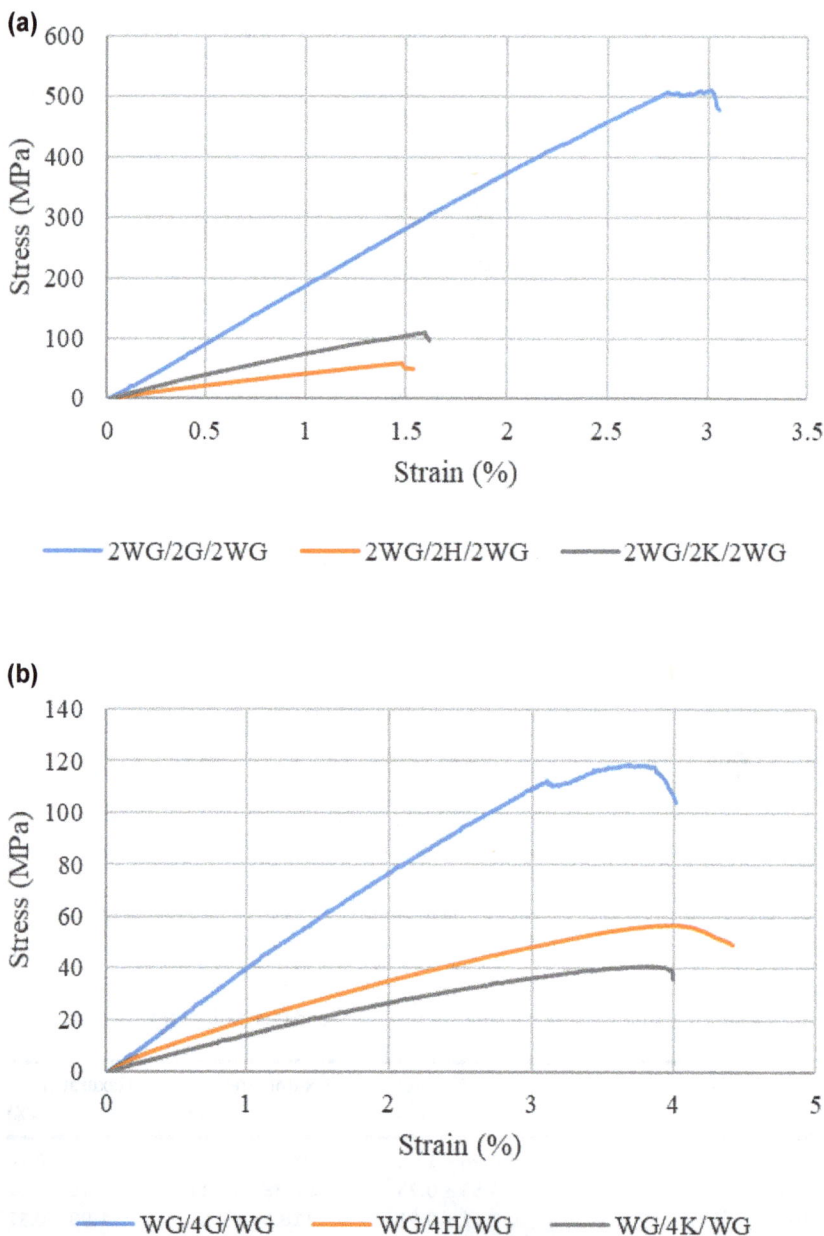

Figure 17.7: Typical stress-strain curves of (a) fibre as core (2:2:2) configuration and (b) fibre as skin (1:4:1) configuration of hybrid composites.

(a)

Flexural Strength (MPa)

(b)

Flexural Modulus (GPa)

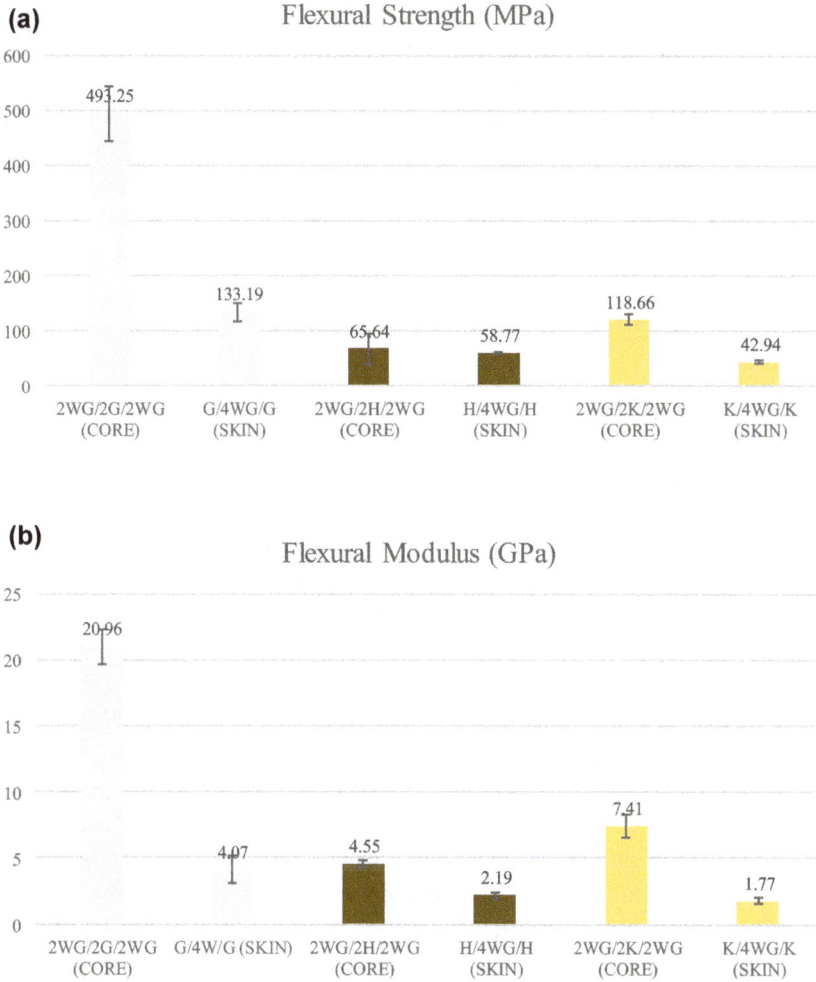

Figure 17.8: (a) Flexural strength and (b) flexural modulus of hybrid composites with system configuration of (2:2:2) and (1:4:1).

that the fibre arrangement of (2:2:2) exhibited higher flexural properties compared to the arrangement (1:4:1) for all types of fibres. Figure 17.8 reveals that the flexural strength and modulus reduced by 80.58 and 73%, respectively for G/4WG/G composite as compared to 2W/2G/2W composite. In the case of comparing the natural fibres, kenaf hybrid composite (2WG/2K/2WG) showed better flexural strength and modulus than hemp hybrid composite (2WG/2H/2WG) as the core. A significant value of 44.68% difference between them really showed that kenaf was much better to be arranged as the core of hybrid laminates. In contrast, hemp hybrid composite (H/4WG/H) showed better flexural strength and modulus than kenaf hybrid composite (K/4WG/K) as skin. It was observed that H/4W/H composite with 58.77 ± 1.93 MPa of flexural strength and

2.19 ± 0.22 GPa of flexural modulus was better than K/4W/K which exhibited flexural strength of 42.94 ± 2.55 MPa and flexural modulus of 1.77 ± 0.25 GPa when subjected to three point bending test. Even with only a slight difference between them, hemp showed better properties as it was arranged as the skin in hybrid composites. A similar trend was reported by V. Dinesh and his colleagues [20]. It is outstanding that the fibre type and the number of fibre layers are primarily in charge of the strength enhancement of the composite. Overall, these cases supported the view of Hunain et al. [29] in which differences in composite strength with different fibre types and the number of loading layers are apparent.

17.4 Conclusions

The effects of the different types of (1) fibre core material; hemp, kenaf or glass mat; (2) fibre core thickness; 1, 2 and 4 plies of hemp, kenaf and glass, and (3) fibre arrangement; hemp, kenaf and glass mat arranged in the middle (core) or outer layer (skin) on flexural response were successfully investigated. It can be concluded that:

- 2WG/G/2WG composite showed the highest flexural strength and modulus compared to 2WG/H/2WG, and 2WG/K/2WG, with about 377.15 ± 48.41 MPa and 16.74 ± 7.15 GPa. Among natural fibres, kenaf fibre (2WG/K/2WG) composite showed better flexural properties compared to hemp fibre (2WG/H/2WG) composite.
- 2WG/2G/2WG composites with two plies of CSM glass showed maximum flexural properties. Meanwhile, the use of four plies of CSM glass (2WG/4G/2WG) composite reduced the flexural properties. As for hemp/woven glass and kenaf/glass hybrid composites, the flexural properties reached maximum value in the system arrangement of (2:1:2) and reduced in the system arrangement of (2:2:2) and (2:4:2).
- hemp, kenaf and glass mat used as core (arrange in the middle; (2:2:2)) showed higher flexural properties as compared to its use as skin (arrange in outer; (1:4:1). (2WG/2K/2WG) showed better flexural properties than (2WG/2H/2WG) as the core, while (H/4WG/H) showed better flexural properties than (K/4WG/K) as skin.

Acknowledgment: The authors thank the Carbon Tech Global Sdn. Bhd. (CTG) Rawang for the technical supports.

References

1. Das SC, Ashek-E-Khoda S, Sayeed MA, Suruzzaman, Paul D, Dhar SA, et al. On the use of wood charcoal filler to improve the properties of natural fiber reinforced polymer composites. Mater Today Proc 2020;44:926–9.

2. Kumar PSS, Allamraju KV. A review of natural fiber composites [jute, sisal, kenaf]. Mater Today Proc 2019;18:2556–62.

3. Vinod A, Sanjay MR, Suchart S, Jyotishkumar P. Renewable and sustainable biobased materials: an assessment on biofibers, biofilms, biopolymers and biocomposites. J Clean Prod 2020;258: 120978.

4. Jawaid M, Abdul Khalil HPS. Cellulosic/synthetic fibre reinforced polymer hybrid composites: a review. Carbohydr Polym 2011;86:1–18.

5. Akter N, Saha J, Das SC, Khan MA. Effect of bitumen emulsion and polyester resin mixture on the physico-mechanical and degradable properties of jute fabrics. Fibers 2018;6:44.

6. Bourmaud A, Beaugrand J, Shah DU, Placet V, Baley C. Towards the design of high-performance plant fibre composites. Prog Mater Sci 2018;97:347–408.

7. Faruk O, Bledzki AK, Fink HP, Sain M. Biocomposites reinforced with natural fibers: 2000–2010. Prog Polym Sci 2012;37:1552–96.

8. Das SC, Paul D, Saha J. Study on the mechanical properties of non-woven glass fiber reinforced polyester composites. In: International conference on computer, communication, chemical, materials and electronic engineering. Higher Education Quality Enhancement Project, Rajshahi, Bangladesh; 2017:180–3 pp.

9. Sathishkumar TP, Naveen J, Satheeshkumar S. Hybrid fiber reinforced polymer composites – a review. J Reinforc Plast Compos 2014;33:454–71.

10. Sood M, Dwivedi G. Effect of fiber treatment on flexural properties of natural fiber reinforced composites: a review. Egypt J Petrol 2018;27:775–83.

11. Manaia JP, Manaia AT, Rodriges L. Industrial hemp fibers: an overview. Fibers 2019;7:106.

12. Sapiai N, Jumahat A, Jawaid M, Midani M, Khan A. Tensile and flexural properties of silica nanoparticles modified unidirectional kenaf and hybrid glass/kenaf epoxy composites. Polymers 2020;12:1–11.

13. Bhoopathi R, Ramesh M, Deepa C. Fabrication and property evaluation of banana-hemp-glass fiber reinforced composites. Procedia Eng 2014;97:2032–41.

14. Castro DO, Ruvolo-Filho A, Frollini E. Materials prepared from biopolyethylene and curaua fibers: composites from biomass. Polym Test 2012;31:880–8.

15. De Fazio D, Boccarusso L, Durante M. Tribological behaviour of hemp, glass and carbon fibre composites. Biotribology 2020;21:100113.

16. Sapiai N, Jumahat A, Mahmud J. Flexural and tensile properties of kenaf/glass fibres hybrid composites filled with carbon nanotubes. J Teknol 2015;76:115–20.

17. Hamidon MH, Sultan MTH, Ariffin AH, Shah AUM. Effects of fibre treatment on mechanical properties of kenaf fibre reinforced composites: a review. J Mater Res Technol 2019;8:3327–37.

18. Sapiai N, Jumahat A, Jawaid M, Khan A. Effect of MWCNT surface functionalisation and distribution on compressive properties of kenaf and hybrid kenaf/glass fibres reinforced polymer composites. Polymers 2020;12:1–18.

19. Sapiai N, Jumahat A, Shaari N, Tahir A. Mechanical properties of nanoclay-filled kenaf and hybrid glass/kenaf fiber composites. Mater Today Proc 2020;46:1787–91.

20. Dinesh V, Shivanand HK, Vidyasagar HN, Chari VS. Investigation of mechanical properties of kenaf, hemp and E-glass fiber reinforced composites. AIP Conf Proc 2018;1943:020117.

21. Väisänen T, Batello P, Lappalainen R, Tomppo L. Modification of hemp fibers (Cannabis Sativa L.) for composite applications. Ind Crop Prod 2018;111:422–9.

22. Spagnuolo S, Rinaldi Z, Donnini J, Nanni A. Physical, mechanical and durability properties of GFRP bars with modified acrylic resin (modar) matrix. Compos Struct 2021;262:113557.

23. Dhakal HN, Zhang ZY, Guthrie R, MacMullen J, Bennett N. Development of flax/carbon fibre hybrid composites for enhanced properties. Carbohydr Polym 2013;96:1–8.

24. Koradiya SB, Patel JP, Parsania PH. The preparation and physicochemical study of glass, jute and hybrid glass-jute bisphenol-C-based epoxy resin composites. Polym Plast Technol Eng 2010;49: 1445–9.
25. Shireesha Y, Nandipati G, Chandaka K. Properties of hybrid composites and its applications: a brief review. Int J Sci Technol Res 2019;8:335–41.
26. Rajak DK, Pagar DD, Menezes PL, Linul E. Fiber-reinforced polymer composites. India: Reinforced Polymer Composites; 2019.
27. Thyavihalli Girijappa YG, Mavinkere Rangappa S, Parameswaranpillai J, Siengchin S. Natural fibers as sustainable and renewable resource for development of eco-friendly composites: a comprehensive review. Front Mater 2019;6:1–14.
28. Akil HM, Omar MF, Mazuki AAM, Safiee S, Ishak ZAM, Abu Bakar A. Kenaf fiber reinforced composites: a review. Mater Des 2011;32:4107–21.
29. Hunain MB, Al-Turaihi AS, Alnomani SN. Tensile and charpy impact behavior of E-glass/ unsaturated polyester laminated composite material at elevated temperature. J Eng Sci Technol 2021;16:1547–60.

Mohd Azrul Jaafar, Shahrul Azam Abdullah*, Aidah Jumahat,
Mohamad Asrofi Muslim, Napisah Sapiai and
Raymond Siew Teng Loy

18 Effect of stacking sequence on tensile properties of glass, hemp and kenaf hybrid composites

Abstract: Natural fibre reinforced polymer composites have the potential to be utilized at various applications due to their non-hazardous effect to the environment, bio-degradable properties as well as enhanced mechanical characteristics. Nevertheless, mechanical properties of these composites are complicated to understand and predicted due to complex interaction between matrix and different type of fibres, fibres architecture and fibres arrangement. Therefore, this paper aims to study the effect of various types of fibres; kenaf mat, hemp mat and Glass Chopped Strand Mat as a core, core thickness; 1, 2 and 4 layers, and fibre arrangements; kenaf mat, hemp mat and Glass Chopped Strand Mat arranged in middle layer (core) or outer layer (skin) on tensile properties of hybrid composites. The hybrid composite specimens were prepared through combinations of hand lay-up and vacuum methods in which both methods are commonly employed techniques in industry. There are four types of fibre arrangement systems involved: (2:1:2), (2:2:2), (2:4:2) and (1:4:1). As expected, the glass fibres hybrid composites had the highest tensile performance compared to other hybrid composites. The fibre arrangement (2-1-2) was the best option for all types of fibres, while the use of 4 layers of kenaf mat, hemp mat and Glass Chopped Strand Mat as core material reduced the tensile properties. In comparison of (2-2-2) and (1-4-1), kenaf mat and Glass Chopped Strand Mat performed better as a core, while hemp mat performed better as skin.

Keywords: hemp fibres; hybridisation; kenaf fibres; polyester resin; tensile properties.

---

*Corresponding author: Shahrul Azam Abdullah, School of Mechanical Engineering, College of Engineering, Universiti Teknologi MARA, 40450, Shah Alam, Selangor, Malaysia,
E-mail: shahrulazam@uitm.edu.my
Mohd Azrul Jaafar, Aidah Jumahat, Mohamad Asrofi Muslim and Napisah Sapiai, School of Mechanical Engineering, College of Engineering, Universiti Teknologi MARA, 40450, Shah Alam, Selangor, Malaysia
Raymond Siew Teng Loy, Carbon Tech Global Sdn Bhd, PT-1361, Jalan Kesidang 5 Kampung Mohd Taib Kawasan Industri Kampung Sungai Choh, 48000, Rawang, Selangor, Malaysia

As per De Gruyter's policy this article has previously been published in the journal Physical Sciences Reviews. Please cite as: "Effect of stacking sequence on tensile properties of glass, hemp and kenaf hybrid composites" *Physical Sciences Reviews* [Online] 2022. DOI: 10.1515/psr-2022-0052 | https://doi.org/10.1515/9783110769227-018

18.1 Introduction

Natural fibres reinforced polymer composites have been gaining great attention in several applications due to its eco-friendly nature, sustainability, stronger and low cost. Natural fibres are segmented based on their origins, i.e., they originated from plants, animals, or minerals. Plant fibres are the most prevalent natural fibres utilized as reinforcement in fibre reinforced composites [1, 2]. Plants that produce cellulose fibres can be divided into several categories, including bast fibres (jute, flax, ramie, hemp and kenaf), seed fibres (cotton, coir and kapok), leaf fibres (sisal, pineapple and abaca), grass and reed fibres (rice, corn and wheat), core fibres (hemp, kenaf and jute), and other kinds (wood and roots) [3]. Bast and core fibres such as flax, jute, hemp, ramie, sisal and kenaf are the most often used plant-based natural fibres in applications [4]. Natural fibres have lower density (1.2–1.6 g/cm^3) than glass fibre (2.4 g/cm^3), allowing for the creation of lightweight composites.

Kenaf (*Hibiscus cannabinus*) in the family of *alvacea* is a plant that grows in southern Asia. Kenaf can grow in a variety of weather conditions. It can reach a height of about 3 m and a base diameter of 3–5 cm. They are unbending, strong and versatile, with amazing pesticide opposition. It is also much planted in Africa, Asia, America and Europe. It is primarily utilized in manufacturing rope and paper, as one of the most significant bast fibres [5]. Kenaf fibre composite is made effectively with different resins and offers better sliding wear obstruction when blended in with polyester than epoxy [6].

Hemp (*Cannabis sativa* plant species) are cultivated mostly in Europe and Asia. It may very well be utilized to make a wide scope of products (paper, rope, textiles, etc.). Along with bamboo, hemp is among the quickest developing plants on Earth. It may reach heights of 1.2–4.5 m and a diameter of 2 cm [7, 8]. It is now utilized to make various composites in recent advancements [9]. Currently, hemp is widely used in the car manufacturing process by the world's leading manufacturers including Audi, BMW, Ford, GM, Chrysler, Honda, Iveco, Lotus, Mercedes, Mitsubishi, Porsche, Saturn, Volkswagen and Volvo.

Glass is a material composed of many extremely fine glass fibres. It is also a synthetic fibre, which is popular and widely used especially in engineering application. There are three types of glass fibres: C-glass, E-glass and S-glass. As we know, C-glass is a material that is used in structural design due to its excellent corrosion resistance. E-glass, on the other hand, is intended for electrical applications. Finally, S-glass is chosen for its high strength and is widely used in polymer matrix composites that require high strength [10].

A hybrid composite is one that contains two or more types of fibres within a single matrix [11]. Environmental concerns have compelled researchers to use hybridization to create novel composites with a variety of natural resource reinforcements. The hybridization of fillers and natural fibres results in improved mechanical properties of the composites. The current study compares the tensile properties of various hybrid glass (woven and chopped strand mat) and natural fibre (hemp and kenaf) architectures with a polyester matrix.

18.2 Material and method

18.2.1 Material

Glass in form of Glass Chopped Strand Mat and woven, hemp mat and kenaf mats were supplied by Innovative Pultrusion Sdn. Bhd., Seremban, Negeri Sembilan. Unsaturated ISO polyester resin was purchased from Carbon Tech Global Sdn. Bhd, Rawang, Selangor, Malaysia. Figure 18.1 show the fibres (Glass Chopped Strand Mat, Kenaf, Woven Glass and Hemp) used in the fabrication of hybrid composite samples.

18.2.2 Preparation of hybrid composite

All the hybrid composite laminates were fabricated by the combination of hand lay-up and vacuum bagging techniques over a silicone mould size of 500 × 500 mm. The polyester resin was prepared with the ratio of 100:2 (Polyester: Catalyst). In this study, the fibres were layered according to the designated layer sequence, i.e. fibre arrangement as shown in Table 18.1. The hybrid composites were prepared in four different fibre arrangements; where the glass, hemp or kenaf mat were arranged in middle layer as core (2-1-2, 2-2-2, 2-4-2) and glass, hemp or kenaf mat were arranged in outer layer as skin (1-4-1). All the glass, hemp and kenaf mat fibres were hybridised with woven type of glass fibres. A detailed description of the fabrication process is illustrated in Figure 18.2.

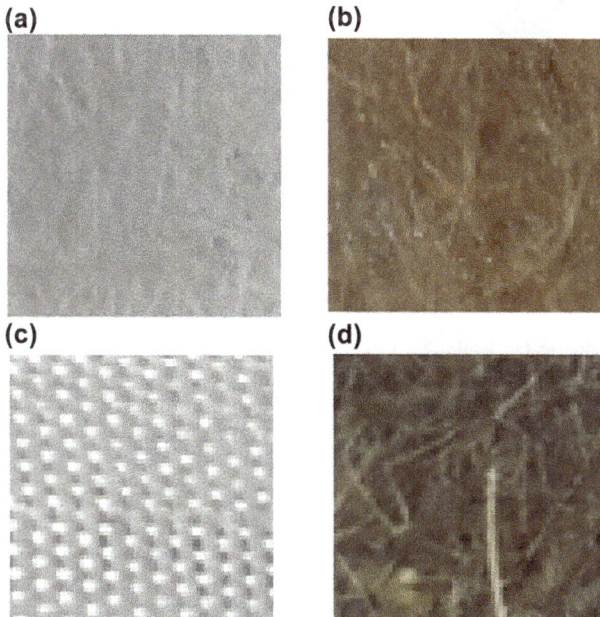

Figure 18.1: Images of fibres. (a) Glass Chopped Strand Mat, (b) Kenaf, (c) Woven Glass and (d) Hemp.

Table 18.1: Designated layer sequences (fibre arrangements) of hybrid composite laminates.

2-1-2	2-2-2	2-4-2	1-4-1
2WG/CG/2WG	2WG/2CG/2WG	2WG/4CG/2WG	CG/4WG/CG
2WG/H/2WG	2WG/2H/2WG	2WG/4H/2WG	H/4WG/H
2WG/K/2WG	2WG/2K/2WG	2WG/4K/2WG	K/4WG/K

Woven Glass (WG)
Glass Chopped Strand Mat (CG)
Hemp (H)
Kenaf (K)

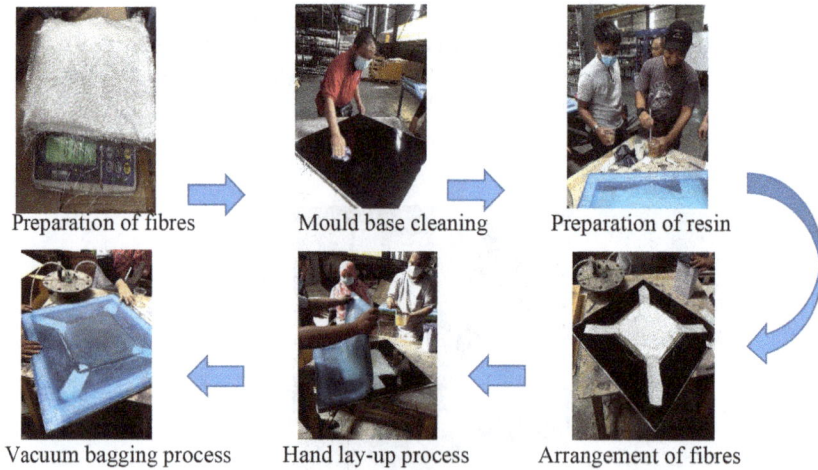

Preparation of fibres — Mould base cleaning — Preparation of resin

Vacuum bagging process — Hand lay-up process — Arrangement of fibres

Figure 18.2: Hybrid composite laminates fabrication.

18.2.3 Tensile test

A universal testing equipment (INSTRON 3382) was used to determine the tensile properties of hybrid composite laminates. The test was conducted in accordance to ASTM D3039 standards. The crosshead speed was set at 2 mm/min. Five identical specimens with dimension of 250 × 25 mm were examined for each system of hybrid composite laminates. Figure 18.3 illustrates the schematic of specimen specifi-cation and Figure 18.4 shows the INSTRON 3382 Universal Testing Machine set-up for tensile test.

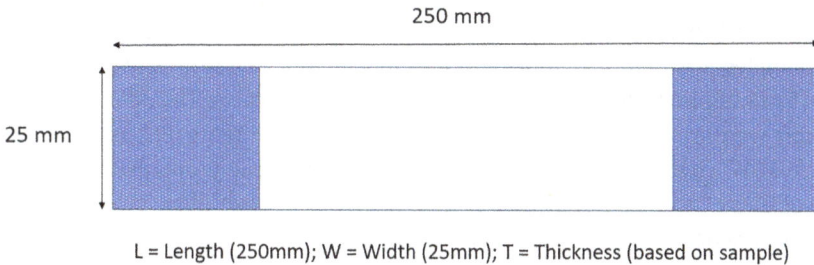

L = Length (250mm); W = Width (25mm); T = Thickness (based on sample)

Figure 18.3: Specimen specification for tensile test.

18.3 Results and discussion

The tensile properties of hybrid composite laminates; kenaf mat/woven glass, hemp mat/woven glass and Glass Chopped Strand Mat glass/woven glass were studied. The finding was thoroughly discussed on the effect of different characters;

(1) core material; which compared the use of hemp, kenaf or glass mat as a core (arranged in the middle layer of hybrid composites).
(2) core thickness; in which different layer of hemp, kenaf and glass mat was used as a core (1, 2 and 4 layers). The 2-1-2, 2-2-2 and 2-4-2 systems arrangements were compared.
(3) fibres arrangement; two systems arrangements were compared which were 2-2-2 and 1-4-1. For 2-2-2 arrangement, hemp, kenaf or glass mat was arranged in the middle as core and woven glass as skin. For 1-4-1 arrangement, hemp, kenaf or glass mat was arranged in outer layer as skin and woven glass as core.

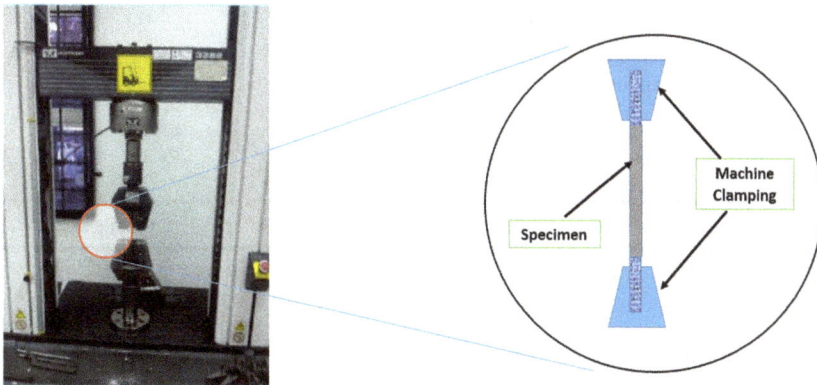

Figure 18.4: Universal testing machine for tensile test.

18.3.1 Effect of different core material on tensile properties of hybrid composite laminates

The tensile properties of 2WG/CG/2WG, 2WG/K/2WG and 2WG/H/2WG hybrid com-posite laminates are shown in Table 18.3 and Figure 18.5. The results are the mean values of five samples tested for each type of composite laminates. The different fibres used as core were arranged in 2/1/2 system arrangement. From Table 18.2, as expected, the 2WG/CG/2WG hybrid composite showed the highest tensile properties. The com-bination of Glass Chopped Strand Mat and woven types of glass fibre, in which both were from synthetic fibres, had higher performance than hybrid kenaf mat/woven glass and hemp mat/woven glass. The tensile strength and modulus for 2WG/CG/2WG hybrid composite were recorded about 230.55 MPa and 9.19 GPa, respectively. Glass Chopped Strand Mat as the core gave a higher tensile value when compared to natural fibres [12]. The result from [13] also showed that Glass gives higher tensile than kenaf. It is equivalent to [14] in which Kenaf and Hemp composites show lower tensile strength than glass fibre composite.

As compared to hemp and kenaf fibres as core, the kenaf fibre hybrid composites of 2WG/K/2WG resulted in higher tensile strength and modulus than hemp fibre hybrid composites of 2WG/H/2WG. The 2WG/K/2WG hybrid composite recorded 46.54%

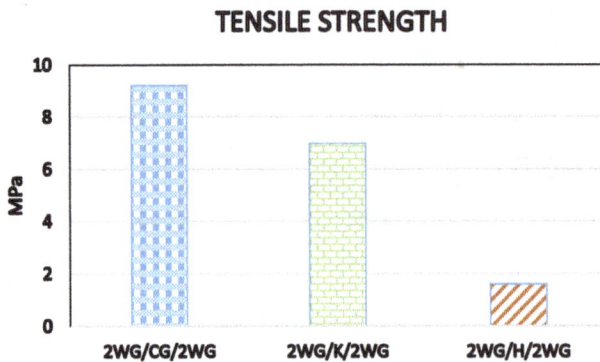

Figure 18.5: Tensile strength for (2:1:2) system hybrid composite.

Table 18.2: Tensile properties of (2:1:2) system hybrid composites.

Systems	Tensile modulus (GPa)	Tensile strength (MPa)	Tensile strain (%)
2WG/CG/2WG	9.19 ± 0.68	230.55 ± 6.44	4.71 ± 0.13
2WG/K/2WG	6.97 ± 3.39	138.57 ± 49.88	3.27 ± 0.44
2WG/H/2WG	2.21 ± 0.06	74.08 ± 7.80	3.19 ± 0.25

higher in strength and 68.29% higher in modulus compared to 2WG/H/2WG hybrid composite. The increase in strength and modulus of 2WG/K/2WG hybrid composites can be ascribed to this effect as the diameter of kenaf fibres is bigger than hemp fibres. The larger diameter of the fibres makes it strong which resulted in higher modulus and strength. Besides that, kenaf outperformed hemp in terms of results, owing to differences in their original properties [1]. Dinesh et al. [15] mentioned that kenaf properties (tensile strength and modulus) were 930 MPa and 53 GPa while hemp properties were 550 MPa and 13 GPa.

18.3.2 Effect of core thickness on tensile properties of hybrid composite laminates

Core thickness refers to laminate ply between 1, 2 and 4 layers of Glass Chopped Strand Mat glass fibre, hemp and kenaf hybridized, which were arranged in the middle of woven glass fibres. In general, the system arrangement of (2-1-2) had the highest tensile properties as compared to (2-2-2) and (2-4-2) for kenaf mat/woven glass and Glass Chopped Strand Mat glass/woven glass hybrid composites (refer Table 18.3). The 2WG/CG/2WG hybrid composites had higher tensile strength and modulus which were about 230.55 MPa and 9.19 ± 0.68 GPa, respectively. The tensile strength reduced 34.69% for 2WG/2CG/2WG and 39.28% for 2WG/4CG/2WG, when increased the layers of Glass Chopped Strand Mat glass. The similar trend was observed for tensile modulus, which reduced 41.46% for 2WG/2CG/2WG and 65.83% for 2WG/4CG/2WG.

For natural fibres hybrid composites, the kenaf mat/woven glass indicated higher tensile properties as compared to hemp mat/woven glass in all systems arrangement. It is the same as reported by Akil et al. [1] where kenaf shows better results than hemp in all conditions. 2WG/K/2WG hybrid composite recorded about 6.97 GPa in modulus and 138.57 MPa in strength.

Table 18.3: Tensile properties of hybrid composites for (2-1-2), (2-2-2) and (2-4-2).

Systems	Tensile modulus (GPa)	Tensile strength (MPa)	Tensile strain (%)
2WG/CG/2WG	9.19 ± 0.68	230.55 ± 6.44	4.71 ± 0.13
2WG/K/2WG	6.97 ± 3.39	138.57 ± 49.88	3.27 ± 0.44
2WG/H/2WG	2.21 ± 0.06	74.08 ± 7.80	3.19 ± 0.25
2WG/2CG/2WG	5.38 ± 2.15	150.57 ± 31.37	2.96 ± 0.36
2WG/2K/2WG	3.22 ± 0.41	87.62 ± 14.33	3.57 ± 0.42
2WG/2H/2WG	2.17 ± 0.42	74.43 ± 9.82	4.33 ± 0.63
2WG/4CG/2WG	3.14 ± 0.60	140.30 ± 16.90	4.08 ± 0.39
2WG/4K/2WG	2.84 ± 0.91	82.11 ± 6.02	3.68 ± 0.35
2WG/4H/2WG	1.61 ± 0.71	44.05 ± 2.38	4.29 ± 0.30

The tensile strength and modulus then decreased when the layers of kenaf mat was increased. This result was different when compared to Hunain et al. [16] where increasing the laminated layers for the same type of fibre of the composite will increase the modulus of elasticity, tensile strength and fracture toughness regardless of fibre orientation. Research by Das et al. [17] also showed opposite result.

For hemp mat/woven glass hybrid composites, the use of 1 or 2 layers hemp fibres had comparable average of tensile strength and modulus, but it decreased when using 4 layers of hemp mat fibre. From the composites specimen observation, the increase of core layers will reduce the efficiency of resin to penetrate inside the hybrid composites. Thus, it resulted in poor adhesive bonding, which later led to reduce the tensile properties.

18.3.3 Effect of fibre arrangement on tensile properties of hybrid composites

In this study, the 2-2-2 and 1-4-1 of fibre arrangement was evaluated on tensile properties of kenaf mat/woven glass, hemp mat/woven glass and Glass Chopped Strand Mat glass/woven glass. In this fibre arrangement, the hemp, kenaf and Glass Chopped Strand Mat were arranged as core (two layers in the middle) for 2-2-2 arrangement and as a skin (one layer in the outer) for 1-4-1 arrangement.

The effect of fibre arrangement on the tensile properties for all hybrid composites is shown in Table 18.4 and Figure 18.6. It can be concluded that in Glass Chopped Strand Mat glass/woven glass and kenaf mat/woven glass hybrid composites, the system arrangement of 2-2-2 resulted in better tensile properties compared to system arrangement of 1-4-1. As for hemp, it shows different results where the 1-4-1 system arrangement has higher tensile properties.

When comparing natural fibre, it was found that hemp mat arranged in skin was better than kenaf. It is the same as reported by Dinesh et al. [15], where hemp showed a slightly higher result than kenaf at skin position. High tensile strength results were shown on the core parts of kenaf (87.62 MPa) as compared to skin (59.00 MPa). This

Table 18.4: Tensile properties of the 2-2-2 and 1-4-1 fibre arrangement of the hybrid composites.

Systems	Tensile modulus (GPa)	Tensile strength (MPa)	Tensile strain (%)
2WG/2CG/2WG	5.38 ± 2.15	150.57 ± 31.37	2.96 ± 0.36
2WG/2K/2WG	3.22 ± 0.41	87.62 ± 14.33	3.57 ± 0.42
2WG/2H/2WG	2.17 ± 0.42	74.43 ± 9.82	4.33 ± 0.63
CG/4WG/CG	4.38 ± 0.73	128.89 ± 30.05	3.49 ± 0.53
H/4WG/H	2.51 ± 0.36	102.29 ± 8.03	5.47 ± 0.12
K/4WG/K	2.24 ± 0.65	59.00 ± 5.07	3.35 ± 0.10

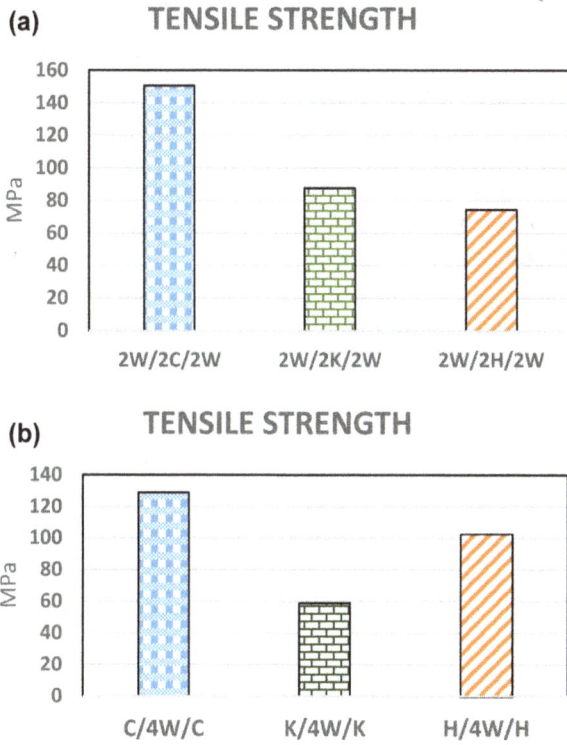

(a) TENSILE STRENGTH

(b) TENSILE STRENGTH

Figure 18.6: Tensile strength of (a) (2:2:2) and (b) (1:4:1) of hybrid composites.

result is supported by Vinoth et al. [18], where the low tensile modulus is shown when kenaf is at skin part due to improper wetting and adhesion of fibre and matrix.

Due to the obvious high tensile property of glass fibre, having the glass fibre type on just the skin part would typically be the best option. It is important to note that when the extreme strength material is being used as the exterior, which would be the principal load-bearing component in tensile strength computation, the tensile strength has become more significant.

18.4 Conclusions

Several hybrid composite laminates were successfully fabricated in this work using Glass Chopped Strand Mat glass, kenaf and hemp mat as reinforcing materials and polyester as the matrix.

– It was observed that the study with different core materials showed that the laminate having kenaf at the core (2WG/K/2GW) was found to have high tensile strength and modulus compared to hemp.

- It was also found that kenaf and hemp showed different properties as the core thickness increased. Kenaf hybrid composite showed a decrease in tensile strength as the thickness increased. The hemp hybrid composites had comparable tensile properties when 1 and 2 layers of hemp were used, but decreased when 4 layers were used.
- Referring to the fibre arrangement, kenaf and hemp showed opposite tensile properties. Kenaf showed a high tensile property when placed as a core, while hemp had a higher tensile property when arranged as a skin.

Acknowledgements: The authors would like to thank Carbon Tech Global Sdn Bhd (CTG) in Rawang, Selangor and Pusat Latihan Pengajar dan Kemahiran Lanjutan (CIAST) Shah Alam, Selangor for the support. The research was conducted at Carbon Tech Global Sdn Bhd (CTG) in Rawang, Selangor.

References

1. Akil HM, Omar MF, Mazuki AAM, Safiee S, Ishak ZAM, Abu Bakar A. Kenaf fiber reinforced composites: a review. Mater Des 2011;32:4107–21.
2. Kumar PSS, Allamraju KV. A review of natural fiber composites [Jute, Sisal, Kenaf]. Mater Today Proc 2019;18:2556–62.
3. Faruk O, Bledzki AK, Fink H, Sain M. Progress in polymer science biocomposites reinforced with natural fibers: 2000–2010. Prog Polym Sci 2012;37:1552–96.
4. Mohammed L, Ansari MNM, Pua G, Jawaid M, Islam MS. A review on natural fiber reinforced polymer composite and its applications. Int J Poly Sci 2015;2015:1–15.
5. Hamidon MH, Sultan MTH, Ariffin AH, Shah AUM. Effects of fibre treatment on mechanical properties of kenaf fibre reinforced composites: a review. J Mater Res Technol 2019;8:3327–37.
6. Matykiewicz D. Hybrid epoxy composites with both powder and thermomechanical fiber filler: a Review of mechanical and thermomechanical properties. Mater 2020;13:1–22.
7. Thiagamani SMK, Krishnasamy S, Muthukumar C, Tengsuthiwat J, Nagarajan R, Siengchin S, et al. Investigation into mechanical, absorption and swelling behaviour of hemp/sisal fibre reinforced bioepoxy hybrid composites: effects of stacking sequences. Int J Biol Macromol 2019;140:637–46.
8. Bhoopathi R, Ramesh M, Deepa C. Fabrication and property evaluation of banana-hemp-glass fiber reinforced composites. Procedia Eng 2014;97:2032–41.
9. Väisänen T, Batello P, Lappalainen R, Tomppo L. Modification of hemp fibers (Cannabis Sativa L.) for composite applications. Ind Crop Prod 2018;111:422–9.
10. Tamrakar S, Kiziltas A, Mielewski D, Zander R. Characterization of kenaf and glass fiber reinforced hybrid composites for underbody shield applications characterization of kenaf and glass fiber reinforced hybrid composites for underbody shield applications. Composites Part B 2021;216: 108805.
11. Sivakumar D, Ng LF, Chew RM, Bapokutty O. Investigation on failure strength of bolted joints woven fabric reinforced hybrid composite. Int Rev Mech Eng 2017;11:138–43.
12. Kennedy ZE, Arul Inigo Raja M. Influence of stacking sequence and hybridization on the mechanical and tribological properties of glass and jute fiber composites. Mater Today Proc 2022;55:220–5.
13. Mishra C, Deo CR, Baskey S. Influence of moisture absorption on mechanical properties of kenaf/ glass reinforced polyester hybrid composite. Mater Today Proc 2021;38:2596–600.

14. Malingam SD, Subramaniam K, Feng NL, Fadzullah SHSMD, Subramonian S. Mechanical properties of plain woven kenaf/glass fiber reinforced polypropylene hybrid composites. Mater Test 2019;61: 1095–100.
15. Dinesh V, Shivanand HK, Vidyasagar HN, Chari VS. Investigation of mechanical properties of kenaf, hemp and E-glass fiber reinforced composites. AIP Conf Proc 2018;1943:1–6.
16. Hunain MB, Al-Turaihi AS, Alnomani SN. Tensile and charpy impact behavior of E-glass/ unsaturated polyester laminated composite material at elevated temperature. J Eng Sci Technol 2021;16:1547–60.
17. Das SC, Paul D, Grammatikos SA, Siddiquee AB, Papatzani S, Vidakis N, et al. Effect of stacking sequence on the performance of hybrid natural/synthetic fi ber reinforced polymer composite laminates. Compos Struct 2021;276:114525.
18. Vinoth KS, Ansari MNM, Begum S, Yahya Z, Atiqah A. Effect of fibre loading on tensile strength of kenaf/glass fibre epoxy hybrid composite for insulator application. Mater Today Proc 2019;29: 123–6.

Mohamad Asrofi Muslim, Aidah Jumahat*, Shahrul Azam Abdullah,
Mohd Azrul Jaafar, Napisah Sapiai and Raymond Siew Teng Loy

19 Investigation on impact properties of different type of fibre form: hybrid hemp/ glass and kenaf/glass composites

Abstract: Natural fibre reinforced polymer composites have high potentials to be used in a variety of applications due to its environmental friendly and biodegradability capabilities. The purpose of this study is to evaluate the effects of core fibre type, core thicknesses, and fibre configurations on the impact behaviour of hybrid natural fibre reinforced polymer (FRP) composites. The samples were made of kenaf, hemp and glass mat fibers, and polyester used as matrix resin. These samples were fabricated using a combination of hand lay-up and vacuum bagging systems. The Instron Dynatup 8250 was used in accordance to ASTM D7136. The results showed that the highest impact properties were in hemp hybrid composites. For fibre arrangement, system (1/4/1) in which kenaf, hemp and glass mat were arranged in outer layer (as skin) resulted a higher energy absorbed compared to system (2/2/2) in which kenaf, hemp and glass mat were arranged in middle layer (as core). The impact properties increased with the increasing of core thickness. These findings are significant for possible applications of natural/synthetic fibre reinforced polymer hybrid composites in the fields of vehicles, biomedical, transportation and other specific application could have benefited for further study in hybrid composite material improvement.

Keywords: hemp fibres; impact properties; kenaf fibres; natural fibre reinforced polymer (NFRP); polyester resin.

19.1 Introduction

Natural fibre composite materials have been extensively used nowadays with the growth of environmental needs and awareness on using eco-friendly materials. It becomes an alternative reinforcement in polymer composites that concerned many researchers' attention due to their advantages over conventional glass and carbon

*Corresponding author: Aidah Jumahat, Faculty of Mechanical Engineering, Universiti Teknologi MARA, 40450, Shah Alam, Selangor, Malaysia, E-mail: aidahjumahat@uitm.edu.my

Mohamad Asrofi Muslim, Shahrul Azam Abdullah, Mohd Azrul Jaafar and Napisah Sapiai, Faculty of Mechanical Engineering, Universiti Teknologi MARA, 40450, Shah Alam, Selangor, Malaysia

Raymond Siew Teng Loy, NCT Global Sdn Bhd, PT-1361, Jalan Kesidang 5, Kampung Mohd Taib, Kawasan Industri, Kampung Sungai Choh, 48000, Rawang, Selangor, Malaysia

As per De Gruyter's policy this article has previously been published in the journal Physical Sciences Reviews. Please cite as:
M. A. Muslim, A. Jumahat, S. A. Abdullah, M. A. Jaafar, N. Sapiai and R. S. Teng Loy "Investigation on impact properties of different type of fibre form: hybrid hemp/glass and kenaf/glass composites" *Physical Sciences Reviews* [Online] 2022.
DOI: 10.1515/psr-2022-0053 | https://doi.org/10.1515/9783110769227-019

fibres (M.F [11]. The natural fibres are renewable and low-price natural resource, which make them widely used in aerospace, leisure, construction, sports, packing and automobile [8, 13, 15, 23]. Moreover, natural fibre composites can be recycled and composed by burning and the fibres' element of re-growing supply is another favourable feature on this kind of fibre [13, 23]. Natural fibres are categorised based on their sources, i.e., they originated from plants, animals or minerals [12, 19, 26]. According to previous studies, plant fibres are the most prevalent natural fibres utilised as reinforcement in fibre reinforced composites [17]. Plant fibres include leaf or stiff fibres, bast fibres, seed, fruit, wood, cereal straw and various grass fibres.

With the recent interest in applying natural fibres, kenaf and hemp fibres are used in this study to achieve high impact performance at a lower cost. Kenaf is one of the natural fibres derived from the *Kenaf- Hibiscus cannabinus* in the family of *alvacea*. Kenaf bast fibre has the property of absorbing oil, high toughness, high aspect ratio and less in weight [4, 16, 18, 27, 29]. Kenaf finds its wide application as oil absorbent, reinforcing fibre in thermoplastic composites, paper production on a limited basis, insulation and packing material [29]. It offers better sliding wear resistance when mixed with polyester than epoxy [16]. Hemp (*Cannabis sativa* plant species) is an annual herbaceous fast-growing plant with a length of up to 2 m. It can be found in the countries such as China, Canada, the United States and France [22]. The properties of hemp that are durable, stiff and lightweight, cheap and natural solid fibres make this fibre as an alternative natural fibre to be considered [6]. The hemp fibres are used in wide industries application such as textiles, construction, paper, composites and animal bedding.

Even though natural fibres have great potential in fibre reinforced polymer composite, the properties of natural fibres will vary considerably depending on the chemical structure such as cellulose, the degree of polymerization, orientation and crystallinity which are affected by condition during growth of plant as well as the age of the plant from which the fibres are extracted [24]. Therefore, the hybridisation with synthetic fibres has been proven to improve the mechanical performance of natural fibres (Muhammad F [11, 22, 26, 27]. In fibre reinforced polymer composites, glass fibre is one of the most commonly utilised synthetic fibres. It has low cost and good mechanical qualities, making it ideal for non-critical structural applications such as the automotive industry, where price and production are crucial [12, 26]. Glass fibre is divided into three categories, such as E-glass, S-glass and C-glass. The E-glass is designated for electrical applications, whereas the S-glass is selected for high strength. C-glass is used in structural design applications because of its excellent corrosion resistance [30]. S-Glass is widely used for fibre reinforced polymer composites that require improved mechanical properties compared to E-glass and C-glass fibres.

Many studies were conducted to investigate the hybridisation of natural fibres and synthetic fibres [2, 25, 27]. N.Sapiai et al. reported that the hybridization of kenaf with glass fibres showed an improvement on the tensile and flexural properties of the kenaf composites. These improvements were a result of hybridization effect as well as the fact that glass fiber is comparatively stiffer, which has a higher modulus than kenaf fiber. Saidane et al. studied the hybridisation effect of flax with glass fibres on kinetic diffusion and tensile mechanical. The authors observed that the hybridisation of flax and glass fibres composites improved the young's modulus, tensile strength and moisture resistance by reducing the water absorption and the diffusion coefficient. Another research by Atiqah et al. on the unsaturated polyester resin-based hybrid composites with kenaf and glass showed higher value of flexural, tensile and impact strength when the mercerization treatment was done on the kenaf fibres.

From the literature, most of the studies were focused on the hybridisation effect of natural and synthetic fibres. Therefore, this research aims to focus on the hybridisation of kenaf mat/woven, hemp mat/woven glass and glass chopped strand mat/woven glass composites. The effect of (1) different core material either hemp, kenaf or glass, (2) core thickness either using 1,2 or 4 plies of fibre and (3) fibre arrangement either as core (arrange in the middle layer) or skin (arrange in outer layer) will be thoroughly studied on the impact properties of hybrid composites.

19.2 Materials and methods

19.2.1 Material

The materials used in this experiment were synthetics fibre; twill weave (GTW) and chopped strand mat types of glass fibre (GCSM), natural fibre; kenaf and hemp mat. All these fibres were purchased from Innovative Pultrusion Sdn.Bhd., Seremban, Negeri Sembilan. The ISO polyester resin and Butanox M60 hardener were used as resin matrix funded by NCT Global Sdn Bhd, Rawang Selangor. Figure 19.1 shows the photographic view of kenaf, hemp and glass fibres used in the fabrication of hybrid composite samples.

19.2.2 Hybrid Composites Fabrications

The fibre reinforced hybrid composites were prepared using a hand lay-up and vacuum techniques, as shown in Figure 19.2. The polyester resin was mixed with a hardener in the ratio of 50:1 by weight as recommended by supplier to make sure the resin fully cured. The fabrication of the hybrid composite

(a)   **(b)**   **(c)**   **(d)**

Figure 19.1: Fibres used in fabrication of hybrid composite (a) kenaf fibres (mat), (b) hemp fibres (mat), (c) glass fibres (chopped strand mat) and (d) glass fibres (twill weave).

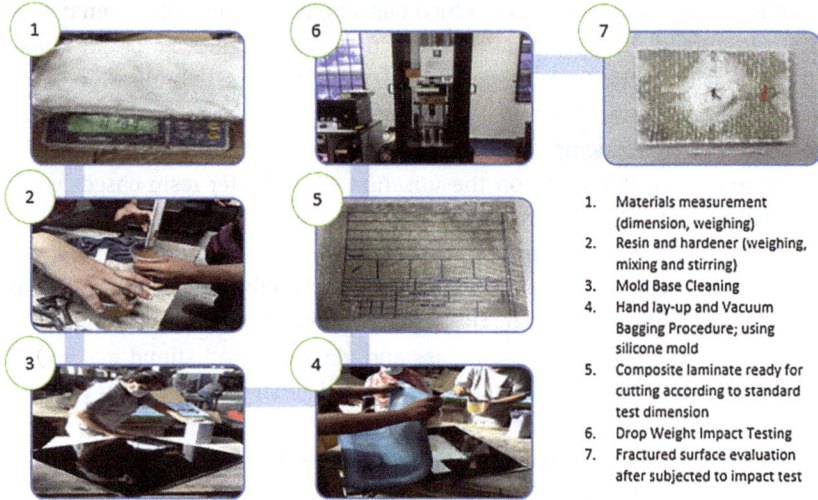

Figure 19.2: Fabrication process of the hybrid composite samples.

1. Materials measurement (dimension, weighing)
2. Resin and hardener (weighing, mixing and stirring)
3. Mold Base Cleaning
4. Hand lay-up and Vacuum Bagging Procedure; using silicone mold
5. Composite laminate ready for cutting according to standard test dimension
6. Drop Weight Impact Testing
7. Fractured surface evaluation after subjected to impact test

specimens started by placing dry fibre fabrics in a pre-defined stacking sequence of an open mould. Once stacking was done, the resin and hardener were poured on the fibres and the roller was rolled over each layer to achieve better bonding between the fibres and resin before sealed using silicon rubber. The roller was used to remove air pockets or bubbles formed inside the hybrid composite specimens. After that, any entrapped air inside the specimens was removed by vacuum as illustrated in Figure 19.3. The hybrid composite specimens were removed from the mold and left for cure at room temperature for 8 h. The hybrid composite specimens were cut to a specific measurement according to ASTM standard for impact test.

The hybrid composites were stacked in four different arrangement systems (1/4/1, 2/4/2, 2/1/2 and 2/2/2); in which hemp mat, kenaf mat or GCSM was hybridised with GTW type of glass fibres for all

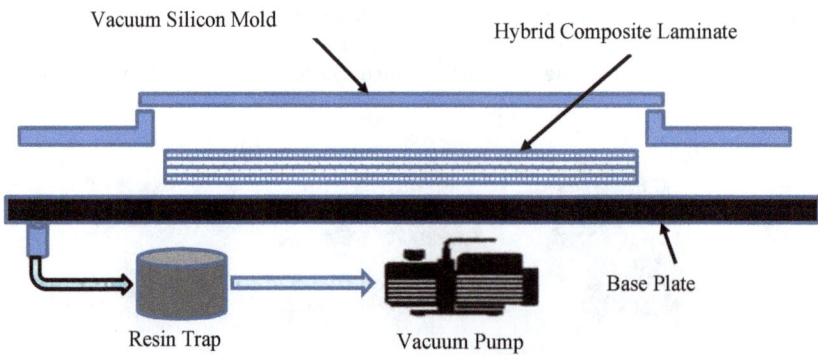

Figure 19.3: Schematic diagram of vacuum silicon mould fabrication process.

arrangement systems. In studying the effect of different core material, the hemp, kenaf or glass mat types were arranged in the middle of the hybrid composite. For the effect of core thickness, 1,2 or 4 plies of hemp, kenaf or GCSM fibre was layered in the middle of the hybrid composites. In the studying the effect of fibre arrangement, the hemp, kenaf or glass mat types were arranged in the middle layer as core or arranged in outer layer as skin. All the hybrid composites (hemp mat/GTW, kenaf mat/GTW and GCSM/GTW) were prepared according to the designated layering sequence as illustrated in Figure 19.4. Meanwhile, Table 19.1 displays the designation of all systems of hybrid composites.

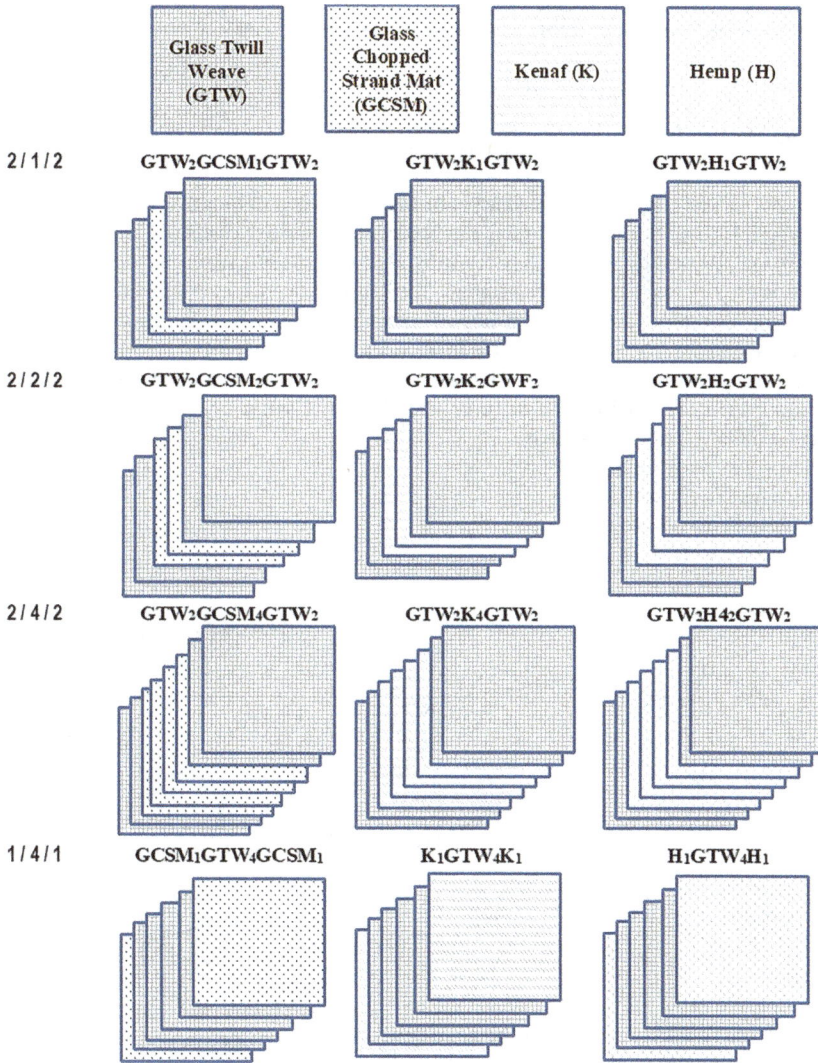

Figure 19.4: Hybrid composite stacking sequence illustrations.

Table 19.1: Stacking system designation for hybrid composites.

Stacking System			
2/1/2	**2/2/2**	**2/4/2**	**1/4/1**
GTW_2 $GCSM_1$ GWF_2	GTW_2 $GCSM_2$ GTW_2	GTW_2 $GCSM_4$ GTW_2	$GCSM_1$ GTW_4 $GCSM_1$
GTW_2 K_1 GTW_2	GTW_2 K_2 GTW_2	GTW_2 K_4 GTW_2	K_1 GTW_4 K_1
GTW_2 H_1 GTW_2	GTW_2 H_2 GTW_2	GTW_2 H_4 GTW_2	H_1 GTW_4 H_1
GCSM = glass chopped strand mat	GTW = glass twill weave	K = kenaf	H = hemp

19.2.3 Drop Weight Impact Test

The drop weight impact test was performed using Instron Dynatup 8250 Drop Weight Impact Tester according to ASTM D7136 [1]. The drop tower is equipped with 13 mm in diameter of hemispherical tip impactor. Five samples were tested for each system of hybrid composite specimens under impact energy of 30 J. Tests were performed at room temperature. The impact properties such as load carrying capabilities, energy absorbed, deflection and impact strength of hybrid composite specimens were measured. The damage surfaces after subjected to impact test were observed under digital camera. Salehi Khojin et al. identified eight types of impact parameters were present during an impact test according to a load–energy–deflection–time curve as in Figure 19.5 below.

The curve utilised the below variables:
A. First damage (damage initiation point)
B. Maximum load (peak load)
C. Damage initiation impact energy
D. Maximum energy absorbed

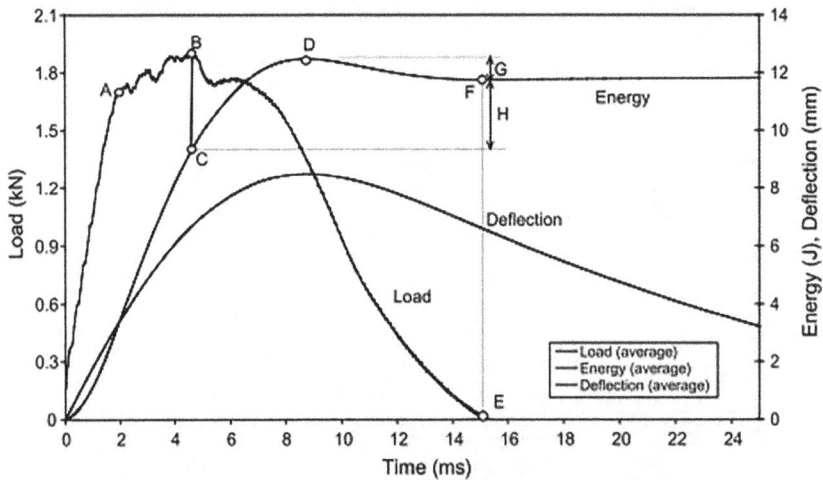

Figure 19.5: Load – energy – deflection – time curve (Salehi Khojin et al. 2006).

E. Energy absorbed (integral area under the force–displacement curve, which occurred from time zero to the time when the load decreased to zero, point E).
F. Total energy
G. Elastic energy
H. Propagation energy

The impact properties obtained from the test were as follows:
(i) Incipient damage load, P_i, was the first failure or damage point that could be identified by the first sudden load drop in the load versus time curve.
(ii) Peak load, PL, is the maximum load value that a composite laminate can withstand in the load versus deflection curve.
(iii) Deflection at peak load, δP_i, is a deflection to a point where the first peak load occurred in the load versus deflection curve.
(iv) Initiation energy (energy at maximum load), E_m and propagation energy (energy absorbed after maximum load), E_p was determined from the area under the load versus deflection curve.
(v) Total energy absorbed, E_t, was the sum of propagation energy, E_p and initiation energy, E_m [14, 20] as demonstrated in Eq. (19.2.1):

$$E_t = E_m + E_p \tag{19.2.1}$$

(vi) Ductility index, DI was defined as the ratio of propagation energy to initiation energy [3, 21] as the following Eq. (19.2.2)

$$DI = E_p / E_m \tag{19.2.2}$$

(vii) Impact strength or toughness value, σ_i in joule per meter square [J/m²] was determined by dividing the initiation energy with the area of an impactor. Where d_i was impactor diameter with a value of 10 mm. Eq. (19.2.3) displays the formula of impact strength value.

$$\sigma_i = E_m / \left(\pi d_i^2 4 \right) \tag{19.2.3}$$

A clearer definition of peak load, initiation energy, propagation energy and total energy absorbed is illustrated in Figure 19.6.

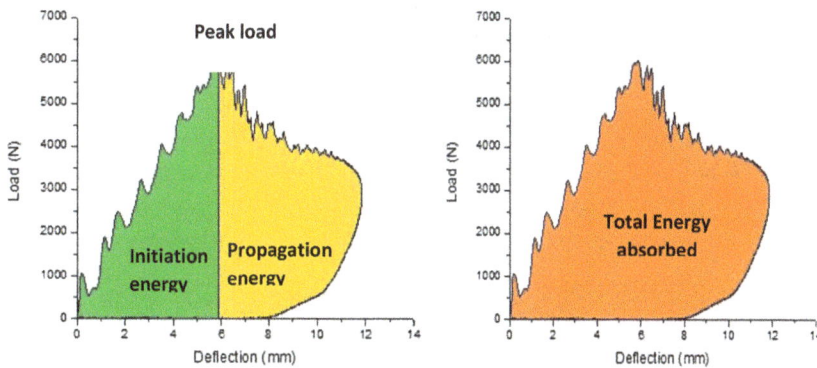

Figure 19.6: The identification of peak load, initiation energy, propagation energy and total energy absorbed from typical load-deflection curve (Salehi Khojin et al. 2006).

19.3 Results and discussion

The impact properties of composites incorporating glass, kenaf, and hemp fibres as reinforcements were studied in different stacking sequences and discussed in the following subsections.

19.3.1 Effect of different core material on impact properties of hybrid composites

Table 19.2 and Figure 19.7 display the results of the impact tests performed on the hybrid composite specimens. In studying the effect of different core materials, three systems of hybrid composite; which are $GTW_2 GCSM_1 GTW_2$, $GTW_2 K_1 GTW_2$ and $GTW_2 H_1 GTW_2$ were compared. These hybrid composites were fabricated in system arrangement of 2/1/2 with one ply hemp, kenaf or glass mat placed in middle layer and two plies of woven glass as outer layer. Based on Table 19.2, $GTW_2 H_1 GTW_2$ owned the highest energy absorbed values followed by $GTW_2 K_1 GTW_2$ and $GTW_2 GCSM_1 GTW_2$. The energy absorbed value for $GTW_2 H_1$ GTW_2 specimen was recorded at about 34.86 J, while the energy absorbed was recorded at about 33.49 J for $GTW_2 K_1 GTW_2$ and 20.7 J for $GTW_2 GCSM_1 GTW_2$.

Table 19.2 shows that $GTW_2 K_1 GTW_2$ had a higher peak load, deflection at peak load and initiation energy compared to $GTW_2 GCSM_1 GTW_2$. The result indicated that the hybridization of woven glass fibers with natural fibers in the core had better load carrying capabilities and resistance to deformation as compared to $GTW_2 GCSM_1 GTW_2$. It was also observed that the propagation or absorbed energy of $GTW_2 H_1 GTW_2$ lami-nates was higher than $GTW_2 GCSM_1 GTW_2$, indicating that $GTW_2 H_1 GTW_2$ can absorb more energy upon impact. The ductility index of $GTW_2 K_1 GTW_2$ was found to be lower than $GTW_2 GCSM_1 GTW_2$ and $GTW_2 H_1 GTW_2$.

According to the calculated impact strength value, it was confirmed that $GTW_2 K_1$ GTW_2 had a higher impact strength value (8.39928 kJ/m^2) compared to $GTW_2 H_1 GTW_2$ (6.87856 kJ/m^2) and $GTW_2 GCSM_1 GTW_2$ (2.88137 kJ/m^2). Usually, an impact failure is initiated as a matrix crack, which propagates towards the interface of laminates and grows as delamination. Kenaf and Hemp as the core give a higher energy absorbed value, higher peak load, deflection at peak load, initiation energy, propagation energy and impact strength when compared to glass fibre. According to the energy conservation principle, an acutely damaged laminate will absorb more energy than one that is less damaged [22].

19.3.2 Effect of core thickness on impact properties of hybrid composites

Core thickness refers to laminate ply between 1, 2 and 4 layers of GCSM, kenaf and hemp hybridised with GTW. The effect of core thickness on impact properties is shown

Table 19.2: Impact Test Result by Different core.

Composites Systems	Impact Properties						
	Peak load (*N*)	Deflection at peak load (mm)	Total energy absorbed, E_t (J)	Initiation energy, E_m (J)	Propagation energy, E_p (J)	Ductility index, E_p/E_m	Impact Strength, σ_i [J/m²]
GTW₂ GCSM₁ GTW₂	2348.50 ± 161.07	5.55 ± 0.43	20.70 ± 3.63	6.12	14.58	2.38	2881.37
GTW₂ K₁ GTW₂	4127.12 ± 756.57	6.61 ± 1.17	33.49 ± 1.69	17.84	15.65	0.88	8399.28
GTW₂ H₁ GTW₂	4002.74 ± 231.69	5.53 ± 1.17	34.86 ± 0.90	14.61	20.25	1.39	6878.56

Energy Absorbed (J)

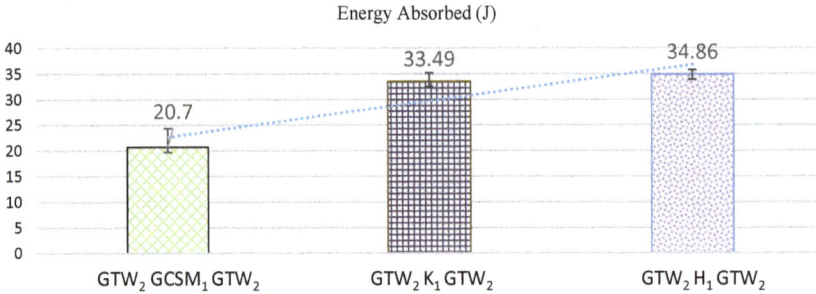

Figure 19.7: Energy absorbed (J) of stacking 2/1/2 for different core material.

in Table 19.3 and Figure 19.8. The table and figure display the results on the impact properties of nine hybrid composites as coded as GTW_2 $GCSM_1$ GTW_2, GTW_2 K_1 GTW_2, GTW_2 H_1 GTW_2, GTW_2 $GCSM_2$ GTW_2, GTW_2 K_2 GTW_2, GTW_2 H_2 GTW_2, GTW_2 $GCSM_4$ GTW_2, GTW_2 K_4 GTW_2 and GTW_2 H_4 GTW_2.

In general, the hemp as core layer had the highest impact energy values, followed by kenaf and GCSM. The tests indicated that hybrid composites with four layers/plies of hemp fibres, GTW_2 H_4 GTW_2 hybrid composites had a maximum energy absorbed of 36.12 J, which increased by 7.55% and 5.86% compared to GTW_2 H_1 GTW_2 and GTW_2 H_2 GTW_2, respectively. The similar trend was found in hybrid kenaf/GTW and GCSM/GTW hybrid composites. All hybrid composites exhibited consistent increasing energy absorbed when the core thickness increased. According to Hunain et al., increasing laminated layers for the same fiber composite will increase the modulus of elasticity, tensile strength, and fracture toughness regardless of fibre orientation.

19.3.3 Effect of fibre arrangement on impact properties of hybrid composites

Fibre arrangement is one of the factors that influence the impact strength of composite fiber [7, 9, 28]. In this study, the comparison between GCSM, kenaf and hemp fibres as core (arrange in the middle) in 2/2/2 fibre arrangement and as a skin (outer layer) in 1/4/1 fibre arrangement was evaluated. Each hybrid composites had 2 plies of GCSM, kenaf or hemp mat which were arranged in two layers in the core and one layer in skin position.

From the data tabulated in Table 19.4 and Figure 19.9, the energy absorbed for GCSM, kenaf and hemp fibres as skin was higher than kenaf and hemp fibres as core. In the skin parts, the energy absorbed for kenaf was 32.06 J, which was higher as compared to kenaf in core (30.57 J). Energy absorbed property difference of 4% was found when kenaf acted as core and skin. As for hemp, H_1 GTW_4 H_1 hybrid composite recorded energy absorbed at about 34.49 J while GTW_2 H_2 GTW_2 recorded a little bit

Table 19.3: Effect of core thickness on impact properties of hybrid composites.

Composites Systems	Peak load (N)	Deflection at peak load (mm)	Total energy absorbed, E_t (J)	Impact Properties Initiation energy, E_m (J)	Propagation energy, E_p (J)	Ductility index, E_p/E_m	Impact Strength, σ_i [J/m²]
GTW₂ GCSM₁ GTW₂ K₁	2348.50 ± 161.07	5.55 ± 0.43	20.70 ± 3.63	6.12	14.58	2.38	2881.37
GTW₂ GTW₂ H₁	4127.12 ± 756.57	6.61 ± 1.17	33.49 ± 1.69	17.84	15.65	0.88	8399.28
GTW₂	4002.74 ± 231.69	5.53 ± 1.17	34.86 ± 0.90	14.61	20.25	1.39	6878.56
GWF₂ GCSM₂ GWF₂	3622.99 ± 254.36	4.74 ± 0.37	28.23 ± 3.71	8.92	19.31	2.16	4199.64
GWF₂ K₂	2852.76 ± 268.10	6.56 ± 1.03	30.57 ± 3.37	12.48	18.09	1.45	5875.73
GWF₂ GWF₂ H₂	7243.74 ± 378.99	4.57 ± 0.36	34.60 ± 2.65	13.71	20.89	1.52	6454.83
GWF₂ GWF₂ GCSM₄ GWF₂	4873.08 ± 326.58	4.68 ± 0.58	36.12 ± 0.50	13.06	23.06	1.77	6148.80
GWF₂ K₄	3497.50 ± 308.05	4.62 ± 1.83	34.12 ± 3.34	10.51	23.61	2.25	4948.23
GWF₂ GWF₂ H₄ GWF₂	5579.24 ± 266.44	2.33 ± 0.06	33.55 ± 1.14	6.21	27.34	4.40	2923.74

Energy Absorbed (J)

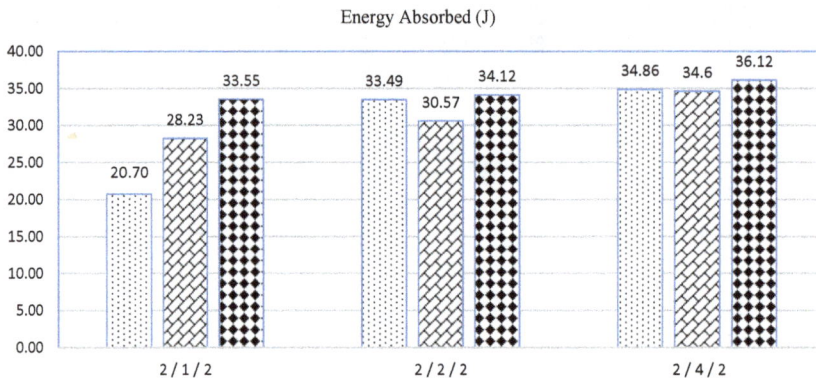

Figure 19.8: Effect of core thickness on impact properties of hybrid composites.

lower compared to skin, which was 30.57 J. Hemp showed very low difference of energy absorbed with 0.3% when compared to core as skin. A similar finding was observed in GCSM/GTW hybrid of $GCSM_1$ GTW_4 $GCSM_1$ in which GCSM fibre arranged as skin resulted a higher energy absorbed as compared to GTW_2 $GCSM_2$ GTW_2, which was arranged as core.

As compared to different materials, the hemp and kenaf fibres had higher energy absorbed for both system arrangements of 2/2/2 and 1/4/1 as compared to GCSM. From here, it was found that the hemp result was higher than kenaf when placed on the skin or core position, which was the same as reported by [5]. It is outstanding that the different fibre type and fibre layers are primarily in charge of the strength enhancement of the composite. Thus, differences in composite strength with different fibre types and the number of layers are apparent [10]. It can be concluded that the natural fibres i.e., kenaf and hemp have potential in replacing synthehic fibres i.e, glass fibres.

19.3.4 Damage Evaluation of hybrid composites

Figure 19.10 depicts the visual inspection of the damaged hybrid composites specimens; front and back views. All specimens appeared to have a nearly identical damage pattern on the back face, with delamination of glass fibres. In contrast, different damage patterns were noticed on the front face. As compared to different core type in the arrangement of 2/1/2, the damage pattern of GTW_2 $GCSM_1$ GTW_2 hybrid composites showed several cracks existed on front face. Meanwhile, for GTW_2 K_1 GTW_2 and GTW_2 H_1 GTW_2 hybrid composites, damage pattern around the impacted surface was observed. This supported that the natural fibres i.e., kenaf and hemp can absorb more energy and endure the load when subjected to impact loading. As for comparison with different core thickness, the hybrid composite specimens had a wide damage area around the struck surface, and the damaged area decreased as the amount of natural

Table 19.4: Effect of fibre arrangement on impact properties of NFRPPC composites.

Composites Systems	Impact Properties						
	Peak load (N)	Deflection at peak load (mm)	Total energy absorbed, E_t (J)	Initiation energy, E_m (J)	Propagation energy, E_p (J)	Ductility index, E_p/E_m	Impact Strength, σ_i [J/m²]
GWF$_2$ GCSM$_2$ GWF$_2$	3622.99 ± 254.36	4.74 ± 0.37	28.23 ± 3.71	8.92	19.31	2.16	4199.64
GWF$_2$ K$_2$ GWF$_2$	2852.76 ± 268.10	6.56 ± 1.03	30.57 ± 3.37	12.48	18.09	1.45	5875.73
GWF$_2$ H$_2$ GWF$_2$	7243.74 ± 378.99	4.57 ± 0.36	34.60 ± 2.65	13.71	20.89	1.52	6454.83
GCSM$_1$ GWF$_4$ GCSM$_1$	3441.31 ± 286.23	6.15 ± 0.40	36.10 ± 1.16	12.29	23.81	1.94	5786.27
K$_1$ GWF$_4$ K$_1$	2982.87 ± 83.94	7.85 ± 0.30	32.06 ± 5.94	15.64	16.42	1.05	7363.49
H$_1$ GWF$_4$ H$_1$	4954.21 ± 218.58	3.98 ± 2.27	34.49 ± 0.21	13.68	20.81	1.52	6440.70

Energy Absorbed (J)

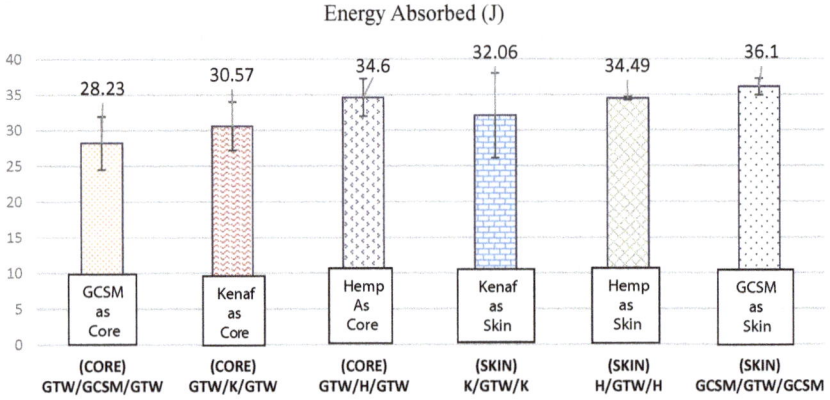

Figure 19.9: Effect of fibre arrangement on impact properties of NFRPPC composites.

Systems	Top/Front	Bottom/Back
GTW$_2$ GCSM$_1$ GTW$_2$		
GTW$_2$ K$_1$ GTW$_2$		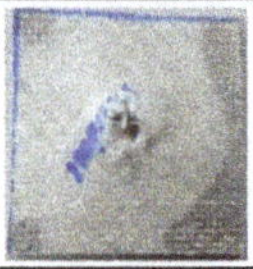
GTW$_2$ H$_1$ GTW$_2$		
GTW$_2$ GCSM$_2$ GTW$_2$	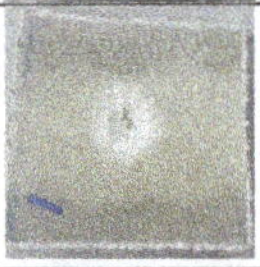	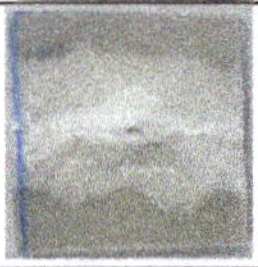
GTW$_2$ K$_2$ GTW$_2$	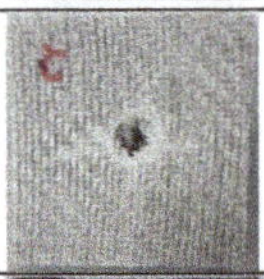	
GTW$_2$ H$_2$ GTW$_2$	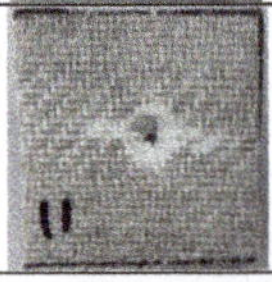	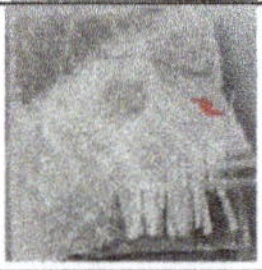

Figure 19.10: Hybrid composite specimens after subjected to impact test.

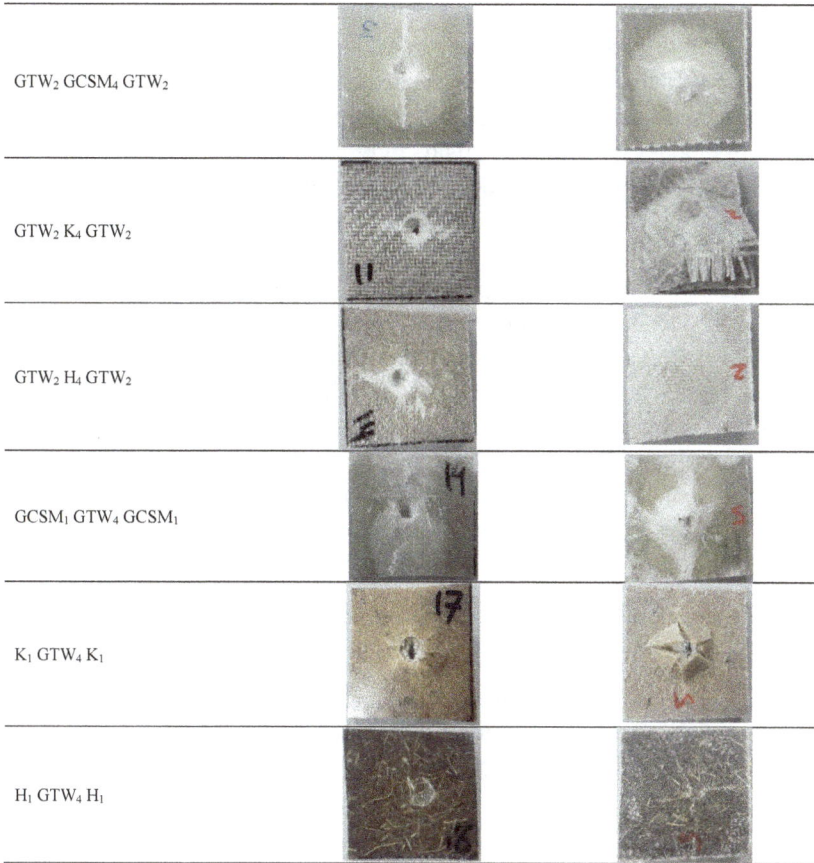

GTW$_2$ GCSM$_4$ GTW$_2$		
GTW$_2$ K$_4$ GTW$_2$		
GTW$_2$ H$_4$ GTW$_2$		
GCSM$_1$ GTW$_4$ GCSM$_1$		
K$_1$ GTW$_4$ K$_1$		
H$_1$ GTW$_4$ H$_1$		

Figure 19.10: Continued.

fibre increased, implying that natural fibres absorbed more energy. The affected surface of the hybrid hemp and hybrid kenaf with glass has a small damaged area; however, the brittle nature of the polyester matrix was obvious by the bright tonalities (cross-shaped cracks) as shown in Figure 19.5. This confirmed that various fibres had different damage or failure modes. As a result, the damage mechanism of hybrid composites explains why natural fibre has a better effect on impact properties.

19.4 Conclusions

In this study, integrating woven glass fibre with hemp and kenaf composites has positive effects on natural fibre, allowing it to be employed in a range of applications. Hemp/woven glass fibre absorbed more energy, followed by kenaf/woven glass and

GCSM/woven glass hybrid composites. Finally, the inclusion of many layers of natural fibre to glass fibre increased the specimen's load-carrying capacity and energy absorption. These findings demonstrated that natural was more resistant to impact loading. As a result of the natural fiber's ability to absorb more energy, hybridising natural fibre to glass fibre laminate tends to lessen the damaged area of the hybrid specimens. Depending on the specific needs of the application, the current composite's impact properties could be useful in a variety of research and industrial applications.

Acknowledgement: The authors appreciate and would like to thank the financial support obtained from the Research Management Institute (RMI) UiTM, the Ministry of Education Malaysia, and the Institute of Graduate Studies (IPSIS). The study was carried out at Malaysia's Universiti Teknologi MARA (UiTM) Faculty of Mechanical Engineering.

References

1. ASTM International. ASTM D 7905-14 - standard test method for determination of the mode II interlaminar fracture toughness of unidirectional fiber-reinforced polymer matrix composites. Annu Book ASTM (Am Soc Test Mater) Stand 2012;i:1–16.
2. Atiqah A, Maleque MA, Jawaid M, Iqbal M. Development of kenaf-glass reinforced unsaturated polyester hybrid composite for structural applications. Compos B Eng 2014;56:68–73.
3. Cantwell WJ, Morton J. The impact resistance of composite materials - a review. Composites 1991; 22:347–62.
4. Castillo-Lara JF, Flores-Johnson EA, Valadez-Gonzalez A, Herrera-Franco PJ, Carrillo JG, Gonzalez-Chi PI, et al. Mechanical properties of natural fiber reinforced foamed concrete. Materials 2020;13: 3060.
5. Dinesh V, Shivanand HK, Vidyasagar HN, Chari VS. Investigation of mechanical properties of kenaf, hemp and E-glass fiber reinforced composites. AIP Conf Proc 2018;1943:20117.
6. Dixit S, Goel R, Dubey A, Bhalavi T. Natural fibre reinforced polymer composite materials-A review. Polym Renew Resour 2017;8:71–8.
7. Elbadry EA, Abdalla A, Aboraia M, Oraby EA. Effect of glass fibers stacking sequence on the mechanical properties of glass fiber/polyester composites. J Mater Sci Eng 2018;07. https://doi.org/10.4172/2169-0022.1000416.
8. Faris M, Khalid S, Jumahat A, Salleh Z, Jawaid M. SCIENCE & TECHNOLOGY flexural properties of random and unidirectional arenga pinnata fibre reinforced epoxy composite. Pertanika J. Sci. & Technol 2017;25:93–102.
9. Hegde SR, Hojjati M. Effect of core and facesheet thickness on mechanical property of composite sandwich structures subjected to thermal fatigue. Int J Fatig 2019;127:16–24.
10. Hunain MB, Al-Turaihi AS, Alnomani SN. Tensile and charpy impact behavior of E-glass/unsaturated polyester laminated composite material at elevated temperature. J Eng Sci Technol 2021;16:1547–60.
11. Ismail MF, Sultan MTH, Hamdan A, Shah AUM, Jawaid M. Low velocity impact behaviour and post-impact characteristics of kenaf/glass hybrid composites with various weight ratios. J Mater Res Technol 2019;8:2662–73.

12. Jeyapragash R, Srinivasan V, Sathiyamurthy S. Mechanical properties of natural fiber/particulate reinforced epoxy composites - a review of the literature. Mater Today Proc 2020;22:1223–7.
13. Khan MZN, Hao Y, Hao H, Shaikh Fuddin A. Mechanical properties and behaviour of high-strength plain and hybrid-fiber reinforced geopolymer composites under dynamic splitting tension. Cement Concr Compos 2019;104:103343.
14. Kounain MA, Al-Sulaiman F, Khan Z. Low velocity impact and post impact tensile properties of plain weave woven GFRP composite laminates. Int J Eng Adv Technol 2015;2:92–9.
15. Mahesh V, Joladarashi S, Kulkarni SM. A comprehensive review on material selection for polymer matrix composites subjected to impact load. Defence Technology 2021;17:257–77.
16. Matykiewicz D. Hybrid epoxy composites with both powder and fiber filler: a review of mechanical and thermomechanical properties. Materials 2020;13:1802.
17. Moustafa H, El-Wakil AEAA, Nour MT, Youssef AM. Kenaf fibre treatment and its impact on the static, dynamic, hydrophobicity and barrier properties of sustainable polystyrene biocomposites. RSC Adv 2020;10:29296–305.
18. Nagalakshmaiah M, Afrin S, Malladi RP, Elkoun S, Robert M. Ansari Ma. Biocomposites: Present trends and challenges for the future. Green Comp Automot App 2019;197–215.
19. Nazim ARM, Ansari MNM. A review on natural fibre polymer composites. Int J Sci Res Eng Technol 2017;6:81–6.
20. Park R, Jang J. Impact behavior of aramid fiber/glass fiber hybrid composite: evaluation of four-layer hybrid composites. J Mater Sci 2001;36:2359–67.
21. Pegoretti A, Fabbri E, Migliaresi C, Pilati F. Intraply and interply hybrid composites based on E-glass and poly(vinyl alcohol) woven fabrics: tensile and impact properties. Polym Int 2004;53:1290–7.
22. Prabhu L, Krishnaraj V, Sathish S, Gokulkumar S, Karthi N, Rajeshkumar L, et al. A review on natural fiber reinforced hybrid composites: chemical treatments, manufacturing methods and potential applications. Mater Today Proc 2021;45:8080–5.
23. Reddy PV, Reddy RVS, Rajendra Prasad P, Mohana Krishnudu D, Reddy RM, Rao HR. Evaluation of mechanical and wear performances of natural fiber reinforced epoxy composites. J Nat Fibers 2020;19:2218–31.
24. Saba N, Jawaid M, Hakeem KR, Paridah MT, Khalina A, Alothman OY. Potential of bioenergy production from industrial kenaf (Hibiscus cannabinus L.) based on Malaysian perspective. Renew Sustain Energy Rev 2015;42:446–59.
25. Saidane EH, Scida D, Assarar M, Sabhi H, Ayad R. Hybridisation effect on diffusion kinetic and tensile mechanical behaviour of epoxy based flax-glass composites. Compos Appl Sci Manuf 2016;87:153–60.
26. Sanjay MR, Arpitha GR, Yogesha B. Study on mechanical properties of natural - glass fibre reinforced polymer hybrid composites: a review. Mater Today Proc 2015;2:2959–67.
27. Sapiai N, Jumahat A, Jawaid M, Khan A. Effect of MWCNT surface functionalisation and distribution on compressive properties of kenaf and hybrid kenaf/glass fibres reinforced polymer composites. Polymers 2020;12:1–18.
28. Setiadi A, Raharjo WW, Triyono T. The effect of core thickness variation of sandwich composite cantala rHDPE on mechanical strength of bending test. In: AIP conference proceedings, 1st ed. AIP Publishing LLC; 2017, 1788:030058 p.
29. Talabari AA, Alaei MH, Shalian HR. Experimental investigation of tensile properties in a glass/epoxy sample manufactured by vacuum infusion, vacuum bag and hand layup process. Rev Compos Matériaux Avancés 2019;29:179–82.
30. Tamrakar S, Kiziltas A, Mielewski D, Zander R. Characterization of kenaf and glass fiber reinforced hybrid composites for underbody shield applications. Compos B Eng 2021;216:108805.

Index

https://doi.org/10.1515/9783110769227-020

www.ingramcontent.com/pod-product-compliance
Lightning Source LLC
Chambersburg PA
CBHW080515220326
41599CB00032B/6085